Principles of Electricity and Magnetism

Principles of Electricity and Magnetism

Saunak Palit

Alpha Science International Ltd.
Harrow, U.K.

Saunak Palit
Department of Physics
St. Xaviers College
Kolkata, India

Alpha Science International Ltd
Hygeia Building, 66 College Road,
Harrow, Middlesex HA1 1BE, U.K.

ISBN 1-84265-205-2

Printed in India

Preface

A couple of years ago, the University Grants Commission (UGC) proposed a Model Curriculum for undergraduates studying for an advanced degree in Physics in India. This book has been written in accordance with this curriculum. It covers Courses P4 (Electricity and Magnetism) and P10 (Electrodynamics, Electromagnetic waves and Relativity). One of the new features of the Model Curriculum is the introduction of Tutorials. I have included tutorials as well as problems with answers at the end of almost all the chapters.

There are six major parts in this book. Part I covers Electrostatics. In chapters 1 and 2, the concepts of electric field and electric potential are explained with examples. The method of images is dealt with in chapter 3. Finally, in chapter 4, polarization of matter under the influence of an electric field is introduced briefly. Part II covers Magnetostatics. The Biot-Savart law and Ampere's law are considered in chapter 5 with standard applications. In chapter 6, magnetization of matter is discussed. Part III is devoted to electric circuits. In chapter 7, the network theorems of Thevenin and Norton are applied to DC circuits. Chapter 8 is a study of AC circuits. The chapter concludes with a survey of some common AC bridges. Chapter 9 deals with generators and rectifiers, including a short introduction to three-phase AC systems.

Part IV of this book is mainly concerned with Maxwell's equations and some of their consequences. Electromagnetic waves in material media are considered in chapter 12. Electromagnetic radiation is just touched upon in chapter 13. Part V deals with Special Relativity. The transformation of electric and magnetic fields is also discussed here.

Part VI is historical in nature. The UGC Model Curriculum includes an account of the contributions of prominent Indian physicists from the ancient times to the time just after independence. In this part of the book, I have given sketches of their academic careers and important accomplishments. In writing this book, I have consulted the well-known books in this subject. The mathematical prerequisite for the student is a knowledge of vector calculus. I shall be happy to receive suggestions for improvement from readers.

Kolkata **Saunak Palit**

Contents

1

The Electric Field

1.1 COULOMB'S LAW

The subject of Electrostatics deals with the effects of charge distributions which are time-independent. It is based on Coulomb's law. In vector notation the law can be stated as follows : the force on a point charge q_2 due to another point charge q_1 is given by

$$\mathbf{F}_{21} = \frac{k_1 q_1 q_2}{r_{12}^2}\,\hat{\mathbf{r}}_{12} \tag{1.1.1}$$

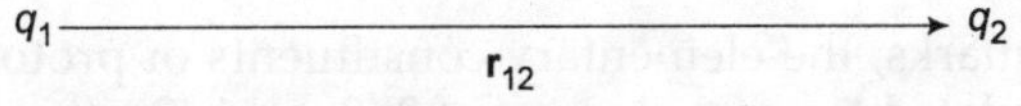

Fig. 1.1

It is a formulation of an action-at-a-distance type of force. Here $\hat{\mathbf{r}}_{12}$ is the unit vector directed from q_1 to q_2 and $r_{12} = |\mathbf{r}_{12}|$ is the distance of separation between them. $\mathbf{F}_{21}$ points away from q_1 if q_1 and q_2 are of the same sign and towards q_1 if they are of opposite sign. The constant of proportionality k_1 is one of the constants which are chosen arbitrarily to define a system of units in electricity and magnetism. Accordingly, the form of the law in some well known systems are: (in free space)

$$\mathbf{F}_{12} = \frac{q_1 q_2}{r_{12}^2}\,\hat{\mathbf{r}}_{12} \text{ (electrostatic and Gaussian)} \tag{1.1.1a}$$

$$\mathbf{F}_{12} = \frac{q_1 q_2}{4\pi r_{12}^2}\,\hat{\mathbf{r}}_{12} \text{ (Heaviside-Lorentz)} \tag{1.1.1b}$$

$$\mathbf{F}_{12} = \frac{q_1 q_2}{4\pi \varepsilon_0 r_{12}^2}\,\hat{\mathbf{r}}_{12} \text{ (rationalized MKS)} \tag{1.1.1c}$$

corresponding to choices $k_1 = 1, 1/4\pi, 1/4\pi\varepsilon_0$ respectively. In eq. (1.1.1c), ε_0 is called the *permittivity* of free space and has a value 8.854×10^{-12} F/m. (F is the Farad) The factor 4π signifies the fact that the electric field around a point charge is spherically symmetric.

The unit of charge in both the electrostatic and Gaussian system of units is defined from eqn. (1.1.1a). Here distance is measured in cm and force in dynes. If we have two equal charges q separated by a distance of 1 cm in air (or vacuum to be more precise) and the force exerted by either one on the other is 1 dyne the charge q is 1 electrostatic unit (esu). In the Heaviside-Lorentz system, distance is measured in cm, force in dynes and so, from eqn. (1.1.1b) the unit of charge is $1/(4\pi)^{1/2}$ times the esu. In the rationalized MKS system, the unit of charge is the Coulomb C but it is not defined from (1.1.1c). Here distance is measured in *m* and force in Newtons. The Coulomb is defined as the charge transported by a steady current of one ampere for one second. (The ampere is the SI unit of current. We will define it later.)

1.2 CONSERVATION AND QUANTIZATION OF CHARGE

The total charge in an isolated system is constant. This result has been experimentally checked and no violation has ever been found. It is true at the fundamental level in the interactions involving elementary particles like electrons, muons, quarks, etc. The first elementary particle to be discovered was the electron, by J.J. Thomson in 1897. Two years later, he measured its charge-to-mass ratio. The American physicist R.A. Millikan, in a series of experiments from 1906 to 1914, made a precise determination of the electronic charge. This value has since been known to be the smallest measurable charge $e^{\oplus}$. Although quarks, the elementary constituents of protons, neutrons, etc. have been postulated to carry charges of 2/3e or 1/3e, they have never been isolated.

1.3 THE ELECTRIC FIELD

The electric field **E** at a point in space can be defined as the ratio of the force $\mathbf{F}_q$ on a charge q placed at that point to the magnitude of the test charge in the limit $q \to 0$.

$$\mathbf{E} = \lim_{q \to 0} \frac{\mathbf{F}_q}{q} \tag{1.3.1}$$

It is important to take this limit so that the test charges does not in any way disturb the charge configuration that gives rise to the electric field. This definition has a limitation because charge is quantized. This definition can be applied only in those cases where the amount of charge involved is large compared to the electronic charge. We can also define the electric field at a

$^{\oplus}e = 1.60217733\ (49) \times 10^{-19}$C. The uncertainty is in the last two digits.

point as the force on a unit positive charge placed at that point. So the force **F** on charge q placed at a point where the electric field is **E** is given by

$$\mathbf{F} = q\mathbf{E} \tag{1.3.2}$$

This description introduces the idea of a region of space permeated by a field where a charge experiences a force. This is to be contrasted with the action-at-a-distance formulation of (1.1.1). The electric field has units of dynes/esu in the Gaussian system and Newton/Coulomb in the rationalized **MKS** system.

In eqn. (1.1.1a), if we put $q_2 = 1$ esu, we get the electric field due to charge q_1 at a distance r_1 from it:

$$\mathbf{E}_1 = \frac{q_1 \hat{\mathbf{r}}}{r_1^2} \tag{1.3.3}$$

where the direction of **E** is radially outward from the location of q_1 if q_1 is positive and inward if q_1 is negative. We find that

$$\nabla_1 \times \mathbf{E}_1 = q_1 \left(\nabla_1 \times \frac{\hat{\mathbf{r}}}{r_1^2} \right) = 0$$

This is a general result for an electrostatic field. The electric field due to a system of charges q_i is given by the principle of superposition: the force with which two charges interact is not changed by the presence of a third charge. So the electric field in this case is the vector sum of the fields due to the individual charges: $\mathbf{E} = \sum \frac{q_i}{r_2^2} \hat{\mathbf{r}}_i$. Here $\hat{\mathbf{r}}_i$ is the unit from the charge q_i to the point where we want to find the field.

Example 1: Calculate the electric field at a point $P(x, y)$ due to the following system of charges : $q_1 = 1$ C, $q_2 = 2$ C. What is the field at the origin?

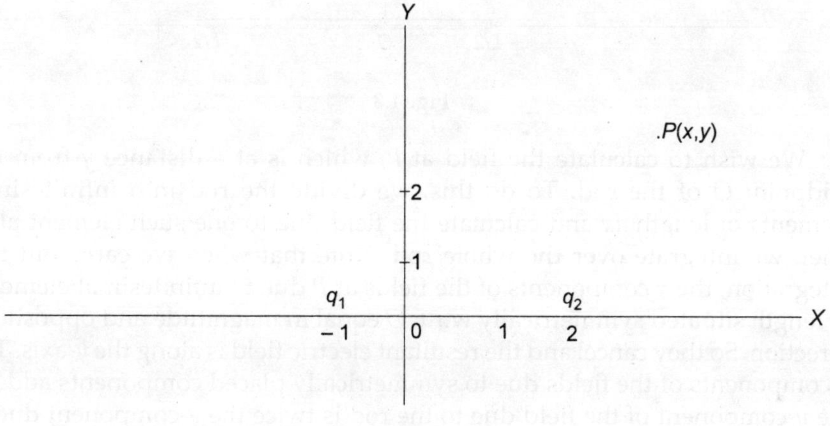

Fig. 1.2

Solution: The electric field at P is given by the vector sum of fields $\mathbf{E}_1$ due to q_1 and $\mathbf{E}_2$ due to q_2 at P. It is,

$$\mathbf{E} = \mathbf{E}_1 + \mathbf{E}_2 = \frac{9\times 10^9\{(x+1)\hat{\mathbf{i}} + y\hat{\mathbf{j}}\}}{\{(x+1)^2 + y^2\}} + \frac{18\times 10^9\{(x-2)\hat{\mathbf{i}} + y\hat{\mathbf{j}}\}}{\{(x-2)^2 + y^2\}}\ \text{N/C} \qquad (1.3.4)$$

where we can take $1/4\,\pi\varepsilon_0 = 9 \times 10^9\ \text{Nm}^2/\text{C}^2$, $\hat{\mathbf{i}}$ and $\hat{\mathbf{j}}$ are the unit vectors along the x and y axes respectively. The electric field at O is obtained by putting $x = 0$, $y = 0$ in (1.3.4) above. The result is

$$E = -\,\mathbf{4.5\,\hat{i} \times 10^9\ N/C.}$$

1.4 THE ELECTRIC FIELD DUE TO A LINE CHARGE

We have so far calculated the electric fields for point charges. Let us now consider continuous charge distributions. The simplest is a line charge, i.e. a uniformly charged nonconducting rod of finite length. Let λ be the charge per unit length and L the length of the rod. Thus the total charge on the rod is $q = \lambda L$. We will calculate the electric field at a point on the perpendicular bisector of the rod. Let us choose the x-axis along the length of the rod, as shown in the figure below.

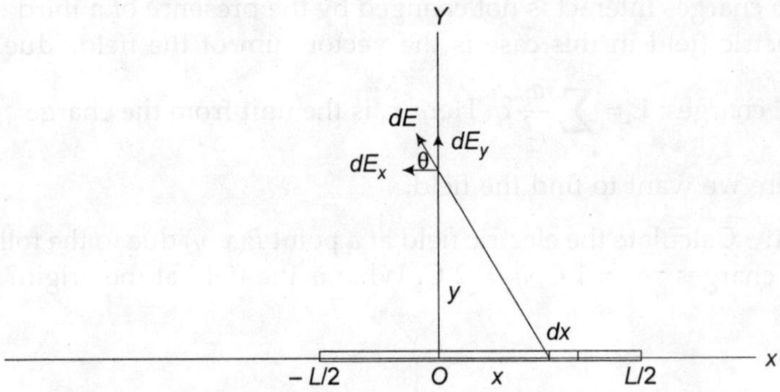

Fig. 1.3

We wish to calculate the field at P, which is at a distance y from the midpoint O of the rod. To do this, we divide the rod into infinitesimal elements of length dx and calculate the field due to one such element at P. Then we integrate over the whole rod. Note that when we carry out this integration, the x components of the fields at P due to infinitesimal elements of length situated symmetrically w.r.t. O equal in magnitude and opposite in direction. So they cancel and the resultant electric field is along the y-axis. The y-components of the fields due to symmetrically placed components add. So the y-component of the field due to the rod is twice the y-component due to half the rod.

The field due to the element dx, which has charge λdx, has magnitude

$$dE = \frac{\lambda dx}{4\pi\varepsilon_0 (x^2 + y^2)} \tag{1.4.1}$$

The y-component of this field is

$$dE_y = dE \sin\theta \tag{1.4.2}$$

From Fig. 1.3,
$$\sin\theta = \frac{y}{(x^2 + y^2)^{1/2}} \tag{1.4.3}$$

So
$$dE = \frac{\lambda y dx}{4\pi\varepsilon_0 (x^2 + y^2)^{3/2}}$$

Thus the electric field at P is

$$\mathbf{E} = \hat{\mathbf{j}}\, 2\int_0^{L/2} \frac{\lambda y dx}{4\pi\varepsilon_0 (x^2 + y^2)^{3/2}}$$

$$= \hat{\mathbf{j}} \frac{\lambda}{2\pi\varepsilon_0 y} \left[\frac{x}{(x^2 + y^2)^{1/2}} \right]_0^{L/2}$$

$$\mathbf{E} = \frac{q}{2\pi\varepsilon_0 (L^2 + 4y^2)^{1/2}} \hat{\mathbf{j}} \tag{1.4.4}$$

For an infinitely long rod, or, for points P very close to the rod (but not near its ends), let us rewrite (1.4.1) as

$$\mathbf{E} = \frac{\lambda}{2\pi\varepsilon_0 y \left(1 + \frac{4y^2}{L^2}\right)^{1/2}} \hat{\mathbf{j}}$$

So as $\mathbf{L} \to \propto$
$$\mathbf{E} = \frac{\lambda}{2\pi\varepsilon_0 y} \hat{\mathbf{j}} \tag{1.4.5}$$

1.5 THE ELECTRIC FIELD DUE TO A SHEET CHARGE

Let us now calculate the electric field due to a uniformly charged disc of radius R. This is an example of a surface charge distribution. Let σ be the surface charge density, i.e. the charge per unit area on the disc.

We will calculate the field at a point P on the axis of the disc at a distance z from the center O. (fig. 1.4) To do this let us divide the disc into rings of infinitesimal thicknesses dr. All points on a ring are equidistant from P and will produce the same field at P (in magnitude). The area of the elementary

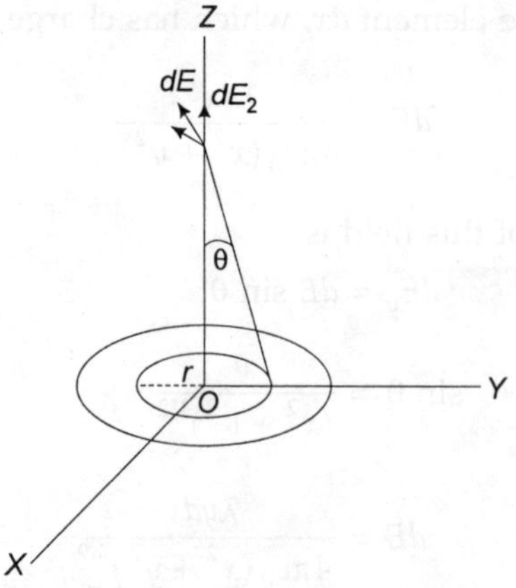

Fig. 1.4

ring is its circumference times its thickness, i.e. $2\pi r dr$ and so it carries a charge $2\pi r dr \sigma$. The horizontal components of the fields due to diametrically opposite segments will cancel each other and the vertical components will add. So the field due to an elementary ring at *P* will be

$$dE = \frac{r\sigma dr \cos\theta}{2\varepsilon_0(r^2 + z^2)} \tag{1.5.1}$$

where $\cos\theta = \dfrac{z}{(r^2 + z^2)^{1/2}}$ from Fig. 1.4

$$dE = \frac{zr\sigma dr}{2\varepsilon_0(r^2 + z^2)^{3/2}}$$

To get the resultant field, which will be along the *z* axis, we have to integrate this from $r = 0$ to $r = R$

$$\mathbf{E} = \hat{\mathbf{k}}\frac{\sigma}{2\varepsilon_0}\int_0^R \frac{rdr}{(r^2 + z^2)^{3/2}}$$

$$\mathbf{E} = \hat{\mathbf{k}}\frac{\sigma}{2\varepsilon_0}\left[1 - \frac{z}{(R^2 + z^2)^{1/2}}\right] \tag{1.5.2}$$

In the limit $R \to \infty$, we get the field due to an infinite sheet of charge. It is

$$\mathbf{E} = \frac{\sigma}{2\varepsilon_0}\hat{\mathbf{k}} \tag{1.5.3}$$

So the electric field is normal to the sheet and points away from it.

1.6 THE ELECTRIC FIELD DUE TO A DIPOLE

An electric dipole is a system of two equal and opposite charges separated by a distance. Let us calculate the electric field due to such a dipole.

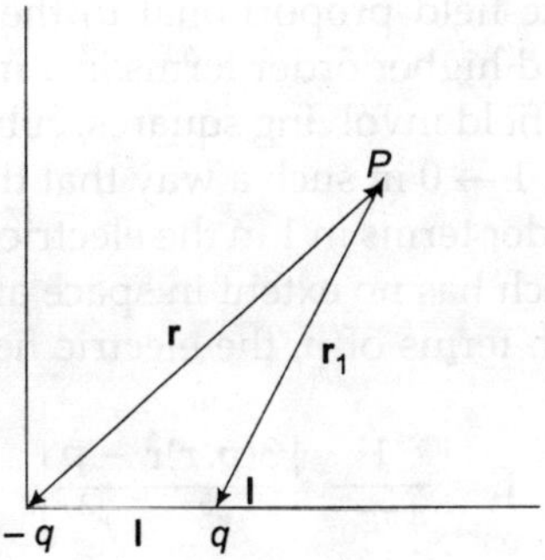

Fig. 1.4

The electric field at P is

$$\mathbf{E} = \frac{1}{4\pi\varepsilon_0}\left\{\frac{q\,\hat{\mathbf{r}}_1}{r_1^2} - \frac{q\,\hat{\mathbf{r}}}{r^2}\right\} = \frac{q}{4\pi\varepsilon_0}\left\{\frac{\mathbf{r}_1}{r_1^3} - \frac{\mathbf{r}\}}{r^3}\right\}$$

From Fig. 1.5 $\quad \mathbf{r} = \mathbf{r}_1 + \mathbf{l}$

$$\mathbf{E} = \frac{q}{4\pi\varepsilon_0}\left\{\frac{\mathbf{r}-\mathbf{l}}{r_3^1} - \frac{\mathbf{r}}{r^3}\right\} = \frac{q}{4\pi\varepsilon_0}\left\{\mathbf{r}\left(\frac{1}{r_1^3} - \frac{1}{r^3}\right) - \frac{\mathbf{l}}{r_1^3}\right\} \tag{1.6.1}$$

$$r_1 = (r^2 + l^2 - 2\mathbf{r}.\mathbf{l})^{1/2} = \mathbf{r}\left(1 + \frac{l^2 - 2\mathbf{r}.\mathbf{l}}{r^2}\right)^{1/2}$$

$$\frac{1}{r_1} = \frac{1}{r}\left\{1 + \frac{l^2 - 2\mathbf{r}.\mathbf{l}}{r^2}\right\}^{-1/2} \tag{1.6.2}$$

We now wish to consider a dipole field, i.e. when 1 is small compared to r. Then, keeping only terms of first order in $\mathbf{l}$, we get, on expanding (1.6.2) binomially,

$$\frac{1}{r_1} = \frac{1}{r}\left\{1 + \frac{\mathbf{r}.\mathbf{l}}{r^2}\right\}$$

$$\frac{1}{r_1^3} = \frac{1}{r^3}\left\{1 + \frac{\mathbf{r}.\mathbf{l}}{r^2}\right\}^3 \cong \frac{1}{r^3}\left\{1 + \frac{3\mathbf{r}.\mathbf{l}}{r^2}\right\}$$

$$\frac{1}{r_1^3} - \frac{1}{r^3} = \frac{3\mathbf{r}.\mathbf{l}}{r^3}$$

Therefore, the electric field is

$$\mathbf{E} = \frac{1}{4\pi\varepsilon_0}\left\{\frac{3q\mathbf{r}(\mathbf{r}.\mathbf{I})}{r^5} - \frac{q\mathbf{I}}{r^3}\right\} \tag{1.6.3}$$

This gives an electric field proportional to the separation between the charges. Had we included higher order terms in 1 in the binomial expansion we would got an electric field involving squares, cubes, etc. in 1. Now we will consider the limit $q \to \infty$, $1 \to 0$ in such a way that the product ql stays finite. In that case, all higher order terms in 1 in the electric field would tend to zero. This is a *point dipole*, which has no extent in space and is characterized by its dipole moment $\mathbf{p} = q\mathbf{I}$. In terms of $\mathbf{p}$, the electric field is

$$\mathbf{E} = \frac{1}{4\pi\varepsilon_0}\left\{\frac{3(\mathbf{p}.\mathbf{r})\hat{\mathbf{r}}}{r^5} - \frac{\mathbf{p}}{r^3}\right\} \tag{1.6.4}$$

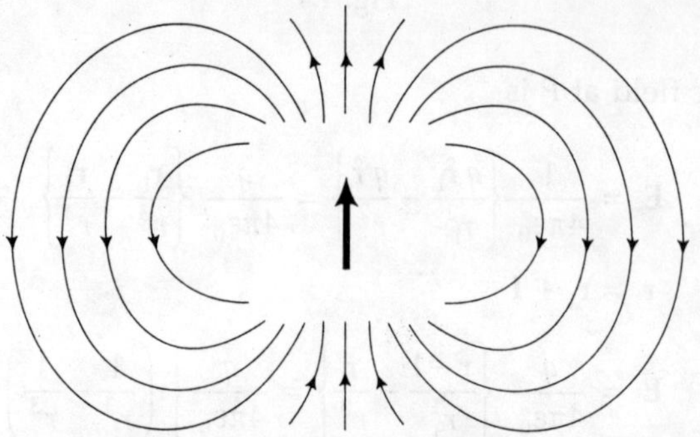

Fig. 1.6

The electric field of a dipole is shown in the Fig. 1.6.

1.7 TORQUE ON A DIPOLE IN AN ELECTRIC FIELD

Let us consider a dipole of moment **p** situated in an electric Field **E**. The dipole makes an angle θ with **E**.

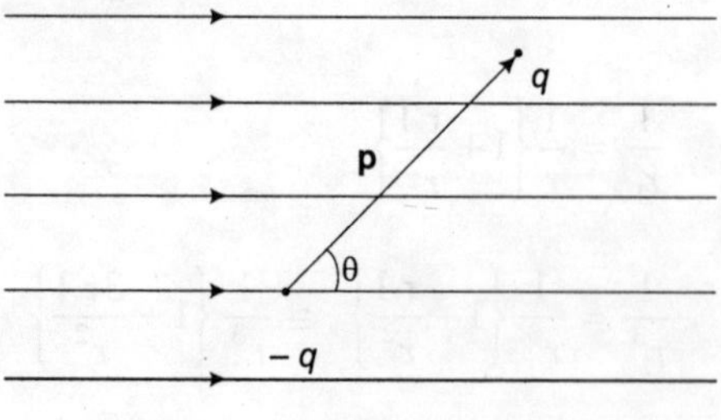

Fig. 1.7(a)

The forces on the two charges are shown below:

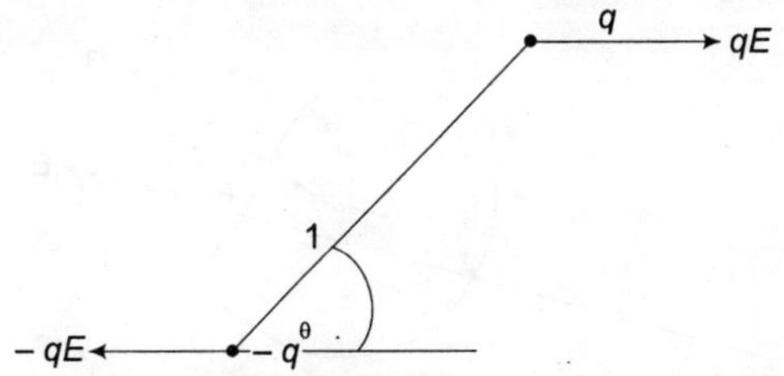

Fig. 1.7(b)

The magnitude of the torque τ is therefore

$$\tau = qEl \sin \theta$$
$$= pE \sin \theta$$

The direction of the torque is perpendicular to the plane of the diagram, pointing downwards. So in vector notation we may write the torque as

$$\boldsymbol{\tau} = \mathbf{p} \times \mathbf{E} \tag{1.7.1}$$

It is clear from this that a dipole parallel to an external field doesn't experience a torque. In general, an electric field exerts a torque on a dipole and tries to align the dipole parallel to itself.

1.8 GAUSS' LAW

Traditionally, electric fields have been represented by lines with arrows on them. The lines are crowded together at places where the field is comparatively larger (as in Fig. 1.6) and the arrow indicates its direction. So if we know the electric field at a point and we want to represent it, we choose a scale and draw a vector whose length is proportional to the magnitude of the field and whose direction is the direction of the field at that point. We can now define the flux ϕ of the electric field through a surface S as the following surface integral:

$$\phi = \int_S \mathbf{E}.d\mathbf{a} \tag{1.8.1}$$

This definition is taken over from Hydrodynamics where the volume of fluid flowing across an area S per second is $\int_S \mathbf{v}.d\mathbf{a}$, $\mathbf{v}$ being the velocity of the fluid at a point.

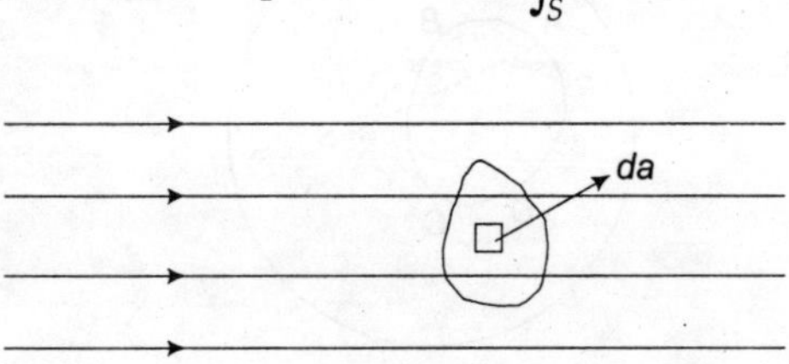

Fig. 1.8

Let us now consider the flux through a surface S due to a point charge q

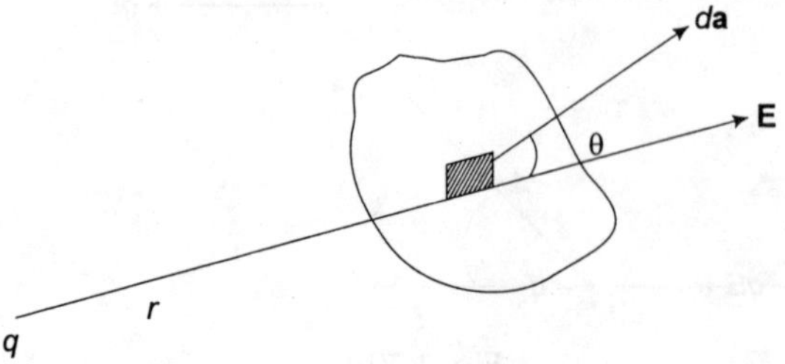

Fig. 1.9

To do this, let us divide the surface S into infinitesimal elements da. Let r be the distance of one such element from q. The electric field due to q at the location of da is

$$\mathbf{E} = \frac{q}{4\pi\varepsilon_0 r^2}\,\hat{\mathbf{r}}$$

The flux of **E** through da is

$$\mathbf{E}.d\mathbf{a} = \frac{q}{4\pi\varepsilon_0 r^2}\,\hat{\mathbf{r}}.d\mathbf{a}$$

$$\mathbf{E}.d\mathbf{a} = \frac{q\,da\cos\theta}{4\pi\varepsilon_0 r^2} \tag{1.8.2}$$

To proceed further, we must define a solid angle. Let me quote from the 'Mathematical Handbook' by M. Vygodshy (English translation, Mir Publishers, 1979)

Solid angle: In geometry, we define a conical surface as one which is generated by the motion of a straight line (called the directrix) which always passes through a fixed point (called the vertex). A conical surface has two sheets, but we usually mean one sheet when we refer to it. A solid angle is a potion of space within one sheet of a conical surface. Just as the

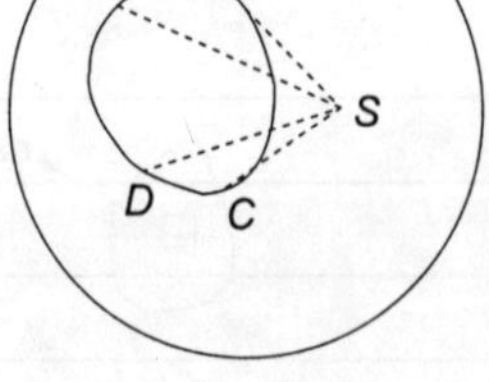

Fig. 1.10

angle between two straight lines is measured by the arc of a circle, a solid angle is measured by a portion of the surface of a sphere. Let us draw a sphere of arbitrary radius with the vertex of a solid angle at its center S.

The surface of a solid angle will cut out of this surface, a portion, say *ABCD* in Fig. 1.10. The area of this portion will vary with the radius of the sphere, but will always *constitute the same fraction of the surface area of the sphere.* So, for the measure of a solid angle, we can take the ratio of the area *ABCD* to the area of the spherical surface, just as the angle between two straight lines may be measured by the ratio of the arc subtended by them (with center at the vertex of the angle) to the perimeter of a circle of the same radius. However, the accepted measure for a solid angle is the ratio of the area *ABCD* to the area of a square constructed on the radius of the sphere which is R^2. Thus, for the measure Ω of a solid angle with vertex S we take the ratio of the area cut out by a solid angle on the surface of a sphere of arbitrary radius with center S to the square of the radius of the sphere:

$$\Omega = \frac{area\ ABCD}{R^2}$$

For the elementary area da which we have been considering, if θ is the angle between the line joining q to da and the normal to da, then the solid angle $d\Omega$ subtended by da at q is the projection of da normal to r, divided by r^2. Thus

$$d\Omega = \frac{da \cos\theta}{r^2} \tag{1.8.3}$$

So the flux of the electric field through da is

$$\mathbf{E}.d\mathbf{a} = \frac{qd\,\Omega}{4\pi\varepsilon_0} \tag{1.8.4}$$

Thus the flux of **E** through the surface S in Fig. 1.9 is

$$\int_S \mathbf{E}.d\mathbf{a} = \frac{q}{4\pi\varepsilon_0}\int d\Omega \tag{1.8.5}$$

where $\int d\Omega$ is the solid angle subtended by S at the location of q. From the definition of Ω it follows that if *ABCD* is the entire surface of the sphere, i.e. $4\pi R^2$, then Ω is given by 4π. So the flux of the above electric field due the any closed surface S' is

$$\int_{S'} \mathbf{E}.d\mathbf{a} = \frac{q}{4\pi\varepsilon_0} \times 4\pi = \frac{q}{\varepsilon_0}$$

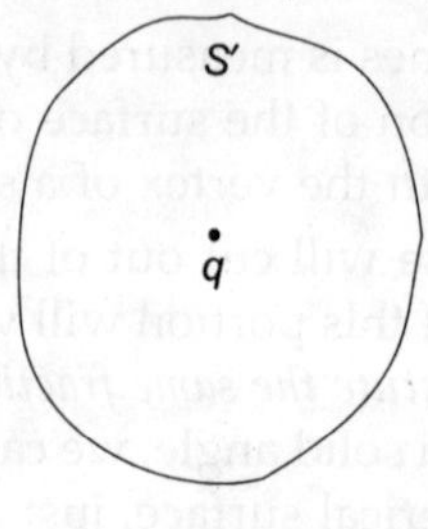

Fig. 1.11

$$\int \mathbf{E}.\mathbf{da} = \frac{q}{\varepsilon_0} \tag{1.8.6}$$

Equation (1.8.6) is the integral form of Gauss' law. If the charge q lies outside S', the total solid angle subtended by S' at q is zero because the normals are in opposite sense w.r.t. the electric field at the nearer end of S'. So

$$\int \mathbf{E}.d\mathbf{a} = 0 \tag{1.8.7}$$

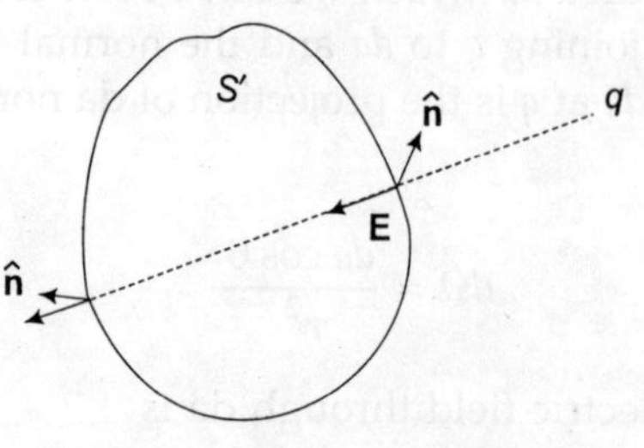

Fig. 1.12

If we have a continuous chage distribution characterized by a charge density ρ, then the total charge Q is given by

$$Q = \int_V \rho dv \tag{1.8.8}$$

For a surface S which encloses Q, we can then write Gauss' law as

$$\int_S \mathbf{E}.d\mathbf{a} = \frac{1}{\varepsilon_0}\int_V \rho dv \tag{1.8.9}$$

The divergence theorem in Vector Calculus states that if $\mathbf{A}(x, y, z)$ is a vector function of co-ordinates with continuous derivatives, the surface integral of $\mathbf{A}$ over a closed surface S equals the volume integral of the divergence of $\mathbf{A}$ over the volume V enclosed by the surface S.

$$\int_S \mathbf{E}.d\mathbf{a} = \int_V \nabla.\mathbf{E}dv \tag{1.8.10}$$

Gauss' law can be rewritten as

$$\int_V \nabla.\mathbf{E}dv = \frac{1}{\varepsilon_0}\int_V \rho dv$$

or
$$\int_V \left(\nabla.\mathbf{E} - \frac{\rho}{\varepsilon_0}\right)dv = 0$$

Since this is true for any volume V, the integrand must be zero, i.e.

$$\nabla.\mathbf{E} = \frac{\rho}{\varepsilon_0} \tag{1.8.11}$$

Eqn. (1.8.11) is the differential form of Gauss' law.

1.9 APPLICATIONS OF GAUSS' LAW

Let us use Gauss' law in the form (1.8.6) to calculate the electric field in three simple cases. Firstly we will consider an infinitely long linear charge distribution with λ as the charge per unit length. From symmetry, the electric field must point away from the line charge. Let us consider a cylinder of length L and radius r around this linear charge with axis along the linear charge. This cylinder is called a Gaussian surface.

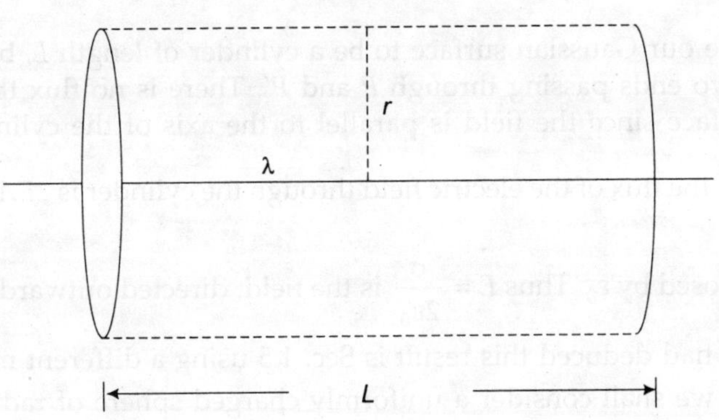

Fig. 1.13

The electric field is radially outward and the vector $d\mathbf{a}$ representing the area of an element da of the curved surface of the cylinder is also radially outward. This problem has cylindrical symmetry. So we expect the field to be a function of r alone. The normal to the base of the cylinder is parallel to its axis. So $\mathbf{E}.d\mathbf{a} = 0$ (since $\mathbf{E}$ and $d\mathbf{a}$ are at right angles) and there is no flux through the base of the Gaussian surface. The flux through the cylinder is then $\int \mathbf{E}.d\mathbf{a} = E.2\pi rL$. The charge enclosed by the cylinder is λL. So from Gauss' law, $E.2\pi rL = \dfrac{\lambda L}{\varepsilon_0}.E = \dfrac{\lambda}{2\pi\varepsilon_0 r}$

We have derived this result earlier in an alternative way is Sec. 1.4

Next we shall consider an infinite sheet of charge characterized by a surface charge density σ. Consider a point P (Fig. 1.14) on one side of the sheet and another point P' on the other side equidistant from the sheet. From the symmetry of the problem, we expect the field to have the same magnitude at P' as at P.

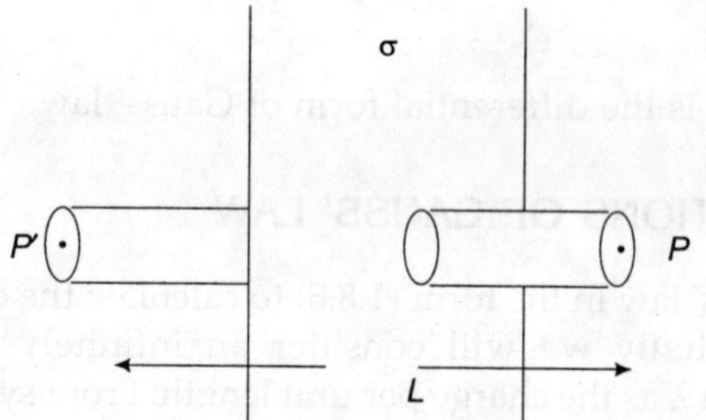

Fig. 1.14

We take our Gaussian surface to be a cylinder of length L, base area A with the two ends passing through P and P'. There is no flux through the curved surface since the field is parallel to the axis of the cylinder. From Gauss' law, the flux of the electric field through the cylinder is $2EA = \dfrac{\sigma A}{\varepsilon_0}$, the charge enclosed by ε_0. Thus $E = \dfrac{\sigma}{2\varepsilon_0}$ is the field, directed outwards both at P and P'. We had deduced this result is Sec. 1.5 using a different method.

Finally we shall consider a uniformly charged sphere of radius R with charge density ρ. The total charge Q on the sphere is given by

$$Q = \frac{4\pi R^3 \rho}{3} \tag{1.9.2}$$

Let us calculate the electric field at a distance r from the center of the sphere using Gauss' law. We take our Gaussian surface to be a sphere of radius $r(< R)$

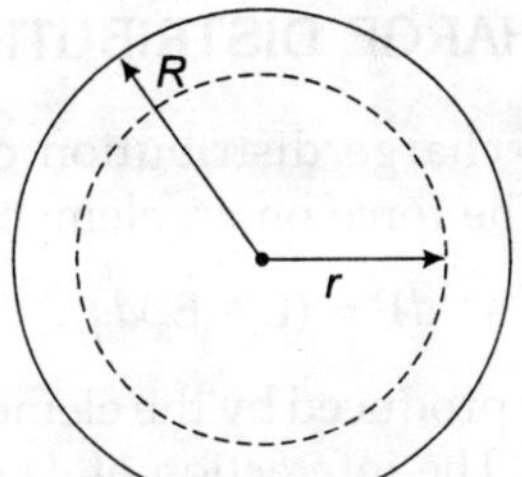

Fig. 1.15

The spherical symmetry of the problem suggests that the electric field is radially outward and a function of r only. The flux of **E** through the dotted sphere is thus $E.4\pi r^2$. Only the charge enclosed by this sphere will contribute to this field E because of (1.8.7). So from Gauss' law, we get,

$$E.4\pi r^2 = \frac{1\rho}{\varepsilon_0}.\frac{4\pi r^3}{3}$$

$$\mathbf{E} = \frac{\rho r}{3\varepsilon_0}\hat{\mathbf{r}} \tag{1.9.3}$$

If we take the Gaussian surface outside our solid sphere so that $r > R$, Gauss' law becomes

$$E.4\pi r^2 = \frac{Q}{\varepsilon_0}$$

because the total charge Q is now enclosed by it. Thus

$$\mathbf{E} = \frac{Q}{4\pi\varepsilon_0 r^2}\hat{\mathbf{r}} \tag{1.9.4}$$

Note that the electric field at points outside the solid sphere has the same magnitude as that due to a point charge Q at the centre of the sphere. Thus for points outside the sphere, we can think of the charge Q as concentrated at its centre. A plot of E vs r is shown below.

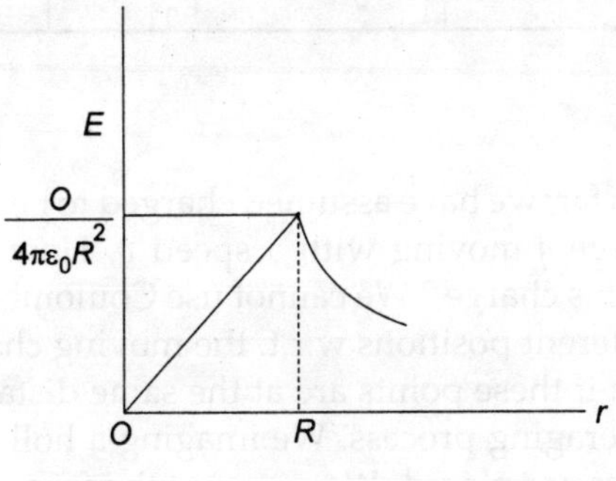

Fig. 1.16

1.10 FORCE ON A CHARGE DISTRIBUTION

Let us consider a surface charge distribution characterized by a surface charge density $\sigma(x, y, z)$. The force on an element of charge dq is given by

$$d\mathbf{F} = (\mathbf{E} - \mathbf{E}_s)dq \tag{1.10.1}$$

where $\mathbf{E}_s$ is the electric field produced by the element dq and $\mathbf{E}$ is the total field (both at the location of dq). The interaction of dq with its own field does not produce any net force on the element and so is excluded in (1.10.1). With $dq = \sigma da$, the force is given by

$$\mathbf{F} = \int (\mathbf{E} - \mathbf{E}_s)\sigma da \tag{1.10.2}$$

The field $\mathbf{E}_s$ is the same as the field due to an infinite sheet of charge because we are referring to the field infinitesimally close to the element of charge σda. Thus at this point the sheet appears to be infinite in extent.

$$\mathbf{E}_s = \frac{\sigma}{2\varepsilon_0}\hat{\mathbf{n}} \tag{1.10.3}$$

where the medium in which the surface charge distribution resides is vacuum, $\hat{\mathbf{n}}$ being along the unit normal to da.

The table below exhibits some important results of this chapter in Gaussian units with the equation nos given for reference.

Physical quantity/law	*MKS*	*Gaussian*
Electric field due to an infinitely long line charge (1.4.5)	$\mathbf{E} = \frac{\lambda}{2\pi\varepsilon_0 y}\hat{\mathbf{j}}$	$\mathbf{E} = \frac{2\lambda}{y}\hat{\mathbf{j}}$
Electric field due to an infinite sheet of charge (1.5.3)	$E = \frac{\sigma}{2\varepsilon_0}\mathbf{k}$	$\mathbf{E} = 2\pi\sigma\hat{\mathbf{k}}$
Gauss' law (1.8.6)	$\int \mathbf{E}.d\mathbf{a} = \frac{q}{\varepsilon_0}$	$\int \mathbf{E}.d\mathbf{a} = 4\pi q$

Tutorial

In our discussions so far, we have assumed charged to be at rest. Suppose now that we have a charge q moving with a speed v. How do we measure the electric field due to this charge ? We cannot use Coulomb' s law because a test charge placed at different positions w.r.t. the moving charge would produce different forces even if these points are at the same distance from the charge q. We can use an averaging process. We imaging a hollow spherical surface on which test charges are placed. We measure the forces on these charges at the instant the moving charges passes through the centre of the sphere. Then we calculate the average force. This is proportional to the surface integral of the electric field and so we can use Gauss' law:

$$q = \varepsilon_0 \int_{S(t)} \mathbf{E}.d\mathbf{a} \tag{1}$$

where $S(t)$ represents the sphere at time t. $S(t)$ can be any surface enclosing the charge q at time t.

The electric field due to a charge q moving with a constant velocity $\mathbf{v}$ is given by

$$\mathbf{E} = \frac{q}{4\pi\varepsilon_0} \frac{\hat{\mathbf{R}}}{R^2} \frac{(1-\beta^2)}{(1-\beta^2 \sin^2 \theta)^{3/2}} \tag{2}$$

where $\beta = v/c$, c being the speed of light in vacuum. $\hat{\mathbf{R}}$ is the unit from the instantaneous location of the charge to the point of observation θ is angle between $\mathbf{v}$ and $\hat{\mathbf{R}}$. Taking a sphere of radius R with charge q at its centre and the direction of velocity to be along the z axis, use spherical polar co-ordinates to calculate the surface integral $\int \mathbf{E}.d\mathbf{a}$ over the surface of the sphere using $\mathbf{E}$ in (2) and show that the value is q/ε_0. The electric field is flattened out in a direction perpendicular ($\theta = 90°$) to the motion and has a magnitude $1/\sqrt{(1-\beta^2)}$ greater than that for charge at rest (Fig. 1.16). In the forward ($\theta = 0$) and backward ($\theta = 180°$) directions, the field is reduced by a factor $(1 - \beta^2)$.

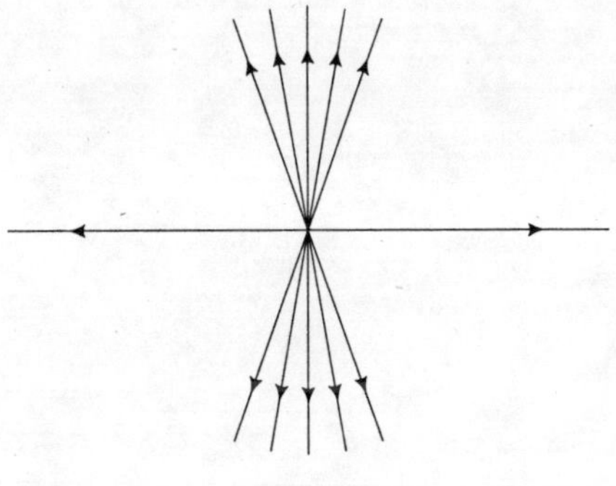

Fig. 1.17

Problems

1. There are three charges: $3e$ located at (2, 0), $2e$ located at (– 2, 0) and $-e$ located at (0, 2) in the xy plane. Calculate the electric field at the origin due to this system of charges. (All distances are in m.)

[**Ans.** $E = 3.6 \times 10^{-10} (-\hat{\mathbf{i}} + \hat{\mathbf{j}})$ N/C]

2. A long cylinder carries a volume charge density which is proportional to the distance from its axis: $\rho = kr$, k being a constant. Find the electric field inside the cylinder at a distance 'd' from the axis.

(University of Calcutta, 2001)

$$\left[\textbf{Ans. E} = \frac{kd^2}{2\varepsilon_0} \text{ radially outward}\right]$$

3. Calculate the electric field at a point on the axis of a ring of radius R, carrying a charge Q, at a distance z from the center of the ring.

$$\left[\textbf{Ans. E} = \frac{Qz}{4\pi\varepsilon_0 (R^2 + z^2)^{3/2}} \text{ away from the ring}\right]$$

4. Calculate the electric field inside and outside a uniformly charged thin spherical shell of radius R, carrying charge Q.

$$\left[\textbf{Ans. } E = 0, \text{ inside; } E = \frac{Q}{4\pi\varepsilon_0 r^2}, \text{ outside}\right]$$

2

The Electric Potential

2.1 DEFINITION

Imagine a region space where there is an electric field. We wish to take a charge q from point A to a point B in this region. Either we have to do work against the field of work is done by the field itself. This work is $-\int_A^B \mathbf{F}.d\mathbf{l}$, where **F** is the force on the charge q. (The $-ve$ sing appears since the work done on the charge is $+ve$ when we move against the field and $-ve$ otherwise.) If **E** is the field at a point then the work done is $-\int_A^B \mathbf{F}.d\mathbf{l}$. So the work done for unit positive charge is $-\int_A^B \mathbf{F}.d\mathbf{l}$. This is defined to be the difference in electric potential V between A and B:

$$V_B - B_A = -\int_A^B \mathbf{F}.d\mathbf{l} \tag{2.1.1}$$

An important property of this work done is that for an electrostatic field, it is independent of the path taken in going from A to B. The value of the integral is the same no matter what path we choose. This allows us to define the electric potential at a point B. Let us take the point A to be at infinity so that, for all practical purposes, the electric field is zero there. We choose the electric potential to be zero at that infinite distance. Then $V_A = 0$ as $A \rightarrow \infty$. So the electric potential V_B at a point B is defined to be

$$V_B = -\int_\infty^B \mathbf{E}.d\mathbf{l} \tag{2.1.2}$$

The MKS unit of potential is the Volt (V). 1 V = 1 J/C. Since the work done is independent of path, it follows from (2.1.1) that when we come back to the same point

$$\oint \mathbf{E}.d\mathbf{l} = 0 \tag{2.1.3}$$

This allows us to define the electric field as the gradient of a potential function. From the definition (2.1.1) this potential function is the electric potential V.

$$\mathbf{E} = -\nabla V \tag{2.1.4}$$

The electrostatic field is an example of a conservative field. From the property of gradient it is clear that the electric field is normal to an equipotential surface, i.e., a surface at all points of which V has the same value. An advantage of using the potential in solving electrostatic problems is that it is a scalar quantity. The potential satisfies the superposition principle. So when we calculate the potential at a point due to a system of charges we have to algebraically add the potentials due to individual source charges.

2.2 EXAMPLES OF CALCULATION OF THE POTENTIAL

Let us calculate the electric potential at a distance r from a point charge q, using (2.1.2). The electric field at a distance r from q is

$$\mathbf{E} = \frac{q}{4\pi\varepsilon_0 r^2}\,\hat{\mathbf{r}}$$

Since the integral in (2.1.2) is independent of path, let us choose a path along a field line, radially inward from infinity. Then (2.1.2) gives

$$V = -\int_{\infty}^{r} \frac{q}{4\pi\varepsilon_0 r^2}\,\hat{\mathbf{r}}.d\hat{\mathbf{r}}$$

$$V = -\frac{q\int_{-\infty}^{r} \frac{dr}{r^2}}{4\pi\varepsilon_0}$$

$$V = \frac{q}{4\pi\varepsilon_0 r} \tag{2.2.1}$$

Clearly then, the electric potential has the same value at all points with the same value of r. So the equipotential surfaces are spheres with centre at the charge q. The field is radially outward everywhere and so is normal to the equipotential surface.

As another example, let us calculate the potential at a point p along the axis of a uniformly charged disc with surface charge density σ and radius R.

The centre of the disc is at the origin O. P is at a distance z from it. All points on the disc on an elementary ring of thickness dr are equidistant from P. The potential at P due to this elementary ring is

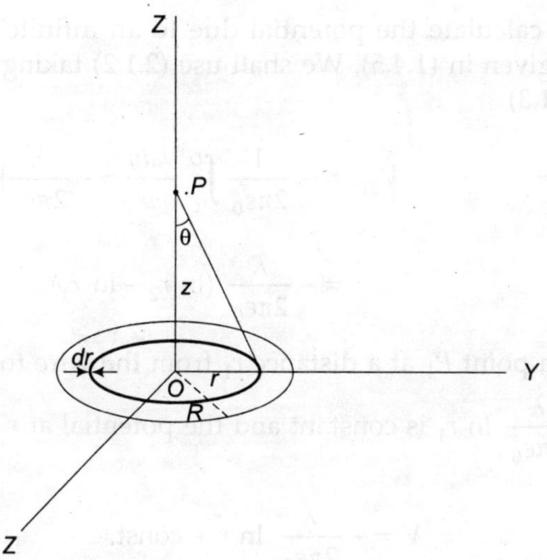

Fig. 2.1

$$dV = \frac{2\pi r dr \sigma}{4\pi\varepsilon_0 \sqrt{(r^2 + z^2)}}$$

$$= \frac{\sigma r dr}{2\varepsilon_0 \sqrt{(r^2 + z^2)}}$$

The potential at P due to the entire disc is

$$V = \int_0^R \frac{r\sigma dr}{2\varepsilon_0 \sqrt{(r^2 + z^2)}}$$

$$= \frac{\sigma}{2\varepsilon_0}\left[\sqrt{(r^2 + z^2)}\right]_0^R$$

$$V = \frac{\sigma}{2\varepsilon_0}\left[\sqrt{(R^2 + z^2)} - z\right] \tag{2.2.2}$$

If we take the derivative of V with respect to z, we get

$$\frac{dV}{dz} = \frac{\sigma}{2\varepsilon_0}\left[\frac{z}{\sqrt{(R^2 + z^2)}} - 1\right]$$

So the electric field at P has magnitude

$$E = \frac{-dV}{dz} = \frac{\sigma}{2\varepsilon_0}\left[1 - \frac{z}{\sqrt{(R^2 + z^2)}}\right]$$

as we had obtained in (1.5.2).

Next let us calculate the potential due to an infinite line charge. The electric field is given in (1.4.5). We shall use (2.1.2) taking a path along the y-axis (see Fig. 1.3)

$$V_{21} = -\frac{1}{2\pi\varepsilon_0}\int_{r_1}^{r_2}\frac{\lambda dy}{y} = -\frac{\lambda}{2\pi\varepsilon_0}[\ln y]_{r_1}^{r_2}$$

$$= -\frac{\lambda}{2\pi\varepsilon_0}(\ln r_2 - \ln r_1)$$

We choose a point P_1 at a distance r_1 from the wire to be our reference point. Then $-\frac{\lambda}{2\pi\varepsilon_0}\ln r_1$ is constant and the potential at r is given by

$$V = -\frac{\lambda}{2\pi\varepsilon_0}\ln r + \text{const.} \tag{2.2.3}$$

As our final example, let us calculate the potential both inside and outside a uniformly charged sphere of radius R. The electric field inside the sphere, at a distance r from the centre is given by (1.9.3) and outside by (1.9.4). Choosing our path to be radially inward, we get, for the potential at a distance r from the centre ($r < R$)

$$V = -\int_{\infty}^{r} E dr$$

$$= \int_{\infty}^{R}\frac{Q\,dr}{4\pi\varepsilon_0 r^2} - \int_{R}^{r}\frac{\rho r\,dr}{3\varepsilon_0}$$

$$= \frac{Q}{4\pi\varepsilon_0}\left[\frac{1}{r}\right]_{\infty}^{R} - \frac{\rho}{3\varepsilon_0}\left[\frac{r^2}{2}\right]_{R}^{r}$$

$$= \frac{Q}{4\pi\varepsilon_0 R} - \frac{\rho(r^2 - R^2)}{6\varepsilon_0}$$

$$= \frac{Q}{4\pi\varepsilon_0 R} - \frac{\rho r^2}{6\varepsilon_0} + \frac{\rho R^2}{6\varepsilon_0}$$

$$= \frac{Q}{4\pi\varepsilon_0 R} - \frac{Qr^2}{8\pi\varepsilon_0 R^3} + \frac{Q}{8\pi\varepsilon_0 R}\left(\text{since } Q = \frac{4\pi R^2\rho}{3}\right)$$

$$= \frac{Q}{8\pi\varepsilon_0 R}\left(3 - \frac{r^2}{R^2}\right) \tag{2.2.4}$$

For a point outside the sphere,

$$V = \frac{Q}{4\pi\varepsilon_0 r} \tag{2.2.5}$$

because of (1.9.4).

2.3 POTENTIAL DUE TO A DIPOLE

Let us consider a dipole, which we defined in Sec. 1.6. Referring to Fig. 1.5, we can calculate the potential due to the dipole at the point P. It is

$$V = \frac{q}{4\pi\varepsilon_0 r_1} - \frac{q}{4\pi\varepsilon_0 r} = \frac{q}{4\pi\varepsilon_0}\left(\frac{1}{r_1} - \frac{1}{r}\right)$$

$$= \frac{q}{4\pi\varepsilon_0}\left\{\frac{1}{r}\left(1 + \frac{\mathbf{l}^2 - 2\mathbf{r.l}}{r^2}\right)^{-1/2} - \frac{1}{r}\right\}$$

Using the binomial expansion and keeping only terms of first order in l, we get

$$V \cong \frac{q}{4\pi\varepsilon_0}\left\{\frac{1}{r}\left(1 + \frac{\mathbf{r.l}}{r^2}\right) - \frac{1}{r}\right\}$$

$$= \frac{q(\mathbf{r.l})}{4\pi\varepsilon_0 r^3} = \frac{\mathbf{p.r}}{4\pi\varepsilon_0 r^3}$$

or

$$V = \frac{\mathbf{p.\hat{r}}}{4\pi\varepsilon_0 r^2} \tag{2.3.1}$$

2.4 ELECTROSTATIC POTENTIAL ENERGY

Let us suppose we have a charge q_1 in a region of free space. We bring a charge q_2 from infinity and place it at a distance r_{12} from q_1. From the definition of potential, the work done in the process is

$$W = q_2 \times (\text{potential } V \text{ at a distance } r_{12} \text{ from } q_2)$$

$$= \frac{q_1 q_2}{4\pi\varepsilon_0 r_{12}}$$

If we now bring another charge q_3 from infinity and place it so that it is at a distance r_{13} from q_1 and r_{23} from q_2, then the work done is

$$\frac{q_1 q_3}{4\pi\varepsilon_0 r_{13}} + \frac{q_2 q_3}{4\pi\varepsilon_0 r_{23}}$$

because the potential at a point distant r_{13} from q_1 is $\frac{q_1}{4\pi\varepsilon_0 r_{13}}$ and that at a distance r_{23} from q_2 is $\frac{q_2}{4\pi\varepsilon_0 r_{23}}$. Thus the total work done in assembling q_1, q_2 and q_3 in this configuration is

$$U = \frac{q_1 q_2}{4\pi\varepsilon_0 r_{12}} + \frac{q_2 q_3}{4\pi\varepsilon_0 r_{23}} + \frac{q_1 q_3}{4\pi\varepsilon_0 r_{13}} \tag{2.4.1}$$

This is the electrostatic potential energy U of the system. This is also the energy required to break up this arrangement, i.e. push the three charges to infinite distances from each other. This energy can be thought of as being stored in the system as a whole. If we have a system of N point charges q_i and the distance between the ith and jth charges is r_{ij}, then the total electrostatic energy can be written as:

$$U = \frac{1}{2}\sum_{i=1}^{N}\sum_{\substack{j=1 \\ (i\neq j)}}^{N} \frac{q_i q_j}{4\pi\varepsilon_0 r_{ij}} \tag{2.4.2}$$

The factor of ½ in front of summation takes into account the fact that each term is counted twice. For example, for $N = 3$,

$$\frac{1}{4\pi\varepsilon_0}\sum_{i=1}^{3}\sum_{j=1}^{3}\frac{q_i q_j}{r_{ij}} = \frac{1}{4\pi\varepsilon_0}\left(\frac{q_1 q_2}{r_{12}} + \frac{q_1 q_3}{r_{13}} + \frac{q_2 q_1}{r_{21}} + \frac{q_2 q_3}{r_{23}} + \frac{q_3 q_1}{r_{31}} + \frac{q_3 q_2}{r_{32}}\right)$$

Since $r_{ij} = r_{ji}$, this is

$$\frac{1}{4\pi\varepsilon_0}\left(\frac{2q_1q_2}{r_{12}} + \frac{2q_1q_3}{r_{13}} + \frac{2q_2q_3}{r_{23}}\right)$$

which is twice the result (2.4.1). The potential V_i at the location of the charge q_i due to the charge q_j at a distance r_{ij} from it is $V_i = \frac{q_i}{4\pi\varepsilon_0 r_{ij}}$. Thus we can rewrite (2.4.2) as

$$U = \frac{1}{2}\sum_{i=1}^{N} q_i V_i \tag{2.4.3}$$

If we have a continuous charge distribution characterized by a charge density ρ then an infinitesimal element of volume dV has charge ρdV. If the electric potential at the location of this element of charge is V then the eqn. (2.4.3) can be replaced by the integral:

$$U = \frac{1}{2}\int V\rho dV \tag{2.4.4}$$

Example 1: Calculate the electrostatic potential energy of a uniformly charged sphere with charge density ρ, total charge Q and radius R.

Solution: Let us consider a spherical volume of radius r (< R) within this sphere. At points outside this volume, the potential due to this sphere is $\frac{q}{4\pi\varepsilon_0 r}$ where $q = \frac{4\pi r^3 \rho}{3}$. So the potential is $\frac{\rho r^2}{3\varepsilon_0}$. Now suppose we add a layer of charge in the form of a spherical shell of thickness dr. The charge it carries is $dq = 4\pi r^2 dr\rho$. ∴ The work done is $dU = \frac{qdq}{4\pi\varepsilon_0 r}$ or, $dU = \frac{\rho r^2}{3\varepsilon_0}.\,4\pi r^2\rho dr = \frac{4\pi r^4\rho^2}{3\varepsilon_0}dr$. In this way, by adding infinitesimally thin spherical shells of charge, we can imagine building up the sphere of radius R. The total work done in this process is the electrostatic potential energy of the sphere. It is obtained by integrating from $r = 0$ to $r = R$.

$$U = \frac{4\pi\rho^2}{3\varepsilon_0}\int_0^R r^4 dr$$

$$= \frac{4\pi\rho^2}{3\varepsilon_0}\frac{R^5}{5} \tag{2.4.5}$$

or, in terms of the total charge Q, we can write

$$U = \frac{3Q^2}{20\pi\varepsilon_0 R} \tag{2.4.6}$$

Let us some back to eqn. (2.4.4). If we use the differential form (1.8.1) of Gauss' law to substitute for ρ, we get

$$U = \frac{\varepsilon_0}{2}\int V(\nabla.\mathbf{E})dv$$

A vector identity involving the divergence reads

$$\nabla.(\alpha\mathbf{A}) = \alpha\nabla.\mathbf{A} + (\nabla\alpha).\mathbf{A}$$

Thus

$$\nabla.(V\mathbf{E}) = V\nabla.\mathbf{E} + (\nabla V).\mathbf{E}$$

$$= V\nabla.\mathbf{E} - E^2$$

We can rewrite the electrostatic potential energy as

$$U = \frac{\varepsilon_0}{2}\left[\int\nabla.(V\mathbf{E})dv + \int E^2 dv\right] \tag{2.4.7}$$

The first integral can be rewritten using the divergence theorem: $\int \nabla.(V\mathbf{E})dv = \int_S \mathbf{E}.d\mathbf{a}$.

$$U = \frac{\varepsilon_0}{2}\left[\int_S V\mathbf{E}.d\mathbf{a} + \int E^2 dv\right] \qquad (2.4.8)$$

where S is the surface enclosing the volume. Let us take a large enough volume which is greater than the volume occupied by the charge distribution. The electric field $\frac{1}{r^2}$ tend to zero at the location of the surface S over which the surface integral is taken. So if we integrate over all space, the first term in (2.4.8) goes to zero and we have

$$U = \frac{\varepsilon_0}{2}\int_{\text{all space}} E^2\, dv \qquad (2.4.9)$$

Note that enlarging the volume beyond the region occupied by the charge distribution has no effect on (2.4.4) because $\rho = 0$ beyond the boundary of the charge distribution. We can think of $\frac{\varepsilon_0}{2}E^2$ as the electrostatic potential energy stored per unit volume.

Let us use eqn. (2.4.9) to calculate the electrostatic potential energy of a sphere of radius R carrying a charge Q with a constant charge density ρ (encountered in Example 1). The electric fields inside and outside the sphere are given in eqns. (1.9.3) and (1.9.4). We have to integrate over all space according to (2.4.9)

$$\int_0^\infty E^2\, dv = \int_0^R E^2\, dv + \int_0^\infty E^2\, dv \qquad (2.4.10)$$

$$\int_0^R E^2\, dv = \int_0^R \frac{\rho^2 r^2}{9\varepsilon_0^2}.r^2.4\pi dr$$

$$= \frac{4\pi\rho^2 R^2}{45\varepsilon_0^2}$$

In terms of the total charge Q, we have

$$\rho = \frac{3Q}{4\pi R^3}$$

$$\therefore \qquad \int_0^R E^2\, dv = \frac{Q^2}{20\pi\varepsilon_0^2 R} \qquad (2.4.11)$$

$$\int_R^\infty E^2\,dv = \int_R^\infty \frac{Q^2}{16\pi^2\varepsilon_0^2 r^4}.4\pi r^2 dr$$

$$= \frac{Q^2}{4\pi\varepsilon_0^2 R} \tag{2.4.12}$$

$$\int_0^\infty E^2\,dv = \frac{3Q^2}{10\pi\varepsilon_0^2 R}$$

$$\therefore \quad U = \frac{3Q^2}{20\pi\varepsilon_0 R} \tag{2.4.13}$$

which agrees with our earlier calculation (2.4.6).

Let us now consider a dipole in an electric field and calculate its potential energy. In Chapter 1, we learnt that when a dipole is oriented parallel to an electric field, it does not experience any torque. Now suppose we wish to orient the dipole at an angle θ_0 relative to an electric field **E**. We have to do work against the field because the field exerts a torque on the dipole, trying to align it parallel to itself. This work goes to increase the potential energy U of the dipole. Let us consider a dipole of moment p initially oriented at an angle θ relative to an electric field **E**. To increase the angle of orientation by an amount $d\theta$, the work required to be done, and thus the increase in potential energy dU of the dipole is

$$dU = \tau d\theta$$

$$= pE \sin\theta d\theta$$

The potential energy U of a dipole in an electric field is thus obtained by integrating the above eqn.

$$U = -pE\cos\theta + C \tag{2.4.14}$$

where C is a constant of integration. It is usually set equal to zero. Then

$$U = -pE\cos\theta$$

$$U = -\mathbf{p}.\mathbf{E} \tag{2.4.15}$$

If we place a point dipole in the electric field due to another point dipole, then we can write down their interaction energy as

$$U = \frac{1}{4\pi\varepsilon_0 r^3}\{\mathbf{p}_1.\mathbf{p}_2 - 3(p_1.\hat{\mathbf{r}})(p_2.\hat{\mathbf{r}})\} \tag{2.4.16}$$

where r is the vector from $\mathbf{p}_2$ to $\mathbf{p}_1$.

2.5 POISSON'S EQUATION AND LAPLACE'S EQUATION

In Section 2.1, we found that the electrostatic field **E** is a conservative field and we can write

$$\mathbf{E} = -\nabla V \tag{2.1.4}$$

so that

$$\nabla \times \mathbf{E} = 0 \tag{2.5.1}$$

If we consider a region of space having a charge density ρ, then the differential form of Gauss' law states that

$$\nabla . \mathbf{E} = \frac{\rho}{\varepsilon_0} \tag{1.8.11}$$

When we combine this with (2.1.4), we get

$$\nabla^2 V = \frac{-\rho}{\varepsilon_0} \tag{2.5.2}$$

This is Poisson's equation. It is partial differential equation. The operator ∇^2 is called the Laplacian. It is obtained by taking the divergence of the gradient of a scalar function of co-ordinates like V. If we are in a region of space permeated by an electric field, but devoid of any charge distribution, then $\rho = 0$ and we get

$$\nabla^2 V = 0 \tag{2.5.3}$$

This is Laplace's equation. The solution of this equation must satisfy either Dirichlet or Neumann boundary conditions. The Dirichlet boundary condition specifies the values of V on boundaries of a region, for example on the surfaces of conductors. The Neumann boundary condition specifies the value of the normal component of the electric field on the boundries of the region. An important property of a solution of Laplace's equation is that it satisfies a uniqueness theorem: if the solution satisfies the equation with given boundary conditions, it is a unique solution of Laplace's equation.

Proof: Let V_1 and V_2 be two solutions of Laplace's equation and consider the function $V = V_1 - V_2$. If they satisfy Dirichlet boundary conditions, then $V = V_1 - V_2 = 0$ on the boundaries. If they satisfy Neumann boundary conditions, then $\hat{\mathbf{n}}.\nabla_1 V_1 = \hat{\mathbf{n}}.\nabla V_2$ on a surface S of the boundary of the region ($\hat{\mathbf{n}}$ is the unit normal vector at a point on S).

$\hat{\mathbf{n}}.\nabla V = 0$ on S. Also $\nabla^2 V = 0$ throughout the region.

Consider $\quad \nabla.(V\nabla V) = (\nabla V)^2 + V\nabla^2 V = (\nabla V)^2$

$$\int \nabla.(V\nabla V)d\tau = \int_S (V\nabla V).d\mathbf{a} \tag{2.5.4}$$

$$\int (\nabla V)^2 d\tau = \int_S (V\nabla V).d\mathbf{a}$$

The surface integral is zero because V satisfies either Dirichlet or Neumann boundary conditions.

$$\int (\nabla V)^2 d\tau = 0 \tag{2.5.5}$$

$(\nabla V)^2$ must be either positive or zero everywhere in the volume. Since its integral is zero, $(\nabla V)^2 = 0$. So V = constant inside the region. For Dirichlet boundary conditions, $V = 0$ on the boundaries, so $V_1 = V_2$ inside the region also. For Neumann boundary conditions, $V_1 - V_2$ = const.

Another important property of the solution of Laplace's equation is that it does not have any local maxima or minima. The extreme values of V occur at the boundaries.

2.6 SOLUTIONS OF LAPLACE'S EQUATION

Let us look at the general solution of Laplace's equation in Cartesian co-ordinates. In three dimensions, the equation reads

$$\frac{\partial^2 V}{\partial x^2} + \frac{\partial^2 V}{\partial y^2} + \frac{\partial^2 V}{\partial z^2} = 0 \tag{2.6.1}$$

We shall solve this equation by the method of separation of variables. Thus we write

$$V(x, y, z) = V_1(x)\, V_2(y)\, V_3(z) \tag{2.6.2}$$

Substituting this in Laplace's equation, we get,

$$V_2 V_3 \frac{d^2 V_1}{dx^2} + V_1 V_3 \frac{d^2 V_2}{dy^2} + V_1 V_2 \frac{d^2 V_3}{dz^2} = 0$$

or, dividing throughout by $V_1 V_2 V_3$ we get

$$\frac{1}{V_1}\frac{d^2 V_1}{dx^2} + \frac{1}{V_2}\frac{d^2 V_2}{dy^2} + \frac{1}{V_3}\frac{d^2 V_3}{dz^2} = 0$$

or,

$$\frac{1}{V_1}\frac{d^2 V_1}{dx^2} + \frac{1}{V_2}\frac{d^2 V_2}{dy^2} = -\frac{1}{V_3}\frac{d^2 V_3}{dz^2}$$

The left hand side of this equation is a function of x and y only while the right hand side is a function of z only. The equation is satisfied provided both are equal to a constant, say k_3, then

$$\frac{1}{V_1}\frac{d^2 V_1}{dx^2} + \frac{1}{V_2}\frac{d^2 V_2}{dy^2} - k_3 \tag{2.6.3a}$$

$$-\frac{1}{V_3}\frac{d^2 V_3}{dz^2} = k_3 \tag{2.6.3b}$$

The equation satisfied by V_3 is thus

$$\frac{d^2V_3}{dz^2} + k_3V_3 = 0$$

whose general solution is

$$V_3(z) = C_1 \cos\{(k_3)^{1/2}z\} + C_2 \sin\{(k_3)^{1/2}z\} \tag{2.6.4}$$

Where C_1 and C_2 are constants.

From eqn. (2.6.3a) we get

$$\frac{1}{V_2}\frac{d^2V_2}{dy^2} = k_3 - \frac{1}{V_1}\frac{d^2V_1}{dx^2}$$

The left hand side of this equation is a function of y only whereas the right hand side is a function of x only. Therefore, this equation can be satisfied if both sides are equal to a constant, say $-k_2$.

$$\frac{1}{V_2}\frac{d^2V_2}{dy^2} = k_3 - \frac{1}{V_1}\frac{d^2V_1}{dx^2} = -k_2$$

So we get two equations:

$$\frac{d^2V_2}{dy^2} + k_2V_2 = 0 \tag{2.6.5a}$$

$$\frac{d^2V_1}{dx^2} - (k_2 + k_3)V_1 = 0 \tag{2.6.5b}$$

The general solutions of these equations are:

$$V_2(y) = B_1 \cos\{(k_2)^{1/2}y\} + B_2 \sin\{(k_2)^{1/2}y\} \tag{2.6.6}$$

$$V_1(x) = A_1 \exp\{(k_2 + k_3)^{1/2}x\} + A_2 \exp\{-(k_2 + k_3)^{1/2}x\} \tag{2.6.7}$$

The boundary conditions appropriate to a given problem are used to find out the constants A_1, A_2, B_1, B_2, C_1, C_2.

We shall consider the solutions of Laplace's equation in cylindrical and spherical polar co-ordinates in the tutorials.

The important results of this chapter are summarized in Gaussian units in the following table:

Physical quantity/equation	*MKS*	*Gaussian*
Potential due to a point charge (2.2.1)	$V = \frac{q}{4\pi\varepsilon_0 r}$	$V = \frac{q}{r}$
Potential due to a charged disc (2.2.2)	$V = \frac{\sigma}{2\varepsilon_0} \times [(R^2 + z^2)^{1/2} - z]$	$V = 2\pi\sigma \times [(R^2 + z^2)^{1/2} - z]$

Potential due to a line charge (2.2.3)	$V = -\dfrac{\lambda}{2\pi\varepsilon_0}\ln r + \text{const.}$	$V - 2\lambda \ln r + \text{const.}$
Potential due to a dipole (2.3.1)	$V = \dfrac{\mathbf{p.r}}{4\pi\varepsilon_0 r^2}$	$V = \dfrac{\mathbf{p.r}}{r^2}$
Elecrostatic potential energy (2.4.9)	$U = \dfrac{\varepsilon_0}{2}\int E^2 dv$ all space	$U = \dfrac{1}{8\pi}\int E^2 dv$ all space
Poisson's equation (2.5.2)	$\nabla^2 V = \dfrac{-\rho}{\varepsilon_0}$	$\nabla^2 V = -4\pi\rho$

Tutorial 1

Let us consider an arbitrary charge distribution characterized by a volume charge density $\rho(\mathbf{r}')$. [See figure below] The potential at P due to an element dv' of this distribution is given by

$$dV = \frac{\rho(\mathbf{r}')dv'}{4\pi\varepsilon_0 r_1}$$

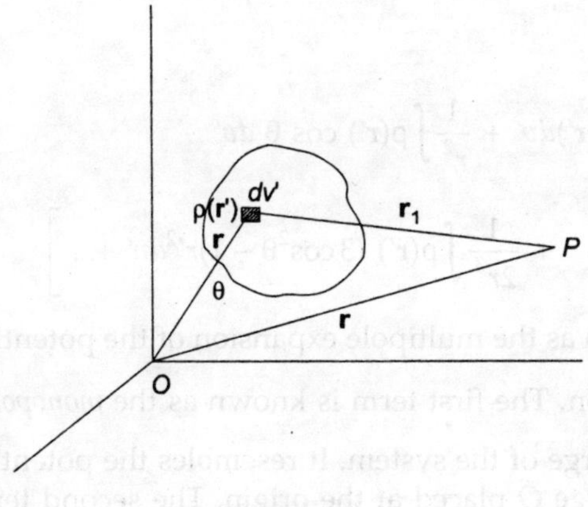

Fig. 2.2

$\therefore$ The potential at P due to the entire charge distribution occupying a volume V is

$$V = \frac{1}{4\pi\varepsilon_0}\int \frac{\rho(\mathbf{r}')dv'}{r_1} \qquad (1.1)$$

r_1 will vary with the location of the element dv'. From the diagram,

$$r_1 = \sqrt{(r^2 + r'^2 - 2rr' \cos\theta)}$$

$$= r\left(1 + \frac{r'^2}{r^2} - \frac{2r'\cos\theta}{r}\right)^{1/2}$$

Since r is a constant for a given P, we can take it outside the integral to rewrite (1.1):

$$V = \frac{1}{4\pi\varepsilon_0} \int \frac{\rho(\mathbf{r}')dv'}{r\left(1 + \frac{r'^2}{r^2} - \frac{2r'\cos\theta}{r}\right)^{1/2}} \tag{1.2}$$

We take the distance $r >> r'$. The binomial expansion $(1 + x)^{1/2} = 1 - \frac{1x}{2} + \frac{3x^2}{8} - \ldots$

Use this to write down the binomial expansion $\left(1 + \frac{r^2}{r^2} - \frac{2r'\cos\theta}{r}\right)^{-1/2}$ keeping only terms upto second order in $\frac{r'}{r}$. Show that the potential V is then given by

$$V = \frac{1}{4\pi\varepsilon_0}\left[\frac{1}{r}\int \rho(\mathbf{r}')dv' + \frac{1}{r^2}\int \rho(\mathbf{r}')\cos\theta\, dv' + \frac{1}{2r^3}\int \rho(\mathbf{r}')\,(3\cos^2\theta - 1)r'^2 dv' + \ldots\right] \tag{1.3}$$

This is known as the multipole expansion of the potential of an arbitrary charge distribution. The first term is known as the *monopole* term. $\int \rho(\mathbf{r}')dv' = Q$, the total charge of the system. It resembles the potential at a distance r from a point charge Q placed at the origin. The second term resembles the potential due to a dipole (2.3.1). This is because, from Fig. 2.2, $r'\cos\theta = \mathbf{r}'.\hat{\mathbf{r}}$ where $\hat{\mathbf{r}}$ is the unit vector along OP. Thus $\frac{1}{r^2}\int \rho r'\cos\theta\, dv' = \frac{1}{r^2}\int \rho\mathbf{r}'.\hat{\mathbf{r}}dv'$ $= \frac{\hat{\mathbf{r}}}{r^2}.\int \rho\mathbf{r}'dv'$. So if we call $\int \rho\mathbf{r}'dv' = \mathbf{p}$, the dipole moment of the charge distribution, the second term agrees with (2.3.1). It is called the *dipole* term.

The third term in (1.3) is called the *quadrupole* term. The quadrupole moment of the charge distribution is defined by

$$\int \rho(\mathbf{r}')(3\cos^2\theta - 1)r'^2 dv'$$

For a spherically symmetric charge distribution, $\rho(\mathbf{r}') = \rho(r')$, show that

$$\int (3\cos^2\theta - 1)d\Omega = 0,$$

where $d\Omega = \sin\theta d\theta d\varphi$ taking the z axis of the co-ordinate system along OP. θ and φ are the azimuthal and polar angles in spherical polar co-ordinates. A charge distribution which has a positive quadrupole moment has a prolate shape (Fig. 2.3a) and one which has a negative quadrupole moment, an oblate shape (Fig. 2.3b) with respect to a reference axis.

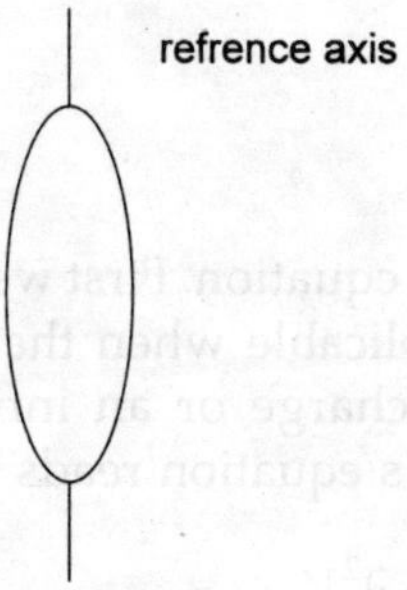

Fig. 2.3a

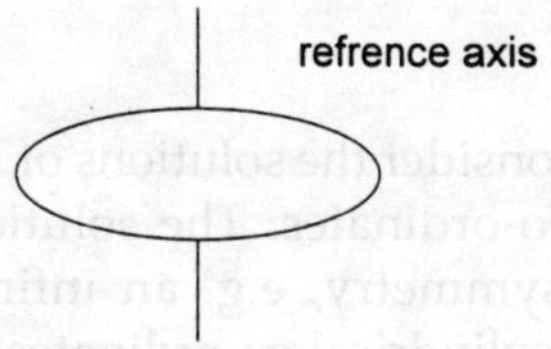

Fig. 2.3b

Tutorial 2

We have seen in Sec. 2.2 that for a point charge, the equipotential surfaces are spheres with centre at the charge. Each sphere has a definite value of V. Plot the equipotential surfaces for a charged disc and a line charge. Draw also the field lines.

Tutorial 3

Let us consider the cubic crystal shown below:

Note that there are 12 pairs of charges ($-e$) separated by a distance b, 12 pairs ($-e$) separated by a distance $b\sqrt{2}$, 8 pairs ($-e$ and $+2e$) separated by $b\sqrt{3}/2$ and 4 pairs ($-e$) by $b\sqrt{2}$. Show that the electrostatic potential energy of this system is

$$U = \frac{0.34e^2}{\varepsilon_0 b}$$

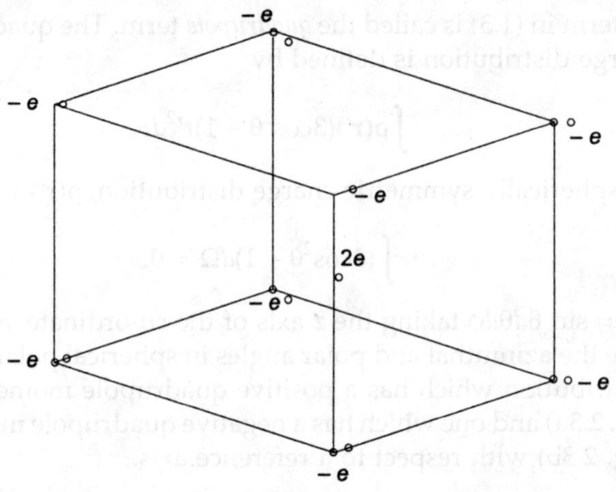

Fig. 2.4

Tutorial 4

Let us now consider the solutions of Laplace's equation. First we shall take up cylindrical co-ordinates. The solution is applicable when the problem has cylindrical symmetry, e.g. an infinite line charge or an infinite charged cylinder. In cylindrical co-ordinates, Laplace's equation reads

$$\frac{1}{r}\frac{\partial}{\partial r}\left(\frac{r\,\partial V}{\partial r}\right)+\frac{1}{r^2}\frac{\partial^2 V}{\partial \theta^2}+\frac{\partial^2 V}{\partial z^2}=0 \tag{4.1}$$

Let us simplify the equation by assuming that V is independent of z. Then (4.1) simplifies to

$$\frac{\partial}{\partial r}\left(\frac{r\,\partial V}{\partial r}\right)+\frac{1}{r}\frac{\partial^2 V}{\partial \theta^2}=0 \tag{4.2}$$

To proceed, we use separation of variables:

$$V(r, \theta) = R(r)f(\theta) \tag{4.3}$$

$$\frac{fd}{dr}\left(\frac{r\,dR}{dr}\right)+\frac{Rd^2 f}{rd\theta^2}=0$$

Dividing throughout by Rf, we get,

$$\frac{1}{R}\frac{d}{dr}\left(\frac{r\,dR}{dr}\right)+\frac{1}{rf}\frac{d^2 f}{d\theta^2}=0$$

or,

$$\frac{rd}{Rdr}\left(\frac{r\,dR}{dr}\right)=-\frac{1}{f}\frac{d^2 f}{d\theta^2}=k,\ \text{say} \tag{4.4}$$

Thus the two equations involving R and f are

$$\frac{d^2 f}{d\theta^2} + kf = 0 \tag{4.5a}$$

$$\frac{rd}{dr}\left(\frac{rdR}{dr}\right) = kR \tag{4.5b}$$

The solutions of (4.5a) are cos $(k^{1/2}\theta)$ and sin $(k^{1/2}\theta)$. For single-valuedness, we must have $k = n^2$ where n is an integer. Then eqn. (4.5b) becomes

$$\frac{rd}{dr}\left(\frac{rdR}{dr}\right) = n^2 R \tag{4.6}$$

Show that $R = r^n$ and $R = r^{-n}$ are solutions of (4.6). So the general solution of (4.2) in this case is

$$V = \sum_n (A_n \cos n\theta + B_n \sin n\theta)(C_n r^n + D_n r^{-n})$$

Consider now an infinite line charge. From cylindrical symmetry, V should be independent of θ and z. Find the solution of Laplace's equation in this case.

Next let us consider the solution of Laplace's equation in spherical polar co-orinates. The equition reads

$$\frac{1}{r^2}\frac{\partial}{\partial r}\left(r^2\frac{\partial V}{\partial r}\right) + \frac{1}{r^2 \sin\theta}\frac{\partial}{\partial \theta}\left(\frac{\sin\theta\, \partial V}{\partial \theta}\right) + \frac{1}{r^2 \sin^2\theta}\frac{\partial^2 V}{\partial \phi^2} = 0 \tag{4.7}$$

where θ and ϕ are polar and azimuthal angles respectively. Again we will restrict ourselves to those solutions which are independent of ϕ. Laplace's equation then reads

$$\frac{\partial}{\partial r}\left(r^2\frac{\partial V}{\partial r}\right) + \frac{1}{\sin\theta}\frac{\partial}{\partial \theta}\left(\frac{\sin\theta\, \partial V}{\partial \theta}\right) = 0 \tag{4.8}$$

To solve this, we will use separation of variables

$$V = R(r)\, f(\theta) \tag{4.9}$$

Show that eqn. (4.8) can then be written as

$$\frac{1}{R}\frac{d}{dr}\left(r^2\frac{dR}{dr}\right) = \frac{-1}{f \sin\theta}\frac{d}{d\theta}\left(\frac{\sin\theta\, df}{d\theta}\right)$$

Since the left hand side is a function of r only and the right hand side is a function of θ only, both must be equal to a constant say k. The equations then read

$$\frac{1}{\sin\theta}\frac{d}{d\theta}\left(\frac{\sin\theta\, df}{d\theta}\right) + kf = 0 \tag{4.10a}$$

and

$$\frac{d}{dr}\left(\frac{r^2\, dR}{dr}\right) = kR \tag{4.10b}$$

The only solutions of (4.10a) which are physically acceptable are those for which $k = n(n + 1)$ where n is an integer. The equation

$$\frac{1}{\sin\theta}\frac{d}{d\theta}\left(\frac{\sin\theta\, df}{d\theta}\right) + n(n+1)\,f = 0 \tag{4.11}$$

is known as *Legendre's* equation and the solutions are known as Legendre polynomials $P_n(\cos\theta)$. For example,

$$P_0(\cos\theta) = 1,\ P_1(\cos\theta) = \cos\theta,\ P_2(\cos\theta) = \frac{1}{2}(3\cos^2\theta - 1) \text{ etc.}$$

The radial equation is

$$\frac{d}{dr}\left(\frac{r^2\, dR}{dr}\right) = n(n+1)R \tag{4.12}$$

Show that $R = r^n$ and $R = r^{-(n+1)}$ are solutions of (4.12). Thus the general solution of (4.8) is

$$V = \sum_n \{A_n r^n + B_n\, r^{-(n+1)}\}\, P_n(\cos\theta) \tag{4.13}$$

Consider the case where V is a function of r alone. Find the solution of Laplace's equation in this case.

Tutorial 5

In this tutorial, we will solve Poisson's equation for a special case when ρ = const. We will look for the solution V inside a uniformly charged sphere of radius R and carrying charge Q. The charge density ρ has a constant value

$$\rho_0 = \frac{3Q}{4\pi R^3} \tag{5.1}$$

From the symmetry of the sphere, we expect V to be a function of r only so that Poisson's equation reads in this case

$$\frac{1}{r^2}\frac{d}{dr}\left(\frac{r^2\, dV}{dr}\right) = -\frac{\rho_0}{\varepsilon_0} \tag{5.2}$$

Solve this equation by integration. To find the two constants of integration, first use the expression for the electric field at a fadius $r(< R)$ [eqn. (1.9.3)] and then use the result that at the surface ($r = R$), the potential must be $V = \dfrac{Q}{4\pi\varepsilon_0 R}$. (Why?) Check your answer with (2.2.4)

Problems

1. Check which of the following function may be possible electrostatic fields:

   (i) $\mathbf{E} = (4y\hat{\mathbf{i}} - 2x\hat{\mathbf{j}} - \hat{\mathbf{k}})A$

   (ii) $\mathbf{E} = (2x\hat{\mathbf{i}} - yz^2\hat{\mathbf{j}} - \hat{\mathbf{k}} - y^2 z\hat{\mathbf{k}})A$

   where A is a constant with suitable dimensions. For the field which you find permissible, find how the charge density changes with position. (Use Poisson's equation.) *(University of Calcutta, 1998)*

   [**Ans.** (ii) is a possible electrostatic field, $\rho(x, y, z) = (y^2 + z^2 - 2)\, A\varepsilon_0$]

2. Find the transverse component of the field due to an electric dipole at a point (r, θ) in space. *(University of Calcutta, 1998)*

   $$\left[\textbf{Ans. } E_\theta = \frac{p \sin\theta}{4\pi\varepsilon_0 r^3}\right]$$

3. What is the volume density of charge in a region of space where the electrostatic potential is given by $V = a - b(x^2 + y^2) - c\log(x^2 + y^2)$, where a, b, c are constants *(University of Calcutta, 2000)*

   [**Ans.** $-4b$]

4. The potential in a medium is given by $V(r) = \dfrac{q\, e^{-r/\lambda}}{4\pi\varepsilon_0\, r}$. Obtain the corresponding electric field and charge density due to $V(r)$. *(University of Calcutta, 2002)*

   $$\left[\textbf{Ans. } \mathbf{E} = \frac{q}{4\pi\varepsilon_0}\left(\frac{1}{\lambda r} + \frac{1}{r^2}\right) e^{-r/\lambda}\hat{\mathbf{r}},\ \rho(r) = \frac{q e^{-r/\lambda}}{4\pi\lambda^2 r}\right]$$

5. Consider two co-axial cylinders with radii 'a' and 'b' ($b > a$). The innerr cylinder is maintained at a potential V_0 and the outer one is grounded. Find the potential and the electric field in the region between the cylinders, at a distance r from the axis.

   $$\left[\textbf{Ans. } V = \frac{\dfrac{-V_0}{b}\ln\dfrac{(r)}{b}}{\ln\dfrac{}{a}},\ E = \frac{-V_0}{r\ln\dfrac{(b)}{a}}\right]$$

6. Calculate the work required to take a proton from (2, –1) to (– 1, 1) in the X-Y plane along the straight path between the two points in a uniform electric field of 2 N/C directed along the positive x-axis.

[Ans. 9.6×10^{-19} J]

7. Show that the dipole moment of a charge distribution is independent of the choice of origin of the co-ordinate system if the total charge of the system is zero.

8. For what relative orientation of two point dipoles $\mathbf{p}_1$ and $\mathbf{p}_2$ is their mutual potential energy zero?

[Ans. $\mathbf{p}_1$ **and** $\mathbf{p}_2$ **at right angles to each other,** $\mathbf{p}_2$ **(say) lying on the radius vector joining the two.]**

9. Consider two concentric spheres of radii '*a*' and '*b*' ($b > a$). The inner sphere has a potential V_0 and the outer one is grounded. Calculate the potential and the electric field in the region between the spheres at a distance *r* from the center of the inner sphere.

$$\left[\textbf{Ans. } V = \frac{abV_0}{(b-a)}\left(\frac{1}{r} - \frac{1}{b}\right), E = \frac{abV_0}{(b-a)r^2}\right]$$

3

Conductors and the Method of Images

3.1 THE CHARGE ON A CONDUCTOR RESIDES ON ITS SURFACE

Let us consider a hollow spherical charged conductor resting on an insulating support (Fig. 3.1). We assume that it has a small hole at the top. Using a proof plane and a gold-leaf elecroscope, we find that the charge resides on the outer

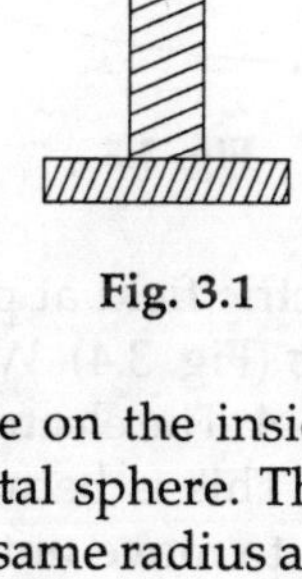

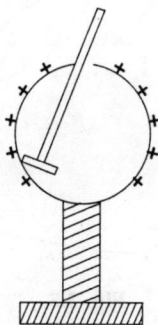

Fig. 3.1

surface and there is no charge on the inside. Another experiment, due to Coulomb, involves a solid metal sphere. The sphere is charged. Two metal hemispheres, which are of the same radius are fitted on it (Fig. 3.2a). They are then removed (Fig. 3.2b). On testing the metal sphere for charge, it is found to be uncharged. All the charge has passed on to the metal hemispheres. Both these experiments demonstrate that the charge on a conductor lies on its surface. We can explain this observation from Gauss' law. In conductors, electrons are free to move under the influence of an applied electric field, unlike an insulator, where they are not free to move. Thus if an electric field

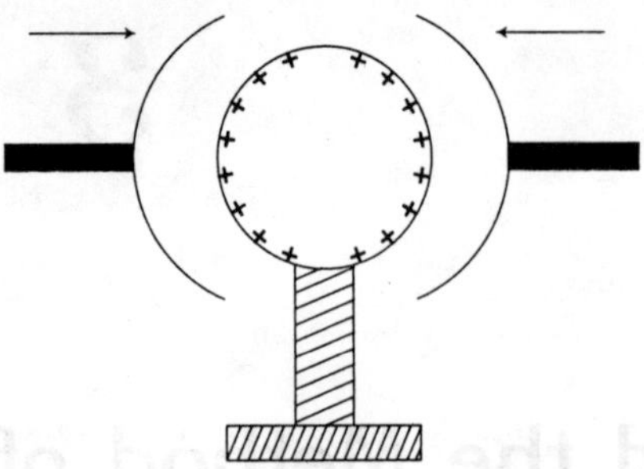

Fig. 3.2a

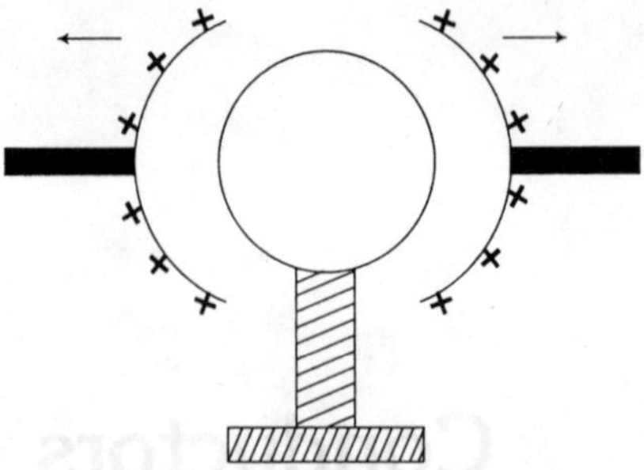

Fig. 3.2b

exists inside a conductor, the electrons will move until that field becomes zero. $E = 0$ inside the conductor. Thus the potential must be the same at all points inside the conductor. Any difference of potential would result in a movement of electrons from points of lower to higher until the potential is the same all throughout. Consider now a Gaussian surface (shown by dotted lines in Fig. 3.3) just inside the conductor, carrying a charge Q. Since $E = 0$ inside, $\int \mathbf{E}.d\mathbf{a} = 0$ and so, from Gauss' law, $Q = 0$ inside the conductor. Thus all the charge must be on the surface.

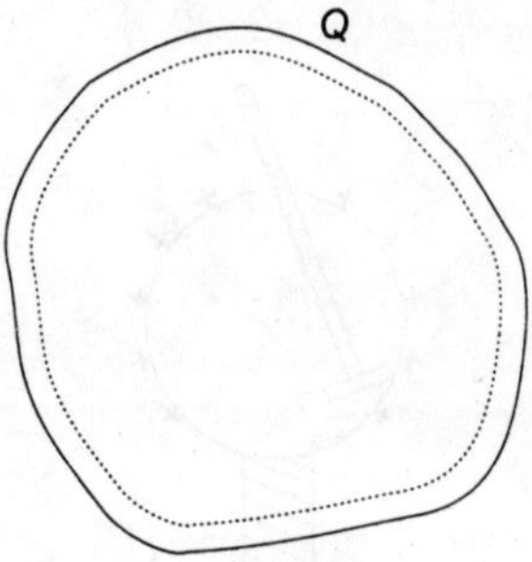

Fig. 3.3

Let us now calculate the electric field at points just outside a conductor having a surface charge density σ (Fig. 3.4). We choose our Gaussian surface to be a small cylinder of base area A. The electric field E is everywhere normal to the surface of the conductor. This is because any component along the surface would cause the charges to move until the surface of the conductor becomes an equipotential surface. So the flux of E through the top face of the cylinder (external) to the conductor is EA. The flux of E through the base of the cylinder is zero because $E = 0$ there. There is no flux through the sides of the cylinder. So Gauss' law gives $EA = \frac{\sigma A}{\varepsilon_0}$ because the charge enclosed by the cylinder is σA. So

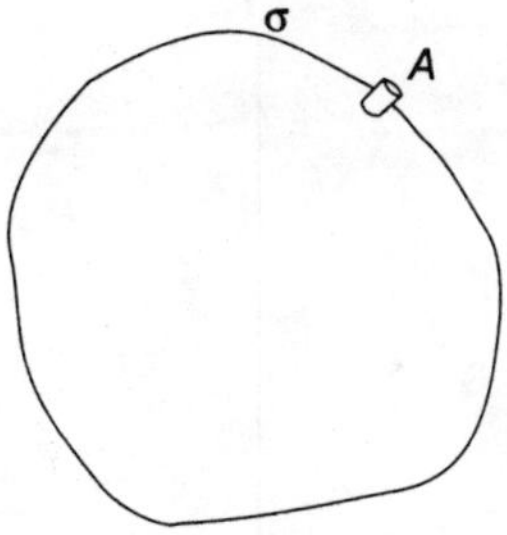

Fig. 3.4

$$\mathbf{E} = \frac{\sigma}{\varepsilon_0}\hat{\mathbf{n}} \tag{3.1.1}$$

where $\hat{\mathbf{n}}$ is the unit vector normal to the surface of the conductor at that point.

3.2 THE METHOD OF IMAGES

Let us suppose we have a point charge q placed at a distance 'a' to the right of an infinite conducting plane (Fig. 3.5). We take the plane to be the Y–Z plane and the charge q to be on the x-axis. We wish to calculate the potential

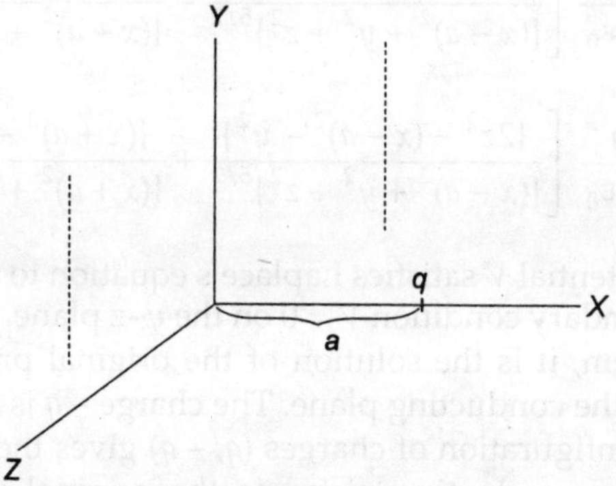

Fig. 3.5

in the region between the charge and the plane. We assume the plane to be grounded. To do this let us suppose that instead of an infinite plane we have a point charge $-q$ at $(-a, 0, 0)$ as shown in Fig. 3.6. The potential at all points on the Y–Z plane is zero because these points are equidistant from the two charges. The potential V at an arbitrary point $P(x, y, z)$ due to these charges is

$$V = \frac{1}{4\pi\varepsilon_0}\left\{\frac{q}{\sqrt{(x-a)^2 + y^2 + z^2}} - \frac{q}{\sqrt{(x+a)^2 + y^2 + z^2}}\right\} \tag{3.2.1}$$

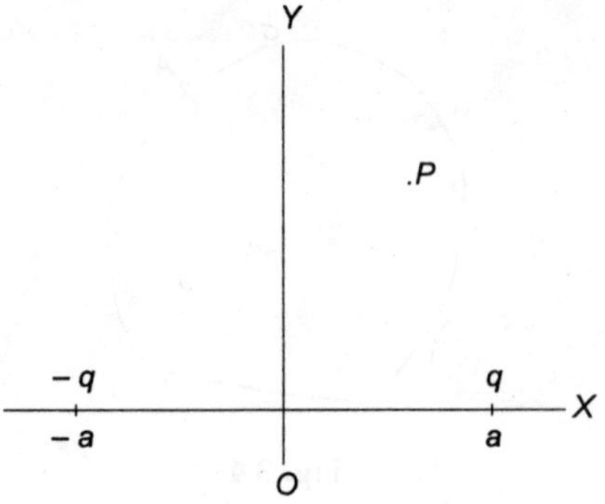

Fig. 3.6

Let us calculate $\frac{\partial^2 V}{\partial x^2}, \frac{\partial^2 V}{\partial y^2}$ and $\frac{\partial^2 V}{\partial z^2}$.

We find

$$\frac{\partial^2 V}{\partial x^2} = \frac{q}{4\pi\varepsilon_0}\left[\frac{\{2(x-a)^2 - y^2 - z^2\}}{\{(x-a)^2 + y^2 + z^2\}^{5/2}} + \frac{\{y^2 + z^2 - 2(x+a)^2\}}{\{(x+a)^2 + y^2 + z^2\}^{5/2}}\right]$$

$$\frac{\partial^2 V}{\partial y^2} = \frac{q}{4\pi\varepsilon_0}\left[\frac{\{2y^2 - (x-a)^2 - z^2\}}{\{(x-a)^2 + y^2 + z^2\}^{5/2}} + \frac{\{(x+a)^2 - 2y^2 + z^2)\}}{\{(x+a)^2 + y^2 + z^2\}^{5/2}}\right]$$

$$\frac{\partial^2 V}{\partial z^2} = \frac{q}{4\pi\varepsilon_0}\left[\frac{\{2z^2 - (x-a)^2 - y^2\}}{\{(x-a)^2 + y^2 + z^2\}^{5/2}} + \frac{\{(x+a)^2 - y^2 - 2z^2\}}{\{(x+a)^2 + y^2 + z^2\}^{5/2}}\right]$$

$\nabla^2 V = 0$. So the potential V satisfies Laplace's equation to the right of the Y–Z plane and the boundary condition $V = 0$ on the y–z plane. So according to the uniqueness theorem, it is the solution of the original problem of the point charge in front of the conducting plane. The charge $-q$ is said to be the *image charge* of q. The configuration of charges $(q, -q)$ gives the same electric field to the right of the conducting plane as the original problem. The only difference is that the field lines terminate on the conducting plane. The induced charge on the plane is calculated using (3.1.1). The surface charge density

$$\sigma = \varepsilon_0 E_x |_{x=0}$$

$$= \frac{-qa}{2\pi(a^2 + y^2 + z^2)^{3/2}} \qquad (3.2.2)$$

The induced charge exerts an attractive force on the charge q. This attractive force is given by the force between the charge q and its image charge $-q$:

$$F = \frac{q^2}{16\pi\varepsilon_0 a^2} \tag{3.2.3}$$

The total induced charge can be obtained by integrating σ in (3.2.2) over the entire Y–Z plane.

$$\int_{-\infty}^{\infty}\int_{-\infty}^{\infty} \sigma\, dydz = -\frac{qa}{2\pi}\int_{-\infty}^{\infty}\int_{-\infty}^{\infty}\frac{dydz}{(a^2+y^2+z^2)^{3/2}}$$

Doing the y-integration first, we get

$$\int_{-\infty}^{\infty}\int_{-\infty}^{\infty}\frac{dydz}{(a^2+y^2+z^2)^{3/2}} = \int_{-\infty}^{\infty}\frac{2dz}{(a^2+z^2)} = \frac{2\pi}{a}$$

$$\int_{-\infty}^{\infty}\int_{-\infty}^{\infty}\sigma\, dydz = \frac{-qa}{2\pi}\frac{2\pi}{a} = -q \tag{3.2.4}$$

Let us now consider another problem, namely a point charge q near a grounded conducting sphere. The radius of the sphere is 'a' and the charge q is at a distance 'd' from the centre of the sphere (Fig. 3.7). Let the image charge q' be at a distance z from the centre of the sphere.

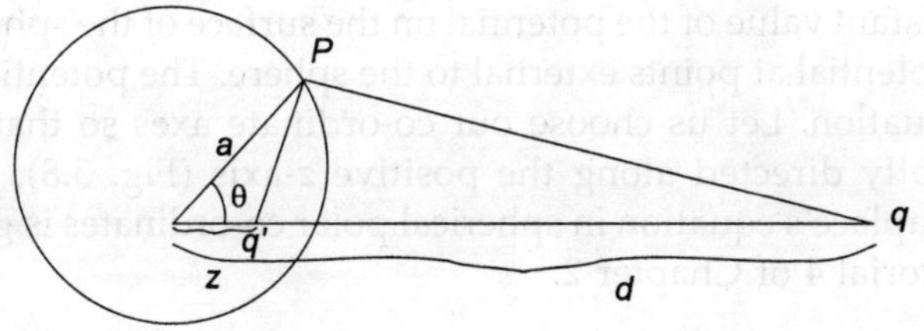

Fig. 3.7

The potential at a point P on the sphere is given by

$$V = \frac{1}{4\pi\varepsilon_0}\left[\frac{q}{\sqrt{(d^2+a^2-2ad\cos\theta)}} - \frac{q'}{\sqrt{(z^2+a^2-2az\cos\theta)}}\right] \tag{3.2.5}$$

The potential must be zero for all θ. If we put $z = a^2/d$ then the denominator in the second term becomes $\frac{a}{d}\sqrt{(a^2+d^2-2ad\cos\theta)}$. So the potential becomes

---

⊕[A point charge q can be represented by a delta function. For example the charge distribution to the right of the Y–Z plane can be written as $q\delta(x-a)$ which is zero everywhere except at $x = a$]

$$V = \frac{1}{4\pi\varepsilon_0 \sqrt{(a^2 + d^2 - 2ad\cos\theta)}}\left(q + \frac{dq'}{a}\right)$$

For **V** = 0, we must have

$$q' = \frac{-aq}{d} \tag{3.2.6}$$

So a charge $-aq/d$ placed at a distance a^2/d from the centre of the sphere satisfies the boundary condition. The potential V at an arbitrary point P outside the sphere at a distance r from the centre is given by

$$V = \frac{q}{4\pi\varepsilon_0}\left[\frac{1}{\sqrt{(r^2 + d^2 - 2rd\cos\theta)}} - \frac{a}{d\sqrt{(r^2 + z^2 - 2rz\cos\theta)}}\right] \tag{3.2.7}$$

3.3 AN UNCHARGED CONDUCTING SPHERE IN AN ELECTRIC FIELD

Let us suppose that in a certain region of space there exists a uniform electric field of magnitude E_0. We place an uncharged conducting sphere of radius 'a' in this region of space. Clearly, the electric field will be distorted near the sphere. This is because the surface of the sphere is an equipotential surface. So the resultant electric field must be normal to the surface at all points. Let V_0 be the constant value of the potential on the surface of the sphere. We wish to find the potential at points external to the sphere. The potential V satisfies Laplace's equation. Let us choose our co-ordinate axes so that the electric field is initially directed along the positive z-axis (Fig. 3.8). The general solution of Laplace's equation in spherical polar co-ordinates is given by eqn. (4.1.3) in Tutorial 4 of Chapter 2.

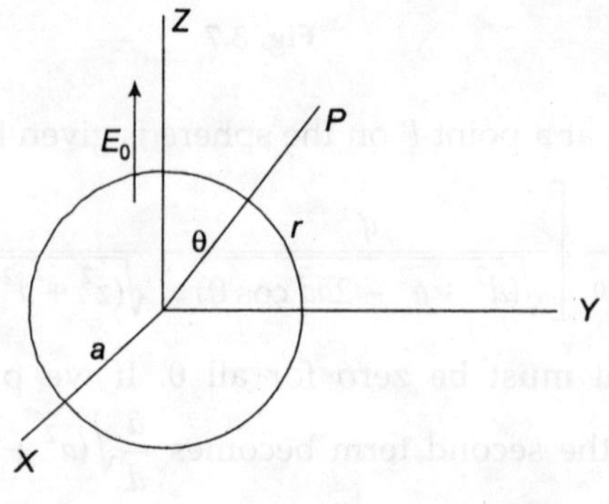

Fig. 3.8

$$V(r, \theta) = \sum_{n=0}^{\infty} \left(A_n r^n + \frac{B_n}{r^{n+1}} \right) P_n(\cos\theta) \qquad (3.3.1)$$

Far from the sphere, the electric field must remain uniform. So as $r \to \infty$, $\mathbf{E} = E_0 \hat{\mathbf{k}}$. In terms of the unit vectors $\hat{\mathbf{r}}$ and $\hat{\theta}$ in spherical polar co-ordinates, $\mathbf{k} = \hat{\mathbf{r}} \cos\theta - \hat{\theta} \sin\theta$. So $\mathbf{E} = \hat{\mathbf{r}} E_0 \cos\theta - \hat{\theta} E_0 \sin\theta$. The electric field is the negative gradient of the potential. Thus, $\mathbf{E} = -\nabla V = -\dfrac{\partial V \hat{\mathbf{r}}}{\partial r} - \dfrac{1}{r}\dfrac{\partial V \hat{\theta}}{\partial \theta}$ because V is independent of the azimuthal angle ϕ. Now $V = -E_0 r \cos\theta + \text{const.}$ as $r \to \infty$. Let us write out the first few terms of the expansion (3.3.1):

$$V(r, \theta) = A_0 + \frac{B_0}{r} + \left(A_1 r + \frac{B_1}{r^2} \right) \cos\theta + \left(A_2 r^2 + \frac{B_2}{r^3} \right) \frac{1}{2} (3\cos^2\theta - 1) + \ldots$$

From the above requirement on the form of V as $r \to \infty$, we must have $A_n = 0$ for $n \geq 2$. Also $A_1 = -E_0$.

$$V(r, \theta) = A_0 + \frac{B_0}{r} + \left(-E_0 r + \frac{B_1}{r^2} \right) \cos\theta + \frac{B_2}{2r^3} (3\cos^2\theta - 1) + \ldots$$

The potential at external points due to a charged sphere is const./r. But our sphere is uncharged. So $B_0 = 0$.

$$V(r, \theta) = A_0 + \left(-E_0 r + \frac{B_1}{r^2} \right) \cos\theta + \frac{B_2}{2r^3} (3\cos^2\theta - 1) + \ldots$$

The other boundary condition is that $V = V_0$ at $r = a$ for all θ. For this to be true, the coefficients of $\cos\theta$, $\cos^2\theta$, etc. must be zero for $r = a$. It is possible to have this for $\cos\theta$, but not possible for higher powers of θ unless all the B_n's are zero for $n \geq 2$. Thus with this condition,

$$V(r, \theta) = A_0 + \left(-E_0 r + \frac{B_1}{r^2} \right) \cos\theta$$

$$V(a, \theta) = V_0$$

$$V_0 = A_0 + \left(-E_0 a + \frac{B_1}{a^2} \right) \cos\theta$$

$$A_0 = V_0, \; -E_0 a + \frac{B_1}{a^2} = 0 \text{ or, } B_1 = E_0 a^3$$

The potential at an arbitrary point P is

$$V(r, \theta) = V_0 + E_0 \left(-r + \frac{a^3}{r^2} \right) \cos\theta \qquad (3.3.2)$$

3.4 SOLUTION OF SIMPLE PROBLEMS USING COMPLEX VARIABLES

From complex analysis, we know that if we have an analysis function $f(z)$ which can be written as a sum of real and imaginary parts $f(z) = u(x, y) + jv(x, y)$, then these functions u and v satisfy the Cauchy-Riemann equations:

$$\frac{\partial u}{\partial x} = \frac{\partial v}{\partial y}, \frac{\partial v}{\partial x} = -\frac{\partial u}{\partial y} \tag{3.4.1}$$

Furthermore, u and v both satisfy Laplace's equation. This can be easily seen if we take the partial derivative of the first equation with respect to x. We get

$$\frac{\partial^2 u}{\partial x^2} = \frac{\partial^2 v}{\partial x \partial y} = \frac{\partial}{\partial y}\left(\frac{\partial v}{\partial x}\right) = -\frac{\partial^2 u}{\partial y^2}$$

using the second equation in (3.4.1). So

$$\frac{\partial^2 u}{\partial x^2} + \frac{\partial^2 u}{\partial y^2} = 0$$

The functions u and v are called *conjugate* functions.

Consider the family of curves represented by

$$u(x, y) = c_1 \tag{3.4.2a}$$

$$v(x, y) = c_2 \tag{3.4.2b}$$

If (x_0, y_0) is a point of intersection of these two curves, the slope of the tangent lines at x_0, y_0, are, respectively,

$$\left.\frac{dy}{dx}\right|_{u = \text{const.}} = -\frac{\dfrac{\partial u}{\partial x}}{\dfrac{\partial u}{\partial y}}$$

and

$$\left.\frac{dy}{dx}\right|_{v = \text{const.}} = -\frac{\dfrac{\partial v}{\partial x}}{\dfrac{\partial v}{\partial y}}$$

Using the Cauchy-Riemann equations (3.4.1), we get

$$\left.\frac{dy}{dx}\right|_{u = \text{const.}} = \frac{\dfrac{\partial v}{\partial y}}{\dfrac{\partial v}{\partial x}} = -\frac{1}{\left.\dfrac{dy}{dx}\right|_{v = \text{const.}}}$$

Hence the slopes are negative reciprocals and therefore the two curves intersect at right angles.[1]

Thus if we can write down an analytic function $f(z)$, we can choose either its real or its imaginary part to be the solution of a problem in electrostatics in two dimensions. For example, consider

$$f(z) = A \ln z + c, \text{ with } z = re^{j\theta}$$

where A and c are constants.

$$f(z) = A \ln r + c + j\theta$$

$$u = A \ln r + c \tag{3.4.3a}$$

$$v = A\theta \tag{3.4.3b}$$

$f(z)$ also represents a mapping from the (x, y) plane to the (u, v) plane. In terms of (x, y), we can rewrite (3.4.3) as:

$$u = A \ln \sqrt{(x^2 + y^2)} + c \tag{3.4.4a}$$

$$v = A \tan^{-1}\frac{(y)}{x} \tag{3.4.4b}$$

Choose u to represent a potential function φ. Then u = constant gives a circle in the $X - Y$ plane. This is the nature of equipotential surfaces in a problem having cylindrical symmetry. If we have two cylinders of radii a and b having potentials φ_1 and φ_2 then, in the region between the cylinders, the potential function φ can be evaluated as follows:

$$\varphi_1 = A \ln a + c$$

$$\varphi_2 = A \ln b + c$$

from which the constants can be evaluated and the solution for φ is

$$\varphi = \frac{(\varphi_2 - \varphi_1)}{\ln\frac{b}{a}} \ln r + \frac{(\varphi_2 \ln a - \varphi_1 \ln b)}{\ln\frac{b}{a}}$$

In general, a mapping from the (x, y) plane to the (u, v) plane is called a conformal mapping. A known solution to a simple problem in the (u, v) plane is used to solve a more complicated problem in the (x, y) plane.

Tutorial

Let us consider the problem of the uncharged conducting sphere of radius a in a uniform electric field, again. Take the analytic function

$$f(z) = E_0\left(-z + \frac{a^3}{z^2}\right) + V_0 \tag{1}$$

with $z = re^{j\theta}$. E_0 is the uniform electric field. What is V_0? Write down the real and imaginary parts of $f(z)$. Choose the real part of this function as the potential V. Show that it is the same as the solution (3.3.2). Calculate the radial and transverse components of the electric field E_r and E_θ.

Problems

1. A point charge +1/2 q is situated at the centre of a spherical cavity (radius r) inside a neutral spherical conductor of radius $4r$. Using Gauss' theorem or otherwise, find the surface density of charge induced on the boundary surface of the cavity. What is the sign of the charge ? *(University of Calcutta, 1998)*

   $$\left[\textbf{Ans. } \sigma = -\frac{q}{8\pi r^2}, \text{ negative}\right]$$

2. A point charge q of mass m is released from rest at a distance d from an infinite grounded conducting plate. Show that the time taken by the charge to hit the plate is $t = \frac{1}{q}\sqrt{(2m\varepsilon_0\pi^3 d^3)}$.

   (University of Calcutta, 2002)

3. Consider the problem of the point charge q near a grounded conducting sphere of radius a. What is the force of attraction between them? Calculate the induced surface charge density σ on the surface of the sphere. Obtain the total induced charge on the sphere.

   $$\left[\textbf{Ans. } F = \frac{aq^2 d}{4\pi\varepsilon_0 (d^2 - a^2)^2}, \sigma = \frac{q(a^2 - d^2)}{4\pi a(d^2 + a^2 - 2ad\cos\theta)^{3/2}}, \frac{-aq}{d}\right]$$

4. Consider an insulated conducting sphere of radius a carrying a charge Q. There is a charge q at a distance d from its centre ($d > a$). Obtain the potential at points external to the sphere. Calculate the force of attraction between the charge q and the sphere.

   $$\left[\textbf{Ans. } V = \frac{1}{4\pi\varepsilon_0}\left\{\frac{q}{\sqrt{(r^2 + d^2 - 2rd\cos\theta)}} - \frac{aq}{d\sqrt{\left(r^2 + \frac{a^4}{d^2} - \frac{2a^2 r\cos\theta}{d}\right)}} + \frac{Q + aq/d}{r}\right\},\right.$$

   $$\left. E = \frac{q}{4\pi\varepsilon_0 d^2}\left\{Q + \frac{a^3 q(a^2 - 3d^2)}{d(d^2 - a^2)^2}\right\}\right]$$

5. Consider a conducting sphere charged to a potential V, having radius a. There is a charge q at a distance 'd' from its centre ($d > a$). Obtain the potential at points external to the sphere. Calculate the force of attraction between the sphere and the charge q.

$$\left[\textbf{Ans. } V = \frac{q}{4\pi\varepsilon_0\sqrt{(r^2 + d^2 - 2dr\cos\theta)}} - \frac{aq}{4\pi\varepsilon_0 d\sqrt{\left(r^2 + \frac{a^4}{d^2} - \frac{2a^2 r\cos\theta}{d}\right)}} + \frac{aV}{r},\ F = \frac{qa}{4\pi\varepsilon_0 d^2}\left\{4\pi\varepsilon_0 V - \frac{qd^3}{(d^2 - a^2)^2}\right\}\right]$$

6. Consider the problem of the uncharged conducting sphere placed in a uniform electric field $\mathbf{E}_0$. Calculate the electric field at an arbitrary point P outside the sphere.

$$\left[\textbf{Ans. } \mathbf{E} = E_0\hat{k} + \frac{a^3E_0}{r^3}(2\cos\theta\,\hat{r} + \sin\theta\,\hat{\theta})\right]$$

4

Electric Fields in Matter

4.1 ATOMIC AND MOLECULAR POLARIZABILITY

In this chapter, we shall study what happens when an insulator, also called a dielectric, is placed in an electric field. In a dielectric, the electrons are rigidly bound to atoms and molecules. They are not free to move as in a conductor. When an electric field acts on an atom or a molecule, there is a force on both the positive and negative charges in them. This force causes a very small displacement of the two types of charge in opposite directions. So the atom or the molecule becomes slightly deformed when the field is present. This relative displacement of positive and negative charges leads to the formation of a dipole. Since the positive charge is displaced in the direction of the field and the electrons against the field direction, the dipole moment points in the field direction. The above description applies to atoms and non-polar molecules like H_2, N_2, etc. which do not have a dipole moment in absence of an electric field. In a non-polar molecule like CO_2, there is an additional complication in that this induced dipole moment is different in different directions. A polar molecule like water, for example, has a permanent dipole moment (in absence of an electric field) of 6.1×10^{-30} Cm. HCl is another example of a polar molecule with dipole moment 3.44×10^{-30} Cm. In a sample of HCl, or any other polar molecule, these dipoles are randomly directed. We learnt in Chapter 1 that when a dipole is placed in an electric field, a torque acts on it. So an HCl molecule is rotated into the direction of the field. All the molecules are not aligned parallel to each other, however, because the thermal motion of the molecules disrupts the alignment at any finite temperature. To sum up, when an atom or non-polar molecule is placed in an electric field, a dipole moment is induced in it. A polar molecule is rotated into the direction of the field. The dielectric sample as a whole is said to be *polarized*. It remains electrically neutral, however. For many materials, the dipole moment **p** of the atom or molecule is proportional to the electric field **E** provided the field is not too large. The constant of proportionality is called the *polarizability* α:

$$\mathbf{p} = \alpha\mathbf{E} \tag{4.1.1}$$

When an atom is involved, α is the atomic polarizability. When a molecule is involved, α is the molecular polarizability. If the polarizability is the same in all directions, the materials is said to be isotropic. In the case of anisotropic materials, one has to define a *polarizability tensor* α_{ij} ($i = 1$ to 3, $j = 1$ to 3) with nine components so that we have

$$p_i = \sum_{j=1}^{3} \alpha_{ij} E_j \quad (i = 1 \text{ to } 3) \tag{4.1.2}$$

We will be discussing only isotropic materials.

4.2 THE POTENTIAL DUE TO POLARIZED MATTER

When a dielectric is placed in an electric field, it becomes polarized. To describe this state of the dielectric, we define the polarization **P** as the dipole moment per unit volume. This will, in general, be a function of position in the material. Let us now calculate the potential φ at an exterior point due to this material. Consider an element of volume $\Delta V'$ in the dielectric. It has a dipole moment $\mathbf{P}\Delta V'$(see Fig. 4.1). So the potential $\Delta\varphi$ it produces at a point A at a distance $|\mathbf{r} - \mathbf{r}'|$ from the element is, from (2.3.1)

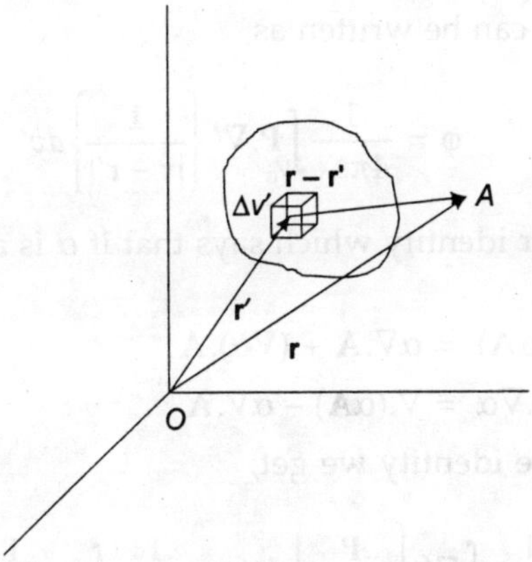

Fig. 4.1

$$\Delta\varphi = \frac{(\mathbf{P}\Delta V').(\mathbf{r} - \mathbf{r}')}{4\pi\varepsilon_0 \, |\mathbf{r} - \mathbf{r}'|^3}$$

The potential φ at A is then given by the integral

$$\varphi = \frac{1}{4\pi\varepsilon_0} \int_{V_0} \frac{\mathbf{P}.(\mathbf{r} - \mathbf{r}')dv'}{|\mathbf{r} - \mathbf{r}'|^3} \tag{4.2.1}$$

over the volume V_0 of the sample. Here $\mathbf{P} = \mathbf{P}(r')$. Let us rewrite this in a different form.

Now $\mathbf{r} - \mathbf{r}' = (x - x')\hat{\mathbf{i}} + (y - y')\hat{\mathbf{j}} + (z - z')\hat{\mathbf{k}}$

Let us take the gradient of $\dfrac{1}{|\mathbf{r} - \mathbf{r}'|}$

$$\frac{\partial}{\partial x}\left\{\frac{1}{|\mathbf{r}-\mathbf{r}'|}\right\} = \frac{\partial}{\partial x}\left\{\frac{1}{\sqrt{(x-x')^2+(y-y')^2+(z-z')^2}}\right\}$$

$$= \frac{-(x-x')}{\left\{(x-x')^2+(y-y)^2+(z-z')^2\right\}^{3/2}}$$

$$\nabla\left\{\frac{1}{|\mathbf{r}-\mathbf{r}'|}\right\} = \frac{-(\mathbf{r}-\mathbf{r}')}{|\mathbf{r}-\mathbf{r}'|^3} \tag{4.2.2}$$

Again
$$\frac{\partial}{\partial x'}\left\{\frac{1}{|\mathbf{r}-\mathbf{r}'|}\right\} = \frac{x-x'}{\{(x-x')^2+(y-y')+(z-z')^2\}^{3/2}}$$

$$\therefore\ \nabla'\left\{\frac{1}{|\mathbf{r}-\mathbf{r}'|}\right\} = \frac{\mathbf{r}-\mathbf{r}'}{|\mathbf{r}-\mathbf{r}'|^3} = -\nabla\left\{\frac{1}{|\mathbf{r}-\mathbf{r}'|}\right\} \tag{4.2.3}$$

The potential φ can be written as

$$\varphi = \frac{1}{4\pi\varepsilon_0}\int_{V_0}\mathbf{P}.\nabla'\left\{\frac{1}{|\mathbf{r}-\mathbf{r}'|}\right\}dv'$$

There is a vector identity which says that if α is a scalar function and $\mathbf{A}$ is a vector function

$$\nabla.(\alpha\mathbf{A}) = \alpha\nabla.\mathbf{A} + (\nabla\alpha).\mathbf{A}$$

$$\mathbf{A}.\nabla\alpha = \nabla.(\alpha\mathbf{A}) - \alpha\nabla.\mathbf{A}$$

Using this above identity we get,

$$\varphi = \frac{1}{4\pi\varepsilon_0}\int_{V_0}\nabla'.\left\{\frac{\mathbf{P}}{\mathbf{r}-\mathbf{r}'}\right\}dv' - \frac{1}{4\pi\varepsilon_0}\int_{V_0}\frac{1}{|r-r'|}\nabla'.\mathbf{P}dv'$$

Applying the divergence theorem, we write the first term as a surface integral over the surface S_0 of the sample.

$$\varphi = \frac{1}{4\pi\varepsilon_0}\oint_{S_0}\frac{\mathbf{P}.d\mathbf{a}}{|\mathbf{r}-\mathbf{r}'|} - \frac{1}{4\pi\varepsilon_0}\int_{V_0}\frac{1}{|\mathbf{r}-\mathbf{r}'|}\nabla'.\mathbf{P}dv' \tag{4.2.4}$$

or
$$\varphi = \frac{1}{4\pi\varepsilon_0}\left[\oint_{S_0}\frac{\sigma_P da}{|\mathbf{r}-\mathbf{r}'|} + \int_{V_0}\frac{\rho_P dv'}{|\mathbf{r}-\mathbf{r}'|}\right] \tag{4.2.5}$$

where $\sigma_P = \mathbf{P}.\hat{\mathbf{n}}, \rho_P = -\nabla'.\mathbf{P}$ (4.2.6)

are polarization charge densities. $\hat{\mathbf{n}}$ is the unit vector normal to the element of surface da. So the potential at an external point A can be written as a sum of two terms, one due to a surface charge σ_P and the other due to a volume change ρ_P. We can think of the polarized dielectric as composed of a large number of elementary dipoles. These dipoles are positioned in horizontal and vertical arrays with the negative end of one dipole near the positive end of the adjacent dipole. If the polarization is uniform, all the dipoles have the same magnitude and $\nabla'.\mathbf{P} = 0$. Thus $\rho_P = 0$. The volume charge density vanishes. This is to be expected since in any finite volume of the material, the positive charge of one dipole neutralizes the effect of the negative charge of the adjacent dipole. On the surface of the sample, however, we have negative charges accumulated at one end and positive charges accumulated at the other end. This gives rise to the surface charge density σ_P. If the polarization is non-uniform, **P** varies from point to point in the dielectric. So the dipole moments of adjacent elementary dipoles are not necessarily equal. The negative charge of one dipole does not cancel the effect of the positive charge of the other. This gives rise to a volume charge density in the material.

4.3 DIELECTRIC CONSTANT

In the last section we introduced the polarization **P** to describe the state of a dielectric when it is placed in an electric field. We shall now restrict our study to those materials for which **P** is proportional to **E**, i.e.

$$\mathbf{P} = \chi_e \mathbf{E} \quad (4.3.1)$$

where χ_e is the *electric susceptibility*. These materials are called linear dielectrics. Let us now suppose that we have a conductor carrying a charge q in a dielectric[(2)] (Fig. 4.2). Let us call q_P the polarization charge. Consider a

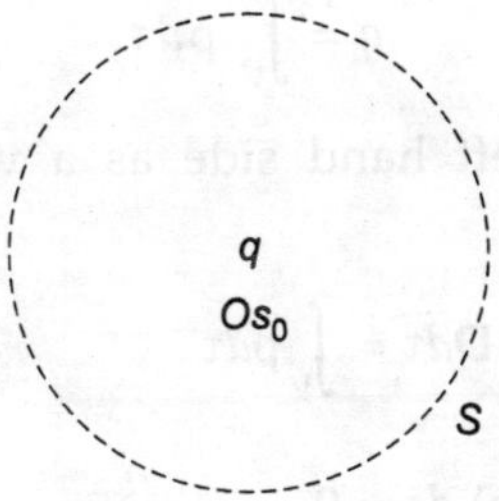

Fig. 4.2

Gaussian surface S. The boundary of the conductor is S_0. From Gauss' law, we have,

$$\varepsilon_0 \int_S \mathbf{E}.d\mathbf{a} = q + q_P \quad (4.3.2)$$

$$q_P = \int_{S_0} \sigma_P da + \int_V \rho_P d\tau \tag{4.3.2}$$

where V is the volume of the dielectric lying between S and S_0. Note that the surface integral is taken over S_0 only because the dielectric does not have any boundary at S.

$$\therefore \quad q_P = \int_{S_0} \mathbf{P}.\hat{\mathbf{n}}\,da - \int_V \nabla.\mathbf{P} d\tau$$

$$= \int_{S_0} \mathbf{P}.\hat{\mathbf{n}}\,da - \int_{S+S_0} \mathbf{P}.\hat{\mathbf{n}}\,da$$

where the divergence theorem has been used to obtain the second term (the volume V is bound S_0 and S).

$$\therefore \quad q_P = -\int_S \mathbf{P}.\hat{\mathbf{n}}\,da \tag{4.3.3}$$

Using (4.3.3) in (4.3.2), we get

$$\varepsilon_0 \int \mathbf{E}.d\mathbf{a} = q - \int_S \mathbf{P}.d\mathbf{a}$$

or

$$\int_S (\varepsilon_0 \mathbf{E} + \mathbf{P}).d\mathbf{a} = q$$

The vector function in the integrand is called the *electric displacement* **D**.

$$\mathbf{D} = \varepsilon_0 \mathbf{E} + \mathbf{P} \tag{4.3.4}$$

So we have

$$\int_S \mathbf{D}.d\mathbf{a} = q \tag{4.3.5}$$

This is the equivalent of Gauss' law for dielectrics in integral form. If the charge q is a volume charge distribution characterized by a charge density ρ,

$$q = \int_V \rho d\tau$$

We can rewrite the left hand side as a volume integral using the divergence theorem

$$\int_V \nabla.\mathbf{D} d\tau = \int_V \rho d\tau$$

or,

$$\int_V (\nabla.\mathbf{D} - \rho)\, d\tau = 0$$

Since this is true for any arbitrary volume V, the integrand must be zero.

$$\nabla.\mathbf{D} = \rho \tag{4.3.6}$$

For linear dielectrics, the electric displacement is

$$\mathbf{D} = (\varepsilon_0 + \chi_e)\mathbf{E}$$

from (4.3.1) and (4.3.4). We may rewrite this as

$$\mathbf{D} = \varepsilon \mathbf{E} \tag{4.3.7}$$

where

$$\varepsilon = \varepsilon_0 + \chi_e \tag{4.3.8}$$

is the *permittivity* of the medium. The *dielectric constant* K of the medium is defined by

$$K = \frac{\varepsilon}{\varepsilon_0} = 1 + \frac{\chi_e}{\varepsilon_0}.$$

4.4 BOUNDARY CONDITIONS ON D AND E

Let us now find out how the electric field and displacement change as we go from one medium to another. Let us consider a small Gaussian pillbox at the interface of two media 1 and 2 (Fig. 4.3). Suppose that charge with surface density σ exists on the interface. This charge is the external charge, not the polarization charge.

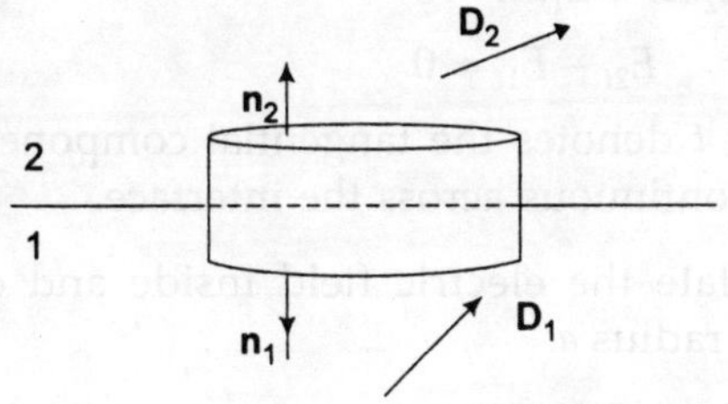

Fig. 4.3

We will use Gauss' law in integral form (4.3.5). Since the curved surface is of small dimensions, we can neglect the value of the integral over this part. Thus we get

$$\mathbf{D}_2.\hat{\mathbf{n}}_2 A + \mathbf{D}_1.\hat{\mathbf{n}}_1 A = \sigma A$$

where A is the area of the top (as well as the bottom) surface of the pillbox. From the diagram it is clear that the two unit normals are oppositely directed, i.e. $\hat{\mathbf{n}}_1 = -\hat{\mathbf{n}}_2$. So

$$(\mathbf{D}_2 - \mathbf{D}_1).\hat{\mathbf{n}}_2 = \sigma$$

or,

$$D_{2n} - D_{1n} = \sigma \tag{4.4.1}$$

where the subscript n denotes the normal component. Thus the normal component of the electric displacement is discontinuous by the amount of surface charge.

The electric field is, as we know, irrotational. $\nabla \times \mathbf{E} = 0$. Or, in integral form,

$$\oint \mathbf{E}.d\mathbf{l} = 0$$

Let us choose a rectangular path *ABCD* (Fig. 4.4) at the interface of two media.

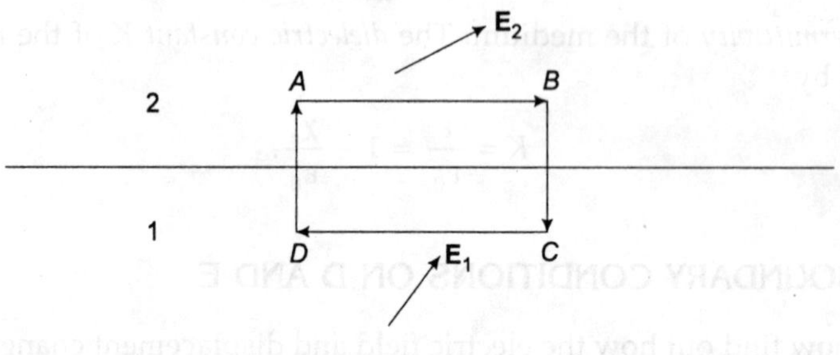

Fig. 4.4

Let Δl be the length of *AB* and *CD*. The edges *AD* and *BC* are infinitesimal. The line integral over the closed loop gives us

$$\mathbf{E}_2.\Delta\mathbf{l} - \mathbf{E}_1.\Delta\mathbf{l} = 0$$

$$E_{2t} - E_{1t} = 0 \tag{4.4.2}$$

where the subscript *t* denotes the tangential component. So the tangential component of **E** is continuous across the interface.

Example 1: Calculate the electric field inside and outside a uniformly polarized sphere of radius *a*.

Solution: We shall take the direction of uniform polarization **P** to be the *z*-axis (Fig. 4.5).

$$\mathbf{P} = P\hat{\mathbf{k}}.$$

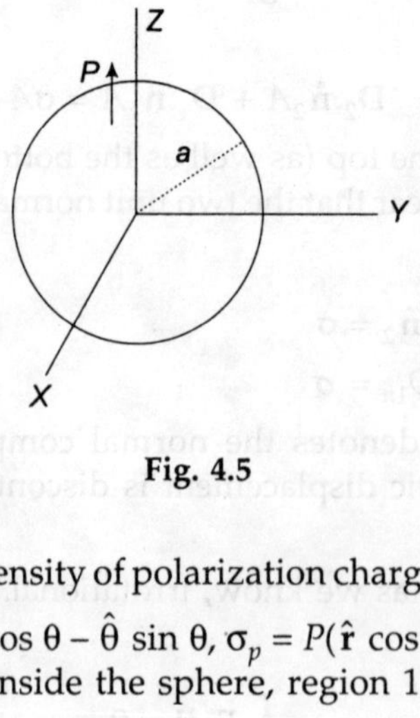

Fig. 4.5

$\therefore$ $\nabla.\mathbf{P} = 0$. The surface density of polarization charge σ_P is given by $\sigma_P = \mathbf{P}.\hat{\mathbf{n}}$.

Here $\hat{\mathbf{n}} = \hat{\mathbf{r}}$. Since $\hat{\mathbf{k}} = \hat{\mathbf{r}}\cos\theta - \hat{\boldsymbol{\theta}}\sin\theta$, $\sigma_p = P(\hat{\mathbf{r}}\cos\theta - \hat{\boldsymbol{\theta}}\sin\theta).\ \hat{\mathbf{r}} = P\cos\theta$. We shall call the region inside the sphere, region 1 and the region outside, region 2. In region 2,

$$\mathbf{D}_2 = \varepsilon_0 \mathbf{E}_2 = -\varepsilon_0 \nabla \mathbf{V}_2$$

where V_2 is the electrostatic potential there. V_2 satisfies Laplace's equation in region 2 because $\nabla.\mathbf{D}_2 = 0$. In region 1, $\mathbf{D}_1 = \varepsilon_0 \mathbf{E}_1 + \mathbf{P}$ or, $\mathbf{D}_1 = -\varepsilon_0 \nabla V_1 + P(\hat{\mathbf{r}} \cos\theta - \hat{\theta} \sin\theta)$ where V_1 is the potential there. Since there is no external charge, eqn. (4.3.5) gives $\nabla.\mathbf{D}_1 = 0$. Since $\nabla.\mathbf{P} = 0$, $\nabla.\mathbf{E}_1 = 0$ and so V_1 also satisfies Laplace's equation. The solution of Laplace's equation in spherical polar co-ordinates appropriate to region 1 is

$$V_1 = \sum_{n=0}^{\infty} A_n r^n P_n(\cos\theta) \tag{4.4.3}$$

from eqn. (4.13) of Tutorial 4, Chapter 2. The coefficients must all be zero for V_1 to remain finite at the origin. In region 2, the solution is

$$V_2 = \sum_{n=0}^{\infty} \frac{B_n}{r^{n+1}} P_n(\cos\theta) \tag{4.4.4}$$

All the coefficients A_n must be zero in this region for V_2 to remain finite at large r. We must now use the boundary condition (4.4.1) on the surface of the sphere remembering that $\sigma = 0$ (there is no external charge on the surface of the sphere; it is only the polarization charge which exists there). Therefore $D_{2r}|_{r=a} = D_{1r}|_{r=a}$. Also, the potential must be continuous at $r = a$. Equating the coefficients of $P_n(\cos\theta)$ on both sides, we get,

$$B_n = A_n a^{2n+1} \tag{4.4.5}$$

$$D_{2r} = \frac{-\varepsilon_0\, \partial V_2}{\partial r},\; D_{1r} = \frac{-\varepsilon_0\, \partial V_1}{\partial r} + P\cos\theta$$

From (4.4.3), $$\frac{\partial V_1}{\partial r} = \sum_{n=0}^{\infty} n A_n r^{n-1} P_n(\cos\theta)$$

From (4.4.4), $$\frac{\partial V_2}{\partial r} = -\sum_{n=0}^{\infty} (n+1)\frac{B_n P_n}{r^{n+2}}(\cos\theta)$$

∴ Using (4.4.5), we get,

$$-\sum_{n=0}^{\infty} (2n+1) A_n a^{n-1} P_n(\cos\theta) = \frac{-P\cos\theta}{\varepsilon_0}$$

$$\frac{A_0}{a} + 3A_1\cos\theta + \frac{5}{2}A_2(3\cos^2\theta - 1) + \ldots = \frac{P\cos\theta}{\varepsilon_0}$$

$\therefore$ All the A_n's except A_1 must be zero. $A_1 = \dfrac{P}{3\varepsilon_0}$, $B_1 = \dfrac{Pa^3}{3\varepsilon_0}$.

$\therefore$ The potentials in regions 1 and 2 are

$$V_1 = \frac{Pr\cos\theta}{3\varepsilon_0}$$

$$V_2 = \frac{Pa^3\cos\theta}{3\varepsilon_0 r^2}$$

The electric field in region 1 is

$$\mathbf{E}_1 = \frac{-P}{3\varepsilon_0}(\hat{\mathbf{r}}\cos\theta - \hat{\boldsymbol{\theta}}\sin\theta) = \frac{-P\hat{\mathbf{k}}}{3\varepsilon_0} \tag{4.4.7}$$

So the electric field inside the sphere is opposite in direction to **P**. It is called a depolarizing field. The electric field in region 2 is

$$\mathbf{E}_2 = \frac{Pa^3}{3\varepsilon_0 r^3}(2\hat{\mathbf{r}}\cos\theta + \hat{\boldsymbol{\theta}}\sin\theta) \tag{4.4.8}$$

This is the electric field due to a dipole of moment $p = \dfrac{4\pi a^3 P}{3}$.

Example 2: Consider an uncharged dielectric sphere of radius a. It is placed in an initially uniform electric field $\mathbf{E}_0$. Calculate the electric field inside and outside the sphere.

Solution: The potential satisfies Laplace's equation both inside and outside the sphere because there are no external charges. Let us call the potential inside, V_1 and outside, V_2. We shall take the direction of the electric field $\mathbf{E}_0$ to be along the z-axis. Far from the sphere, the electric field should be uniform. So $\mathbf{E} = E_0\hat{\mathbf{k}} = E_0(\hat{\mathbf{r}}\cos\theta - \hat{\boldsymbol{\theta}}\sin\theta)$.

$\therefore$ As $r \to \infty$, $V_2 \to -E_0 r\cos\theta$. The general solution for V_2 is

$$V_2 = \sum_{n=0}^{\infty}\left(A_n r^n + \frac{B_n}{r^{n+1}}\right)P_n(\cos\theta)$$

From the above condition, $A_1 = -E_0$. A_0 is a constant term which we can set to zero. All the remaining A_n's are zero. Since the sphere is uncharged, the potential cannot have a $1/r$ term. So $B_n = 0$. Thus

$$V_2 = -E_0 r\cos\theta + \sum_{n=0}^{\infty}\frac{B_n}{r^{n+1}}P_n(\cos\theta) \tag{4.4.9}$$

Since the potential must remain finite as $r \to 0$, we must have

$$V_1 = \sum_{n=0}^{\infty} C_n r^n P_n(\cos\theta) \qquad (4.4.10)$$

Now let us apply the boundary conditions

$$D_{2r}\big|_{r=a} = D_{1r}\big|_{r=a} \qquad (4.4.11)$$

$$E_{2\theta}\big|_{r=a} = E_{1\theta}\big|_{r=a} \qquad (4.4.12)$$

$$D_{1r} = \frac{-\varepsilon\partial V_1}{\partial r} = -\,\varepsilon\sum_{n=1}^{\infty} nC_n r^{n-1} P_n(\cos\theta)$$

$$D_{2r} = \frac{-\varepsilon_0\partial V_2}{\partial r} = \varepsilon_0 E_0\cos\theta + \varepsilon_0\sum_{n=1}^{\infty}(n+1)\frac{B_n}{r^{n+2}}P_n(\cos\theta)$$

The first boundary condition at $r = a$ gives:

$$-\,\varepsilon C_1\cos\theta - \varepsilon\sum_{n=2}^{\infty} nC_n a^{n-1}P_n(\cos\theta) = \varepsilon_0\left(E_0 + \frac{2B_1}{a^3}\right)\cos\theta$$

$$+\,\varepsilon_0\sum_{n=2}^{\infty}(n+1)\frac{B_n}{a^{n+2}}P_n(\cos\theta)$$

Equating the coefficient of $\cos\theta$ on both sides, we get:

$$\frac{2B_1}{a^3} + KC_1 = -\,E_0 \qquad (4.4.13)$$

where $K = \varepsilon/\varepsilon_0$. Equating the coefficients of $P_n(\cos\theta)$ on both sides (for $n \geq 2$), we get:

$$B_n = \frac{-\,nKC_n a^{2n+1}}{(n+1)} \qquad (4.4.14)$$

The second boundary condition at $r = a$ gives:

$$\left(E_0 - \frac{B_1}{a^3}\right)\sin\theta + \sum_{n=2}^{\infty}\frac{B_n}{a^{n+2}}\frac{dP_n}{d\theta} = -\,C_1\sin\theta + \sum_{n=2}^{\infty} Ca^{n-1}\frac{dP_n}{d\theta}$$

Equating the coefficients of $\sin\theta$ on both sides, we get,

$$\frac{B_1}{a^3} - C_1 = E_0 \qquad (4.4.15)$$

Equating the coefficients of $\frac{dP_n}{d\theta}$ on both sides, we get,

$$B_n = C_n a^{2n+1} \tag{4.4.16}$$

From (4.4.14) and (4.4.16), $B_n = 0$, $C_n = 0$ for $n \geq 2$. From the other two equations, we get

$$B_1 = \frac{(K-1)a^3 E_0}{(K+2)} \tag{4.4.17}$$

$$C_1 = \frac{-3E_0}{(K+2)} \tag{4.4.18}$$

So the potentials inside and outside the sphere are:

$$V_1 = \frac{-3E_0 r \cos\theta}{(K+2)} \tag{4.4.19}$$

$$V_2 = -E_0 r\cos\theta + \frac{(K-1)a^3 E_0 \cos\theta}{(K+2)r^2} \tag{4.4.20}$$

The electric fields inside and outside the sphere are

$$\mathbf{E}_1 = \frac{3E_0}{(K+2)}\hat{\mathbf{k}} \tag{4.4.21}$$

and $$\mathbf{E}_2 = \mathbf{E}_0\hat{\mathbf{k}} + \frac{(K-1)E_0 a^3}{(K+2)r^3}(2\cos\theta\,\hat{\mathbf{r}} + \sin\theta\,\hat{\boldsymbol{\theta}}) \tag{4.4.22}$$

The electric field inside the sphere is in the same direction as $\mathbf{E}_0$. Outside, the field is a vector sum of the field $\mathbf{E}_0$ and that due to a dipole of moment

$$\mathbf{p} = 4\pi a^3 \varepsilon_0 E_0 \frac{(K-1)\,\hat{\mathbf{k}}}{(K+2)}$$

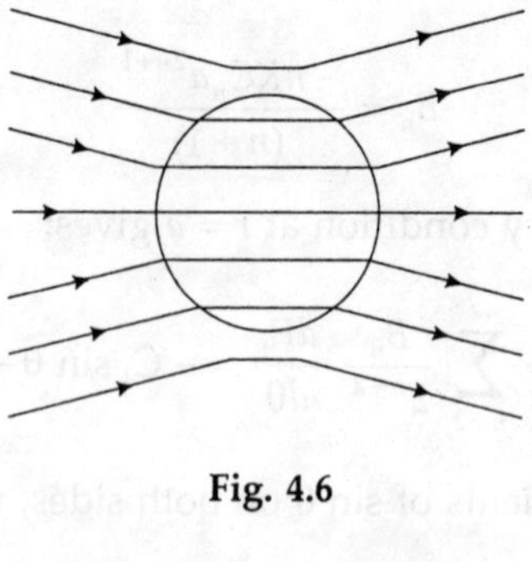

Fig. 4.6

As a limiting case, we may regard a conductor as a substance for which $K \to \infty$. Then, we get

$$\mathbf{E}_1 = 0$$

and
$$\mathbf{E}_2 = E_0\,\hat{\mathbf{k}} + \frac{E_0 a^3}{r^3}(2\cos\theta\,\hat{\mathbf{r}} + \sin\theta\hat{\boldsymbol{\theta}})$$

This is the field obtained from the potential $V(r, \theta)$ in Sec. 3.3 where we discussed an uncharged conducting sphere in an electric field.

4.5 THE CLAUSIUS—MOSSOTTI RELATION

In Section 4.3, we learnt that linear dielectrics are those for which the polarization is proportional to the electric field. This field is the macroscopic field. In Section 4.1, we had taken the induced dipole moment in an atom or molecule to be proportional to the electric field. This latter field is the field which acts on the individual atom or molecule. This is not in general the same as the macroscopic field, because of polarization. Let us call the macroscopic field **E** and the field which acts on an individual atom or molecule, $\mathbf{E}_m$. We can write

$$\mathbf{E}_m = \mathbf{E} + \mathbf{E}_1 - \mathbf{E}_2 \tag{4.5.1}$$

where $\mathbf{E}_1$ is the field produced by neighbouring molecules and $\mathbf{E}_2$, the average field due to the dielectric[(3)], treated as a continuum and described by **P**.

To calculate $\mathbf{E}_2$, let us consider a charge distribution characterized by a volume charge density $\rho(\mathbf{r}')$. We will take a spherical region of radius a and calculate the volume integral of the electric field **E** over this sphere.

$$\int_V \mathbf{E}dv = -\int_V \nabla\varphi dv$$

$$= -\int_S \varphi\hat{\mathbf{n}}da$$

where φ is the potential function, V is the volume of the sphere and S is its surface. The second line follows from a vector identity. We take the origin of our co-ordinate system at the centre of the sphere; $\hat{\mathbf{n}}$ is the unit outward normal, $\hat{\mathbf{n}} = \dfrac{\mathbf{r}}{a}$. Also $da = a^2 d\Omega$, where $d\Omega$ is an element of solid angle.

$$\therefore \qquad \int_V \mathbf{E}dv = -\int_S \varphi\hat{\mathbf{n}}a^2 d\Omega$$

We may write, for a volume charge distribution,

$$\therefore \qquad Q(\mathbf{r}) = \frac{1}{4\pi\varepsilon_0}\int \frac{\rho(\mathbf{r}')dv'}{|\mathbf{r} - \mathbf{r}'|}$$

where the primed co-ordinates refer to the source and the unprimed co-ordinates to the field point.

$$\therefore \qquad \int_V \mathbf{E} dv = \frac{-a^2}{4\pi\varepsilon_0} \iint_S \frac{\rho(\mathbf{r}')\hat{\mathbf{n}} d\Omega dv'}{|\mathbf{r}-\mathbf{r}'|} \tag{4.5.2}$$

In terms of polar and azimuthal angles, we may write $\hat{\mathbf{n}} = \hat{\mathbf{i}} \sin\theta \cos\phi + \hat{\mathbf{j}} \sin\theta \sin\phi + \hat{\mathbf{k}} \cos\theta$.

$$\frac{1}{|\mathbf{r}-\mathbf{r}'|} = \sum_{m=0}^{\infty} \frac{r'^m}{r^{m+1}} P_m(\cos\gamma) \text{ if } r > r'$$

$$= \sum_{m=0}^{\infty} \frac{r^m}{(r')^{m+1}} P_m(\cos\gamma) \text{ if } r' > r$$

$\hat{\mathbf{n}}.\hat{\mathbf{n}}' = \sin\theta \cos\phi \sin\theta' \cos\phi' + \sin\theta \sin\phi \sin\theta' \sin\phi' + \cos\theta \cos\theta'$
or, $\cos\gamma = \cos\theta \cos\theta' + \sin\theta \sin\theta' \cos(\phi - \phi')$

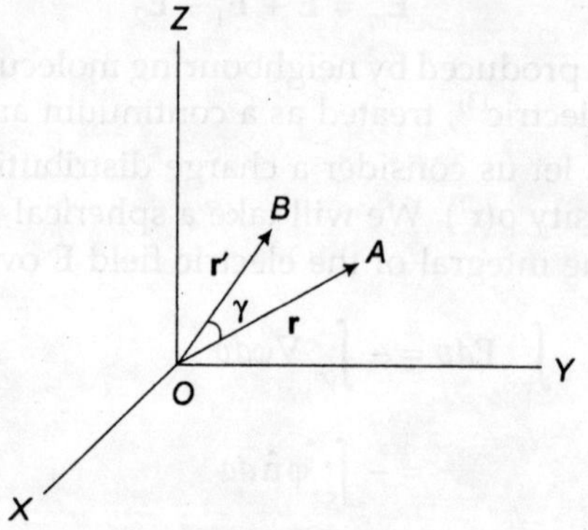

Fig. 4.7

Only the $m = 1$ term in the above series will give nonzero result in the integration over Ω in (4.5.2).

$$\int \frac{\hat{\mathbf{n}} d\Omega}{|\mathbf{r}-\mathbf{r}'|} = \frac{\mathbf{r}'}{\mathbf{r}^2} \int_S \hat{\mathbf{n}} \cos\gamma \, d\Omega$$

$$= \frac{\mathbf{r}'}{\mathbf{r}^2} \Bigg[\hat{\mathbf{i}} \int_0^{2\pi} \int_0^{\pi} \sin^2\theta \cos\phi \cos\gamma \, d\theta d\phi + \hat{\mathbf{j}} \int_0^{2\pi} \int_0^{\pi} \sin^2\theta \sin\phi \cos\gamma \, d\theta d\phi + \hat{\mathbf{k}} \int_0^{2\pi} \int_0^{\pi} \cos\theta \sin\theta \cos\gamma d\theta \, d\phi \Bigg]$$

$$\int_0^{2\pi}\int_0^{\pi} \sin^2\theta \cos\phi \cos\gamma d\theta d\phi = \frac{4\pi}{3}\sin\theta'\cos\phi'$$

$$\int_0^{2\pi}\int_0^{\pi} \sin\theta \sin\phi \cos\gamma d\theta d\phi = \frac{4\pi}{3}\sin\theta'\sin\phi'$$

$$\int_0^{2\pi}\int_0^{\pi} \cos\theta \sin\theta \cos\gamma d\theta d\phi = \frac{4\pi}{3}\cos\theta'$$

$$\therefore \quad \int_S \frac{\hat{\mathbf{n}}d\Omega}{|\mathbf{r}-\mathbf{r}'|} = \frac{r'}{r^2}\left[\hat{\mathbf{i}}\frac{4\pi}{3}\sin\theta'\cos\phi' + \hat{\mathbf{j}}\frac{4\pi}{3}\sin\theta'\sin\phi' + \hat{\mathbf{k}}\frac{4\pi}{3}\cos\theta'\right]$$

$$= \frac{r'}{r^2}.\frac{4\pi\hat{\mathbf{n}}'}{3}$$

where $\hat{\mathbf{n}}' = \dfrac{\mathbf{r}'}{r'}$

$$\therefore \quad \int_V \mathbf{E}dv = \frac{1}{4\pi\varepsilon_0}\left[\frac{-4\pi a^2}{3}\int \frac{r'\hat{\mathbf{n}}'}{r^2}\rho(\mathbf{r}')dv'\right]$$

If the sphere of radius '*a*' completely encloses the distribution, $r = a$ and so

$$\int_V \mathbf{E}dv = \frac{1}{4\pi\varepsilon_0}\left[\frac{-4\pi}{3}\int \mathbf{r}'\rho(\mathbf{r}')dv'\right]$$

or,
$$\int_V \mathbf{E}dv = \frac{-\mathbf{p}}{3\varepsilon_0} \tag{4.5.3}$$

where $\mathbf{p} = \int \mathbf{r}'\rho(\mathbf{r}')dv'$ is the dipole moment of the charge distribution with respect to the centre of the sphere.

Coming back to the dielectric, the average field inside a spherical volume of radius '*a*' within it is

$$\mathbf{E}_2 = \frac{3}{4\pi a^3}\int_S \mathbf{E}dv = \frac{3}{4\pi a^3}\frac{(-\mathbf{p})}{3\varepsilon_0} = \frac{-\mathbf{p}}{4\pi a^3\varepsilon_0}$$

The dipole moment of this region is

$$\mathbf{p} = \frac{4\pi a^3\mathbf{P}}{3} \tag{4.5.4}$$

assuming the spherical volume to be small enough so that **P** is constant inside it.

$$\therefore \qquad \mathbf{E}_2 = -\frac{\mathbf{P}}{3\varepsilon_0} \tag{4.5.5}$$

The electric field is more difficult to calculate but it was shown by Lorentz that for a simple cubic lattice, it is zero. In this case we may write

$$\mathbf{E}_{\mathrm{m}} = \mathbf{E} + \frac{\mathbf{P}}{3\varepsilon_0} \tag{4.5.6}$$

If there are N molecules per unit volume in a material, then we may write

$$\mathbf{P} = N\mathbf{p} = N\alpha\mathbf{E}_m$$

where α is the molecular polarizability

or,
$$\mathbf{P} = N\alpha\left(\mathbf{E} + \frac{\mathbf{P}}{3\varepsilon_0}\right)$$

or,
$$\mathbf{P} = \frac{N\alpha\,\mathbf{E}}{\left(1 - \dfrac{N\alpha}{3\varepsilon_0}\right)}$$

The electric susceptibility χ_e is given by

$$\chi_e = \frac{N\alpha}{\left(1 - \dfrac{N\alpha}{3\varepsilon_0}\right)}$$

$$\varepsilon = \frac{\varepsilon_0 + \dfrac{2N\alpha}{3}}{1 - \dfrac{N\alpha}{3\varepsilon_0}}$$

In terms of the dielectric constant K of a material, we have,

$$\alpha = \frac{3\varepsilon_0}{N}\frac{(K-1)}{(K+2)} \tag{4.5.7}$$

Equation (4.5.7) is called the *Clausis–Mossotti* relation. It is a relation between a microscopic quantity (α) and a macroscopic quantity (K). It is true for gases and only approximately valid for liquids and solids.

4.6 ENERGY IN DIELECTRIC SYSTEMS

The electrostatic energy of a charge distribution characterized by a charge density $\rho(\mathbf{r})$ is given by

$$U = \frac{1}{2}\int \rho\varphi dv$$

where φ is the electric potential. In the case of dielectrics, we have to do work to bring the true charges in from infinity and also to achieve the state of polarization. Let us suppose we increase the charge density by a small amount δρ. Then the change in the energy is given by

$$\delta U = \int \delta\rho\varphi dv \tag{4.6.1}$$

This equation follows from the elementary result that work done = potential × change in charge at that point. Let us rewrite this in terms of fields. We will use eqn. (4.3.6): $\nabla.\mathbf{D} = \rho$. So

$$\nabla.(\delta\mathbf{D}) = \delta\rho$$

$$\delta U = \int \nabla.(\delta\mathbf{D})\varphi dv \tag{4.6.2}$$

A vector identity is

$$\nabla.(\varphi\delta\mathbf{D}) = \varphi\nabla.(\delta\mathbf{D}) + \delta\mathbf{D}.\nabla\varphi$$

$$\delta U = \int \nabla.(\varphi\delta\mathbf{D})dv - \int \delta\mathbf{D}.\nabla\varphi dv$$

$$= \int_S \varphi\delta\mathbf{D}.dv + \int \delta\mathbf{D}.\mathbf{E}dv$$

where the first term has been written using the divergence theorem with S being the surface bounding the volume V of integration. We take the charge distribution ρ to be localized. Then, taking the volume V to be large so that the surface S is far away, $\varphi \to 0$ there, and the first term goes to zero.

$$\therefore \qquad \delta U = \int \mathbf{E}.\delta\mathbf{D}dv \tag{4.6.3}$$

The total electrostatic energy is obtained by integrating from 0 to D.

$$U = \int\int_0^D \mathbf{E}.\delta\mathbf{D}dv$$

If the medium is linear, then,

$$\delta(\mathbf{E.D}) = \delta\mathbf{E.D} + \mathbf{E}.\delta\mathbf{D} = 2\varepsilon\mathbf{E}.\delta\mathbf{E}$$

$$\frac{1}{2}\delta(\mathbf{E.D}) = \varepsilon\mathbf{E}.\delta\mathbf{E} = \mathbf{E}.\delta\mathbf{D} \tag{4.6.4}$$

$$\therefore \qquad U = \frac{1}{2}\int \mathbf{E.D}dv \tag{4.6.5}$$

4.7 DISPERSION IN DIELECTRICS: A SIMPLE MODEL

In our discussion of polarization in the earlier sections of this chapter, we have considered electrostatic fields. In this section, we will discuss briefly what happens when a dielectric is placed in a time-varying field, e.g. an electromagnetic wave. The nature of the response to a time-varying electric field depends on whether it consists of polar or non-polar molecules. If the dielectric is non-polar, dipole moments are induced in the molecules. As the external electric field changes with time, so does the dipole moment. The motion of the electrons involves times of the order of 10^{-16}s.[4] So as the frequency of the field is increased the dipole moments change in step with the field. This goes on for frequencies from zero (dc) up to that of visible light ($\sim 10^{14}$ Hz). Thus the dielectric constant does not change for this range of frequencies. A polar molecule, on the other hand, has a permanent dipole moment. The external field exerts a torque on the dipole. When the frequency of the field is increased, the alignment of the dipole gradually becomes out of step because the moment of inertia of the molecules is quite large. The time required for a polar molecule like water to respond to a change in the field is about 10^{-11}s. Experimentally, we find that the dielectric constant of water remains at about 80 for up to $\sim 10^{10}$ Hz and then falls rapidly to a value for a non-polar liquid at $\sim 10^{11}$ Hz.

We will take as a model of an atom an electron of charge e and mass m bound to a nucleus by a force proportional to the displacement (force constant = k) from the equilibrium position. In addition we will assume a damping force proportional to velocity (proportionality constant = b) to take into account the fact that as the electron accelerates, it loses energy by radiation. The electric field, in this case, is the molecular field E_m, which varies sinusoidally with time, with a frequency ω. The equation of motion of the electron is then

$$\frac{m\, d^2x}{dt^2} + \frac{b\, dx}{dt} + kx = eE_m$$

Dividing the above equation throughout by m and calling $\gamma = b/m$, we get

$$\frac{d^2x}{dt^2} + \frac{\gamma dx}{dt} + \omega_0{}^2 x = \frac{eE_m}{m}$$

where $\omega_0 = \sqrt{(k/m)}$ is the natural angular frequency of the electron's motion. We take $E_m = E_0 e^{-j\omega t}$. Then we get,

$$\frac{d^2x}{dt^2} + \frac{\gamma dx}{dt} + \omega_0{}^2 x = \frac{eE_0 e^{-j\omega t}}{m} \tag{4.7.1}$$

Let us choose $x = Ae^{-j\omega t}$ as our solution. So we get,

$$[(\omega_0^2 - \omega^2) - j\gamma\omega]A = \frac{eE_0}{m} \tag{4.7.2}$$

The polarization P then has an amplitude given by $P = NeA = \dfrac{Ne^2 E_0/m}{(\omega_0^2 - \omega^2) - \gamma j\omega}$. If we assume that $P = \chi_e E$, where E is the peak macroscopic field, we have, from the above equation,

$$\chi_e E = \frac{Ne^2 E_0/m}{(\omega_0^2 - \omega^2) - \gamma j\omega} \tag{4.7.3}$$

Since $E_0 = E + \dfrac{P}{3\varepsilon_0}$, from (4.5.6), we find, on substituting $P = \chi_e E$,

$$E = \frac{E_0}{1 + \dfrac{\chi_e}{3\varepsilon_0}} \tag{4.7.4}$$

Putting this on the L.H.S. of (4.7.3), we get

$$\frac{\chi_e}{1 + \dfrac{\chi_e}{3\varepsilon_0}} = \frac{Ne^2}{m\{(\omega_0^2 - \omega^2) - \gamma j\omega\}}$$

In terms of the dielectrics constant K, $\chi_e = (K - 1)\varepsilon_0$, so that we get

$$\frac{K-1}{K+2} = \frac{Ne^2}{3m\varepsilon_0} \cdot \frac{1}{\{(\omega_0^2 - \omega^2) - \gamma j\omega\}}$$

In an atom, there would be a number of electrons, each with its own natural frequency ω_{0n}. If f_n is the fraction of electrons with this frequency, then we may write

$$\frac{K-1}{K+2} = \frac{Ne^2}{3m\varepsilon_0} \cdot \sum_n \frac{f_n}{\{(\omega_{0n}^2 - \omega^2) - \gamma j_n \omega\}} \tag{4.7.5}$$

On the basis of this model, we can say that the dielectric constant is dependent on the frequency of the external field and it is complex. This is a classical model. In quantum mechanics, we would interpret ω_{0n} as proportional to the difference in energy between two levels of an atom, and f_n is called the *oscillator strength*. Equation (4.7.5) is a generalization of the Clausius–Mossotti relation for time-varying fields.

The table below summarizes some of the important results of this chapter in Gaussian units:

Physical quantity/equation	*MKS*	*Gaussian*
Electric displacement (4.3.4)	$\mathbf{D} = \varepsilon_0\mathbf{E} + \mathbf{P}$	$\mathbf{D} = \mathbf{E} + 4\pi\mathbf{P}$
Gauss' law (4.3.6)	$\nabla.\mathbf{D} = \rho$	$\nabla.\mathbf{D} = 4\pi\rho$
Clausius–Mossotti relation (4.5.7)	$\alpha = \frac{3\varepsilon_0}{N}\frac{(K-1)}{(K+2)}$	$\alpha = \frac{3}{4\pi N}\frac{(K-1)}{(K+2)}$
Energy in dielectric Systems (4.6.5)	$U = \frac{1}{2}\int \mathbf{E.D}dv$	$U = \frac{1}{8\pi}\int \mathbf{E.D}dv$

Tutorial

Consider a parallel-plate capacitor with vacuum between its plates. The area of the plates is A and separation between them is d. The plates are connected to a battery with potential difference V_0. Write down an expression for the charge q_0 on the positive plate and the capacitance C_0. Now suppose that with the battery connected, a dielectric slab having dielectric constant K is introduced between the plates. Using Gauss' law for dielectrics (4.3.5) calculate the charge on the capacitor plates and its capacitance. Next consider what happens if instead of keeping the battery connected when the dielectric is introduced, we first disconnect the battery and then introduce the dielectric slab. Again, using (4.3.5), calculate the electric displacement, capacitance and charge on the capacitor plates.

Problems

1. A dielectric sphere of radius 'a' has a polarization $\mathbf{P} = k\mathbf{r}$, where k is a constant and the origin is at the centre. Where do the polarization charges appear as far as this sphere is concerned ? For this sphere, calculate the surface density of the polarization charges and their volume density. What is the field intensity outside the sphere due to this polarization? Also find the field intensity just inside the polarized sphere. *(University of Calcutta, 1998)*

   [Ans. $\sigma_P = ka$, $\rho_P = -3k$, $\mathbf{E}_{out} = 0$, $\mathbf{E}_{in} = -k\mathbf{r}/\varepsilon_0$**]**

2. A dielectric sphere of radius 'a' and permittivity ε_1 is placed in a medium ε_2. Show that the polarization induced in the sphere is given by

$$\mathbf{P} = \frac{3\varepsilon_2\,(\varepsilon_1 - \varepsilon_2)\,E_0\hat{\mathbf{k}}}{(\varepsilon_1 + 2\varepsilon_2)}$$

   $\hat{\mathbf{k}}$ being the unit vector along the z-axis. *(University of Calcutta,2001)*

3. The distance between the plates of a parallel plate air capacitor is d. A dielectric slab of thickness x is introduced in the air gap. Show that the capacitance will be doubled if the dielectric constant of the slab is $k = \dfrac{2x}{2x - d}$. *(University of Calcutta, 2002)*

4. A spherical conductor of radius 'a', with no other conductor in its neighbourhood, is coated with a uniform thickness d of shellac of dielectric constant k. Show that the capacitance of the conductor in increased in the ratio $\dfrac{k(a + d)}{ka + d}$.

5

The Magnetic Field

5.1 DEFINITION

Consider a charge q moving with velocity $\mathbf{v}$ in a certain region of space. No electric field exists in this region. Suppose that it experiences a sideways force at right angles to $\mathbf{v}$. This force is a maximum in a certain direction and decreases to zero when $\mathbf{v}$ is perpendicular to this direction. Furthermore the force goes to zero if the particle is at rest. Then the region of space is said to be permeated by a magnetic field $\mathbf{B}$. The magnitude of the field is given by

$$\mathbf{B} = \frac{F}{qv_\perp}$$

where $v_\perp$ is the component of $\mathbf{v}$ perpendicular to $\mathbf{B}$. If θ is the angle between $\mathbf{v}$ and $\mathbf{B}$, then $v_\perp = v \sin\theta$.

$$\text{B} = \frac{F}{qv \sin\theta} \tag{5.1.1}$$

The direction of $\boldsymbol{B}$ is that along which F becomes zero. The force is thus given by the cross product of $\mathbf{v}$ and $\mathbf{B}$

$$\mathbf{F} = q(\mathbf{v} \times \mathbf{B}) \tag{5.1.2}$$

The SI unit of B is Tesla (T). It is defined in the next section. 1 T = 10^4 gauss, where gauss is the unit of B in the Gaussian system. In the presence of the both electric field and magnetic fields, the force is

$$\mathbf{F} = q(\mathbf{E} + \mathbf{v} \times \mathbf{B}) \tag{5.1.3}$$

Equation (5.1.3) is called the Lorentz force law.

5.2 FORCE A CURRENT-CARRYING IN A MAGNETIC FIELD

An electric current in a conductor is a flow of charges in it. Let us suppose that there are N charges each of magnitude q, per unit volume, all of which are moving with the same velocity **v**. Let the area of cross section of the conductor be A, assumed uniform. The charge in a volume element Adl (where dl is an element of length) of the conductor is then $NqAdl$. If the conductor is placed in a magnetic field **B**, the Lorentz force $d\mathbf{F}$ on this charge is

$$d\mathbf{F} = NqAdl\mathbf{v} \times \mathbf{B} \tag{5.2.1}$$

The current **I** is given by $\mathbf{I} = NqA\mathbf{v}$ (charge flowing per unit time). So we have

$$d\mathbf{F} = \mathbf{I}dl \times \mathbf{B} \tag{5.2.2}$$

Often, the vector $d\mathbf{l}$ is taken in the direction of current I and so the above equation is written as $d\mathbf{F} = Id\mathbf{l} \times \mathbf{B}$. Thus **B** has units of force per unit current per unit length. In the SI system, the unit of force is Newton an the unit of current is ampere (A). So

$$1\ T = 1\ \mathrm{N/Am} \tag{5.2.3}$$

The total force on a current loop C is obtained by integrating (5.2.2):

$$\mathbf{F} = I\oint_C d\mathbf{l} \times \mathbf{B} \tag{5.2.4}$$

5.3 TORQUE ON A CURRENT LOOP IN A MAGNETIC FIELD

Using eqn. (5.2.2) let us now consider the forces on a rectangular current-carrying conductor in a magnetic field. We will assume that the dimensions of the loop are 'a' and 'b' (see Fig. 5.1b) perpendicular to and along the z-axis, respectively. We take the field to be along the y-axis, making an angle θ with the normal to the plane of the coil. There will be a force $F_1 = IbB$ in the direction of the negative x-axis on the side BA and an equal force on the side DC along the positive x-axis. There are no forces on the sides AD and BC because the lengths AD and BC parallel to **B**. These two forces constitute a couple of moment F_1 a sin θ = $IabB \sin\theta$. Now the area of the loop is $A = ab$. So the torque is given by

$$\tau = IAB \sin\theta$$

The quantity IA is called the magnetic moment m of the loop. So

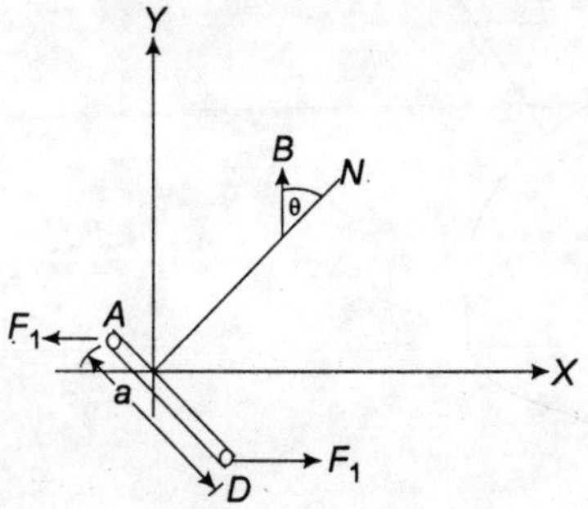

Fig. 5.1a

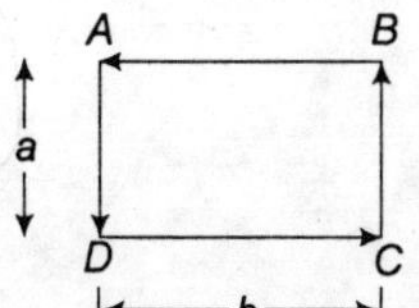

Fig. 5.1b

$$\tau = mB \sin\theta \quad (m = IA)$$

or, in vector notation,

$$\boldsymbol{\tau} = \mathbf{m} \times \mathbf{B} \tag{5.3.1}$$

The direction of **m** is perpendicular to the plane of the rectangular loop and is related to the direction of the current by the right-hand screw rule. Thus, in Fig. 5.1a, **m** is along *ON*. The torque is zero when $\theta = 0$, i.e. **m** is parallel to **B**. The plane of the coil is perpendicular to **B**. Therefore, when a current-carrying conducting loop is placed in a magnetic field, it experiences a torque. The magnetic field tries to align the magnetic moment of the loop parallel to itself.

In atoms we have electrons moving around a nucleus. These circulating electrons constitute current loops. So an atom has a magnetic moment associated with it. The electrons also have momentum about the nucleus, for example. The ratio of the magnetic moment m of an atom to its angular momentum L is called its gyromagnetic ratio

$$g = \frac{m}{L} \tag{5.3.2}$$

This definition is also used for a current distribution.

5.4 MAGNETIC FIELDS DUE TO CURRENTS: BIOT–SAVART'S LAW

The pioneering experiment on the magnetic fields due to current-carrying conductors was done by Oersted in 1819. However, it was Biot, Savart and Ampere who established the foundation of the subject in their detailed experiments in the years 1820–1825. The magnetic field due to an element dl of a conductor carrying a current I at a point distant r from the element is given by

$$d\mathbf{B} = \frac{\mu_0}{4\pi} \frac{I d\mathbf{l} \times \mathbf{r}}{r^3} \tag{5.4.1}$$

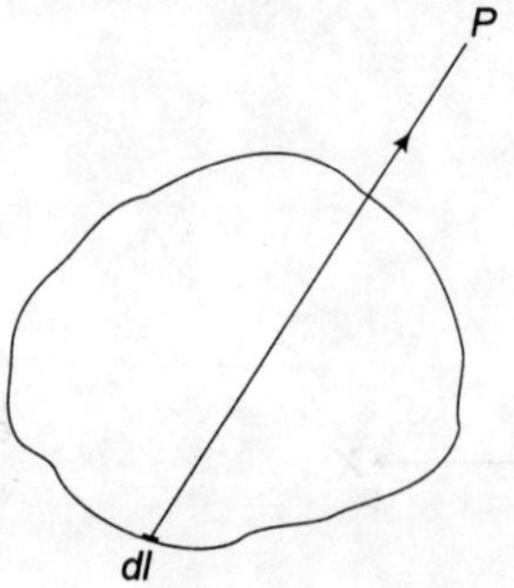

Fig. 5.2

in the SI system of units. By definition, the permeability of free space μ_0 is

$$\mu_0 = 4\pi \times 10^{-7}\ \text{N/A}^2 \tag{5.4.2}$$

Let us use (5.4.1) to find **B** in some simple cases.

First we shall calculate **B** due to a straight conductor carrying a current I (Fig. 5.3). Consider an element of length dx of the wire, which lies along the x-axis. The current flows in the direction of positive x. Therefore, the field dB at P on the y-axis is given by

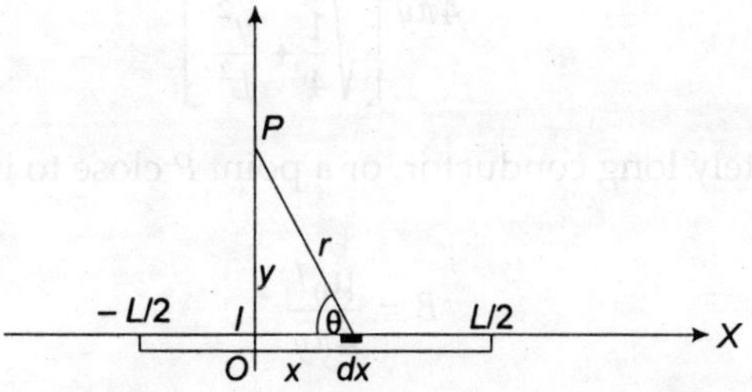

Fig. 5.3

$$dB = \frac{\mu_0}{4\pi}\frac{Idx\sin\theta}{(x^2+y^2)} \tag{5.4.3}$$

[since $\sin(\pi-\theta) = \sin\theta$]

It is directed along the positive z-axis. From the diagram,

$$\sin\theta = \frac{y}{\sqrt{x^2+y^2}}$$

Putting this in (5.4.3), we get

$$dB = \frac{\mu_0}{4\pi}\frac{Iydx}{(x^2+y^2)^{3/2}}$$

To find the field due to the length L of the wire we have to integrate over x. Note that the field at P due to all the elements is in same direction. So they add.

$$\therefore \quad B = \frac{\mu_0 Iy}{4\pi}\int_{-L/2}^{L/2}\frac{dx}{(x^2+y^2)^{3/2}}$$

$$= \frac{\mu_0 I}{4\pi}\left[\frac{x}{\sqrt{x^2+y^2}}\right]_{-L/2}^{L/2}$$

$$B = \frac{\mu_0 I}{4\pi y} \frac{[L]}{\sqrt{y^2 + \frac{L^2}{4}}} \tag{5.4.4}$$

Dividing the numerator and denominator by L, we get

$$B = \frac{\mu_0 I}{4\pi y} \left[\frac{1}{\sqrt{\frac{1}{4} + \frac{y^2}{L^2}}} \right]$$

For an infinitely long conductor, or a point P close to it, we take the limit $L \to \propto$:

$$B = \frac{\mu_0 I}{4\pi y} \tag{5.4.5}$$

The field is directed along the positive z-axis. If the point P were on the negative y-axis, the field would have been along the negative z-axis. The field lines are thus concentric circles with the wire lying along the axis of the circle.

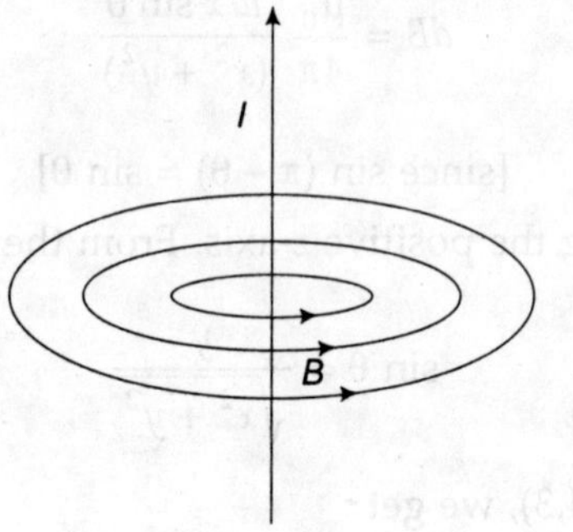

Fig. 5.4

Next let us calculate the field at a point on the axis of a circular loop of radius 'a'. We consider the loop to lie in the XY plane with its center at the origin O (Fig. 5.5). The axis of the loop is the z-axis. An element of the arc dl subtends an angle $d\theta$ at O and the vector $\mathbf{r}$ points from the element to $\mathbf{P}$. The field $d\mathbf{B}$ due to the element is perpendicular to the plane containing $d\mathbf{l}$ and $\mathbf{r}$, making an angle $90°\text{-}\phi$ with respect to the z-axis. To calculate the field due to the entire loop, we have to integrate over it. Only the components along the z-axis will add and the components perpendicular to the z-axis will cancel for symmetrically placed pairs of elements.

$$dB = \frac{\mu_0}{4\pi} \frac{Idl}{(a^2 + z^2)}$$

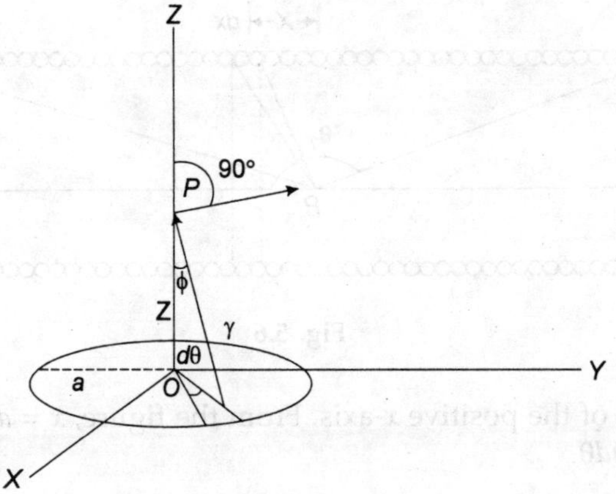

Fig. 5.5

$$= \frac{\mu_0}{4\pi} \frac{Iad\theta}{(a^2 + z^2)}$$

$$B = \int dB \sin\phi$$

$$= \frac{\mu_0 Ia}{4\pi} \int_0^{2\pi} \frac{d\theta}{(a^2 + z^2)} \cdot \frac{a}{\sqrt{a^2 + z^2}}$$

$$= \frac{\mu_0 Ia^2}{4\pi(a^2 + z^2)^{3/2}} \int_0^{2\pi} d\theta$$

$$B = \frac{\mu_0 Ia^2}{2(a^2 + z^2)^{3/2}} \tag{5.4.6}$$

The magnetic field is along the positive z-axis. At the center of the circle (z = 0),

$$B = \frac{\mu_0 I}{2a} \tag{5.4.7}$$

We will now use the result (5.4.6) to calculate the field due to a solenoid having N turns and length L, carrying a current I. Each turn has a radius 'a'. Consider an element of length dx along the solenoid. It has N/Ldx turns. It is at a horizontal distance x from the point P where we wish to calculate the field. θ is the angle subtended by the element at P. The field at P due to this element is

$$dB = \frac{\mu_0 Ia^2}{2} \frac{N/L\,dx}{(a^2 + x^2)^{3/2}} \tag{5.4.8}$$

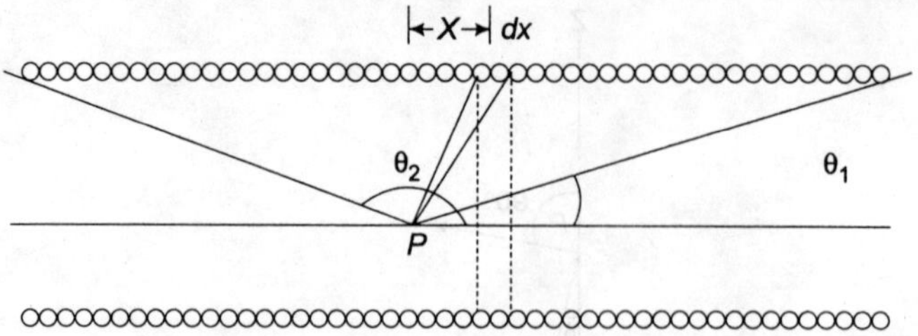

Fig. 5.6

in the direction of the positive x-axis. From the figure, $x = a \cot \theta$. Therefore $dx = -a \operatorname{cosec}^2\theta d\theta$.

$$\therefore \qquad dB = \frac{-\mu_0 NI}{2L} \sin \theta d\theta$$

The total field due to the entire solenoid is obtained by integrating the above expression from $\theta = \theta_2$ to $\theta = \theta_1$, corresponding to the two ends of the solenoid.

$$\boldsymbol{B} = \frac{-\mu_0 NI}{2L} \int_{\theta_2}^{\theta_1} \sin\theta d\theta$$

$$\therefore \qquad B = \frac{\mu_0 NI}{2L} (\cos \theta_1 - \cos \theta_2) \qquad (5.4.9)$$

For an infinitely long solenoid, $\theta_1 \to 0$ and $\theta_2 \to \pi$. Thus

$$B = \frac{\mu_0 NI}{L} \qquad (5.4.10)$$

This field is along the axis of the solenoid. Outside a finite solenoid, the field is very small compared to that inside.

In all the above examples, we have looked at currents flowing along wires. We can have something more general – a current distribution. This could be a current flowing along a sheet (Fig. 5.7) or a volume of material (Fig. 5.8). A current along a sheet by a surface current density **K** whose magnitude is current per unit length-perpendicular-to-flow. So if $dl_\perp$ is an element of length perpendicular to **K**, the current is $dI = \mathbf{K} dl_\perp$. A current

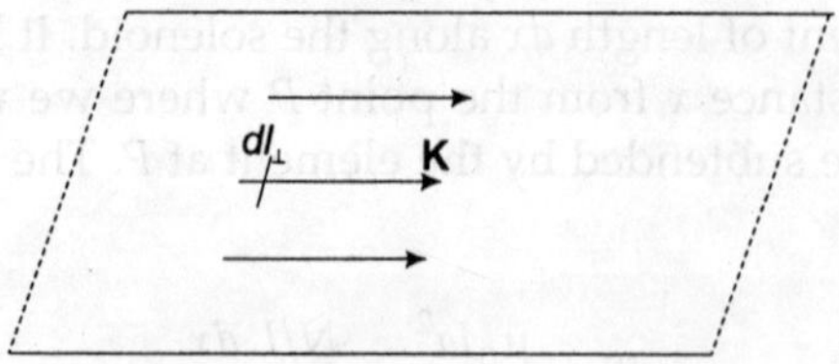

Fig. 5.7

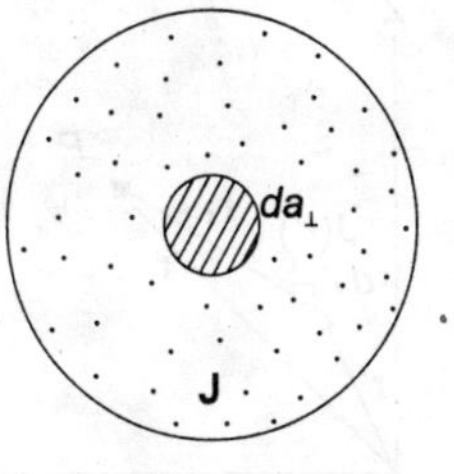

Fig. 5.8

flowing perpendicular to a cross section of a tube is described by a volume current density **J**. If $da_\perp$ is an element of area normal to **J**, the current flowing through it is $d\mathbf{I} = \mathbf{J}da_\perp$.

For a surface current density, Biot-Savart's law is

$$\mathbf{B} = \frac{\mu_0}{4\pi}\int \frac{\mathbf{K}\times\mathbf{r}}{r^3}da \tag{5.4.11}$$

For a volume current density, Biot- Savart's law is

$$\mathbf{B} = \frac{\mu_0}{4\pi}\int \frac{\mathbf{J}\times\mathbf{r}}{r^3}dv \tag{5.4.12}$$

5.5 ABSENCE OF MAGNETIC MONOPOLES AND AMPERE'S LAW

In out study of electrostatics, we came across isolated positive and negative charges. We also encountered and electric dipole. Contrary to this, we do not find isolated north and south poles. They always occur in pairs. In the early twentieth century such a pair was called a dipole. The modern view pictures the dipole as a current loop. The direction of the magnetic moment and the direction of current are related by the right hand screw rule.

We will now express this absence of magnetic monopoles in mathematical form. We shall first rewrite (5.4.12), with the functional dependences on co-ordinates explicitly shown.

$$\mathbf{B}(\mathbf{r}) = \frac{\mu_0}{4\pi}\int \frac{\mathbf{J}(\mathbf{r}')\times\mathbf{r}_1 dv'}{r_1^3} \tag{5.4.12a}$$

Here $\mathbf{J}(\mathbf{r}')$ is a function of the primed co-ordinates (x', y', z') referred to as the source co-ordinates. The point where we wish to calculate the field **B** has co-ordinates (x, y, z). These are referred to as the field co-ordinates. They are independent of the source co-ordinates. Also $\mathbf{r}_1 = \mathbf{r} - \mathbf{r}'$ (Fig. 5.9).

Let us take the divergence of both sides of (5.4.12a) with respect to the field co-ordinates.

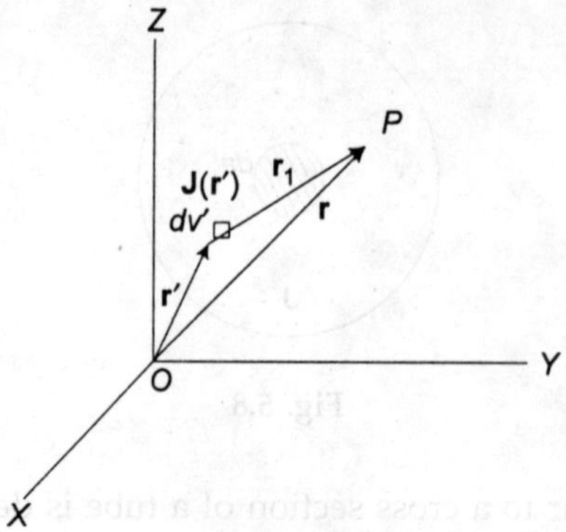

Fig. 5.9

$$\nabla.\mathbf{B}(\mathbf{r}) = \int \nabla.\left\{\frac{\mathbf{J}(\mathbf{r}') \times \mathbf{r}_1}{r_1^3}\right\} dv'$$

Notice that we have been able to take the divergence operator inside the integral because the derivatives are with respect to the field co-ordinates whereas the integration is with respect to the source co-ordinates. We shall now use a vector identity

$$\nabla.(\mathbf{A}_1 \times \mathbf{A}_2) = \mathbf{A}_2.(\nabla \times \mathbf{A}_1) - \mathbf{A}_1.(\nabla \times \mathbf{A}_2)$$

$$\nabla.\left\{\frac{\mathbf{J}(\mathbf{r}') \times \mathbf{r}_1}{r_1^3}\right\} = \frac{\mathbf{r}_1}{r_1^3}.\{\nabla \times \mathbf{J}(\mathbf{r}')\} - \mathbf{J}(\mathbf{r}').\left\{\nabla \times \left(\frac{\mathbf{r}_1}{r_1^3}\right)\right\}$$

The first term is zero because of the reason stated above. To evaluate the second term, we note that

$$\nabla \times \left(\frac{\mathbf{r}_1}{r_1^3}\right) = \begin{vmatrix} \mathbf{i} & \mathbf{j} & \mathbf{k} \\ \dfrac{\partial}{\partial x} & \dfrac{\partial}{\partial y} & \dfrac{\partial}{\partial z} \\ \dfrac{x - x'}{r_1^3} & \dfrac{y - y'}{r_1^3} & \dfrac{z - z'}{r_1^3} \end{vmatrix}$$

Now, $$\frac{\partial}{\partial y}\frac{(z - z')}{r_1^3} - \frac{\partial}{\partial z}\frac{(y - y')}{r_1^3} = \frac{-3(y - y')(z - z') + 3(y - y')(z - z')}{r_1^5} = 0$$

$$\nabla \times \frac{\mathbf{r}_1}{r_1^3} = 0$$

$$\therefore \qquad \nabla.\mathbf{B} = 0 \qquad (5.5.1)$$

This implies that there are no sources or sinks of **B**. In other words, there are no isolated poles. Let us now recall (5.4.5). The magnetic field at a distance r from the wire is given

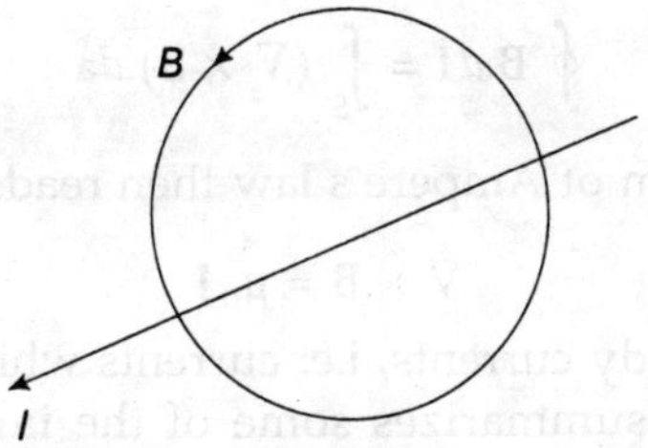

Fig. 5.10

by
$$\mathbf{B} = \frac{u_0 I \hat{\phi}}{2\pi r}$$

The magnetic field line is a circle with the wire passing through the centre. $\hat{\phi}$ is the unit vector along the tangent to the circle. If we now consider the line integral of **B** around the circle, we get

$$\oint \mathbf{B}.d\mathbf{l} = \int_0^{2\pi} \frac{u_0 I}{2\pi r} r d\phi = \frac{u_0 I}{2\pi} \int_0^{2\pi} d\phi$$

$$\therefore \qquad \oint \mathbf{B}.d\mathbf{l} = \mu_0 I$$

This result is true in general. If the path of integration is a closed loop of arbitrary shape, we still have

$$\oint \mathbf{B}.d\mathbf{l} = \mu_0 I_{\text{enc.}} \tag{5.5.2}$$

where I_{enc}. is the current enclosed by the loop. The direction of current and the magnetic field are related by the right hand screw rule. If the loop does not enclose current, the integral is zero. (See Fig. 5.11, curve C)

Equation (5.5.2) is known as Ampere's law. We can rewrite it recalling the definition of current in terms of **J**.

$$I_{\text{enc.}} = \int_S \mathbf{J}.d\mathbf{a} \tag{5.5.3}$$

where S is the surface enclosed by the loop. We can use Stokes' theorem to write

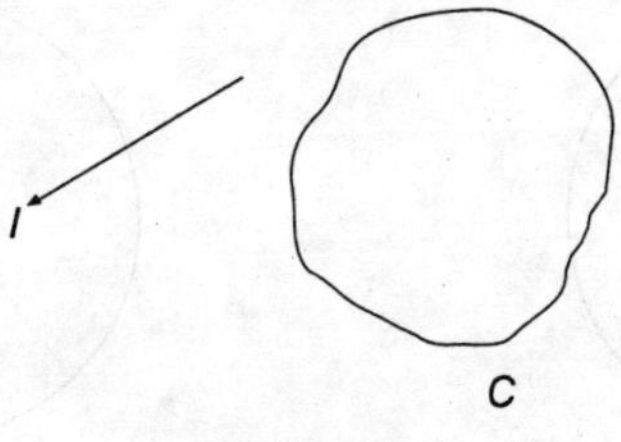

Fig. 5.11

$$\oint \mathbf{B}.d\mathbf{l} = \int_S (\nabla \times \mathbf{B}).d\mathbf{a}$$

The differential form of Ampere's law then reads

$$\nabla \times \mathbf{B} = \mu_0 \mathbf{J} \tag{5.5.4}$$

This is true for steady currents, i.e. currents which are not functions of time. The table below summarizes some of the important results of this chapter in Gaussian units.

Physically quantity/equation	*SI*	*Gaussian*
Lorentz force (5.1.3)	$\mathbf{F} = q(\mathbf{E} + \mathbf{v} \times \mathbf{B})$	$\mathbf{F} = q\left(\mathbf{E} + \frac{\mathbf{v}}{c} \times \mathbf{B}\right)$
Biot-Savart's law (5.4.1)	$d\mathbf{B} = \frac{\mu_0 I}{4\mu} \frac{d\mathbf{l} \times \mathbf{r}}{r^3}$	$d\mathbf{B} = \frac{I}{c} \frac{d\mathbf{l} \times \mathbf{r}}{r^3}$
Ampere's law (5.5.2)	$\oint \mathbf{B}.d\mathbf{l} = \mu_0 \mathbf{I}_{enc.}$	$\oint \mathbf{B}.d\mathbf{l} = \frac{4\pi I_{enc}}{c}$

Tutorial 1

We come across three types of materials in nature: diamegnetic, paramegnetic and ferromagnetic materials. When placed in a non-uniform magnetic field, diamagnetic materials move from the stronger to the weaker region of the field. An example is bismuth. The effort is very weak and a proper explanation requires quantum mechanics. A paramagnetic material, of the other hand, is attracted to the stronger region of the field. Examples are the ions of transition elements like $Mn^{2+}$and Gd^{3+}. The third category are ferromagnetic materials of which iron, cobalt and nickel are examples. These materials have a nonzero magnetic moment even the absence of a magnetic field.

We shall consider a simple classical model of diamagnetism here [(5)]. In a diamagnetic material, an atom has no net magnetic moment. Consider two electrons in an atom of the material, circulating in opposite directions around a positively charged nucleus (not shown) (Figs. 5.12a and 5.12b). Now

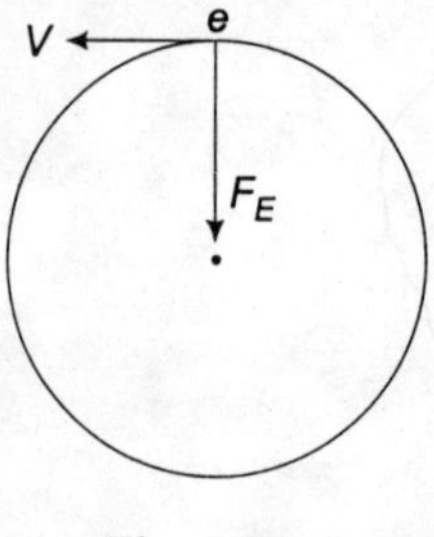

Fig. 5.12a

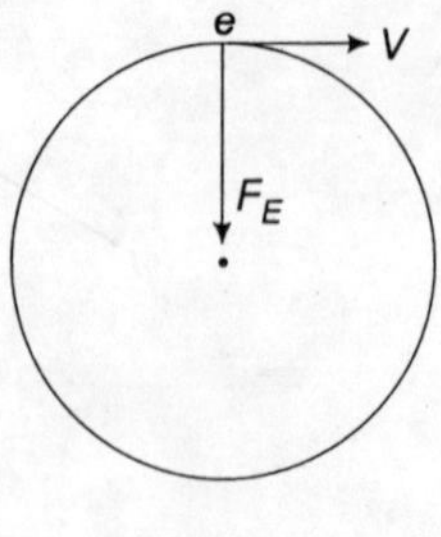

Fig. 5.12b

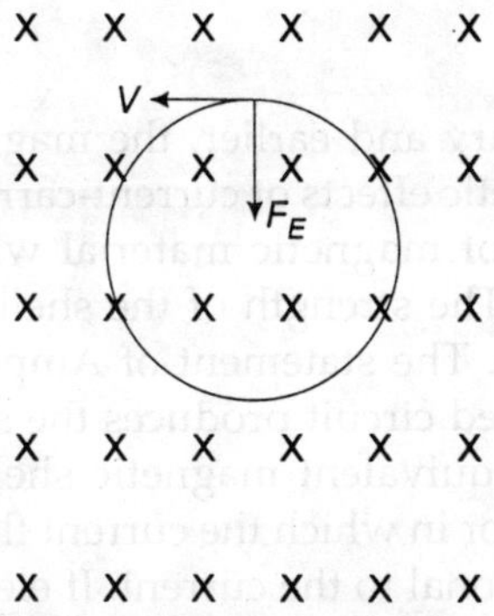

Fig. 5.13a

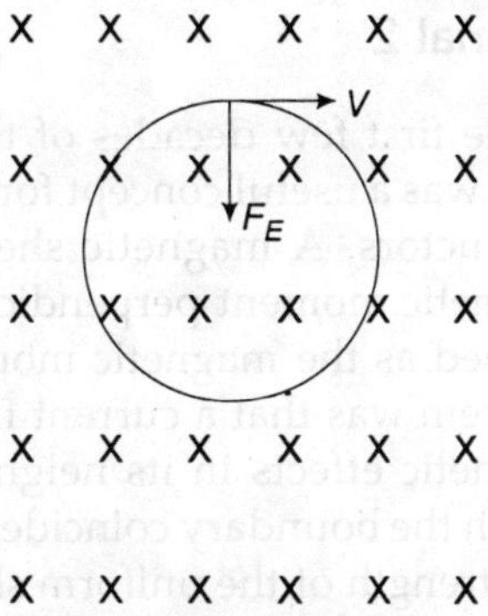

Fig. 5.13b

suppose the atom is placed in a magnetic field B (Fig. 5.13a and Fig. 5.13b), perpendicular to the plane of the diagram. The electrostatic force is F_E. The radius of the path is r, m_e is the mass of the electron and ω_0 is its angular frequency (in absence of the field). What is the direction of the force on the electron due to the magnetic field B in (a) and (b)? If the angular frequency of rotation charges to = $\omega = \omega_0 + \Delta\omega$ in presence of the field, which r remaining constant, show that

$$\Delta\omega = \pm \frac{eB}{2m_e}$$

if $\Delta\omega << \omega_0$. What is the charge is magnetic moment Δm? In absence of the field, the magnetic moments of the two electrons in an atom in Figs. 5.12a and 5.12b would cancel. Show that in the presence of the field, there would be a net magnetic moment $2\Delta\mathbf{m}$ opposite to $\mathbf{B}$. Thus, when a piece of diamagnetic material is placed near the north pole of a powerful magnet, a north pole is induced in the region of the material near it. So it is repelled from the pole of the magnet.

A paramagnetic material, on the other hand, has atoms with magnetic moments. When such a material is placed in a magnetic field, a torque acts on the moments, trying to align them with the field (Sec. 5.3). This alignment is disturbed by the thermal motion of the atoms. However, a fraction of the moments are sligned. So the material acquireds a moment parallel to the field. When a piece of paramagnetic material is placed near the north pole of a powerful magnet, a south pole a induced in the region near it. So the material is attracted towards the pole of the magnet.

Ferromagnetic materials are characterzed by *domains*. These are groups of atoms with aligned magnetic moments. Different groups are not aligned parallel to each other, however. When a ferromagnetic material is placed in a magnetic field, two processes occur: domain rotation and domain growth. The latter is an increase in size of domains oriented favourably with respect to the field at the expense of those oriented unfavourably. This results in a large magnetic moment of the sample. We shall not go into any further details of these.

Tutorial 2

In the first few decades of the twentieth century and earlier, the magnetic shell was a useful concept for calculating magnetic effects of current-carrying conductors. A magnetic shell is a thin sheet of magnetic material with a magnetic moment perpendicular to the sheet. The strength of the shell σ is defined as the magnetic moment per unit area. The statement of Ampere's theorem was that a current flowing in any closed circuit produces the same magnetic effects in its neighbourhood as an equivalent magnetic shell, of which the boundary coincides with the conductor in which the current flows, the strength of the uniform shell being proportional to the current. If electric current is measured in cgs electromagnetic units (emu), then the corresponding magnetic shell will produce the same field in the surrounding space if the strength of the shell in numerically equal to the current. This provides an alternative way of defining 1 emu of current (1 emu = 10 A). In those days, there was a Coulomb's law for magnetism and a potential. Let us calculate the potential and field due to a plane circular shell (of strength σ) or a current loop, at a point P on the axis (Fig. 5.14a). Consider an element of area A_1 formed between two rings of radii r and $r + \delta r$. It is $r\delta\phi\delta r$ (Fig. 5.14b). The solid angle $\delta\omega$ subtended by it at P is $\dfrac{r\delta\phi\delta r\cos\theta}{r^2 + x^2}$. The total solid angle subtended by the whole disc at P is

$$\omega = \int_0^a \int_0^{2\pi} \frac{r d\phi dr \cos\theta}{r^2 + x^2}$$

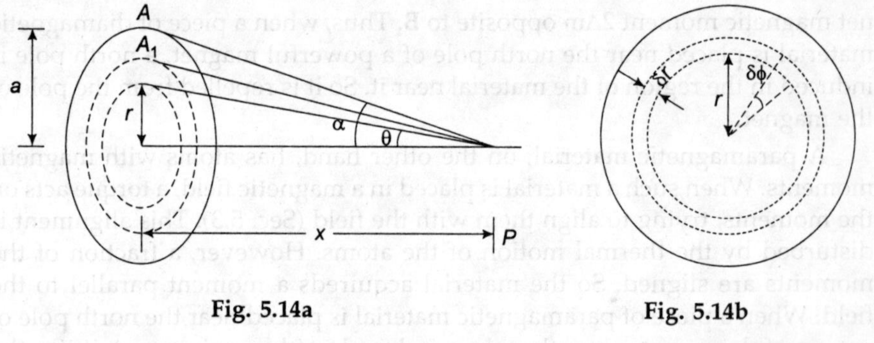

Fig. 5.14a Fig. 5.14b

Complete the calculation. The potential V_P at P is $\omega\sigma$. Therefore, in this case, write down V_P. The magnetic field intensity H in Oersted is

$$H = -\frac{\partial V_P}{\partial x}$$

You will find that the result is proportional to the field on the axis of a circular coil found in Sec. 5.4 (eqn. 5.4.6). The concept of a magnetic shell is obsolete today.

Tutorial 3

Consider two wires carrying currents I_1 and I_2, separated by a distance d in vacuum (Fig. 5.15). What is the direction and magnitude of the magnetic field B_1 due to current I_1 at the location of the other wire? If the length of this wire is 1, show that it is attracted to the first wire (carrying current I_1) with a force

$$F_2 = \frac{\mu_0 I_1 I_2 I}{2\pi d}$$

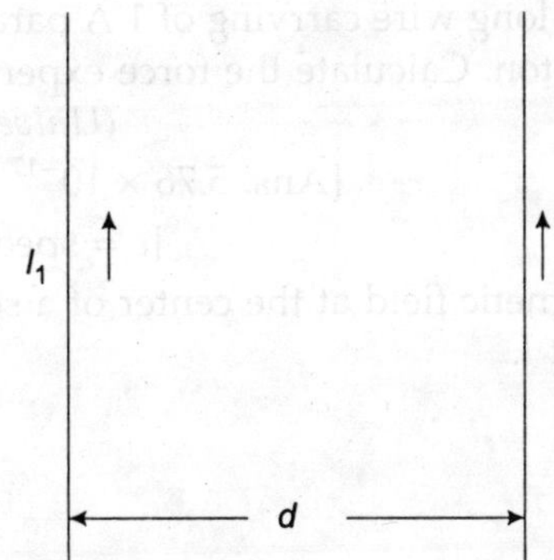

Fig. 5.15

The first wire is similarly attracted to the second wire with an equal force. If we take $I_1 = I_2 = 1$ A and $d = 1m$, then

$$\frac{F_2}{L} = \frac{\mu_0}{2\pi} = 2 \times 10^{-7}\,\text{N/m}$$

This is how the unit of current in the SI system – the ampere (A) is defined. It is that steady current, which, flowing in each of two infinitely long parallel conductors, placed 1 m apart in vacuum, causes a force of attraction of 2×10^{-7} Newton per metre of their length. What is the direction of the force on the wires if I_1 and I_2 are in opposite directions?

Problems

1. Two parallel wires each of length 1m are separated by a distance of 0.2 cm. Current 5A flows through each conductor in opposite directions. Calculate the mutual force between the conductors. Is it attractive or repulsive? *(University of Calcutta, 2002)*

   [Ans. 2.5×10^{-3} N; repulsive]

2. A current I is uniformly distributed over a wire of circular cross section, with radius R. What is the volume current density?-If the

current density is proportianal to distance from the axis, what will be the total current in the wire? *(University of Calcutta, 2001)*

$$\left[\textbf{Ans. } J = \frac{I}{\pi R^2}, \text{ total current} = \frac{2\pi k R^3}{3},\right.$$

$$\left.\text{where } k \text{ is the constant of proportionality}\right]$$

3. A proton moving with a velocity 0.6 c is placed at a distance of 10 cm from an infinitely long wire carrying of 1 A parallel to the direction of motion of the proton. Calculate the force experienced by the proton. *(University of Calcutta, 2000)*

   [**Ans.** 5.76×10^{-17} N, towards the wire.)

   [c = speed of light in vacuum]

4. Calculate the magnetic field at the center of a square loop of side $2R$, carrying a current I.

$$\left[\textbf{Ans. } \frac{\mu_0 I\sqrt{2}}{\pi R}\right]$$

5. A method for producing a uniform magnetic field is the Helmholtz coil arrangement. Consider two sets of circular coils, each set having N turns, radius a and carrying current I, separated by a distance $2b$. They are placed coaxially. Take the common axis to be the z-axis. One of the coils lies in the X-Y plane. Write down the field at a point of the axis, distant z from the centre of the coil. Find the value of z for which $\frac{dB}{dz} = 0$. What is the relation between a and b for $\frac{d^2B}{dz^2} = 0$ at $z = b$? Write down the value of B at $z = b$ under these conditions, using Taylor expansion upto second derivatives about $z = b$.

$$\left[\textbf{Ans. } B = \frac{8\mu_0 NI}{5^{3/2}a}\right]$$

6

Magnetic Fields in Matter

6.1 MAGNETIZATION CURRENTS

When a piece of matter becomes magnetized, we describe its state by a vector **M** – the magnetization vector. It is defined as the magnetic dipole moment per unit volume. If the magnetization is uniform, **M** is constant. If it is non-uniform, **M** is a function of co-ordinates $\mathbf{M}(x, y, z)$. Let us first consider the former. Suppose we have a slab of material of uniform thickness t, magnetized perpendicular to the slab (Fig. 6.1a). We can think of this material as composed of a number of elementary dipoles. The modern picture of a dipole is a current loop. So we look upon the slab as composed of a number of current loops. Each such dipole is thus an elementary block of material (Fig. 6.2) of area A, say, and thickness t. A current I goes around the periphery of the block. This is the sum of elementary currents due to circulating electrons in the atoms of the block. The magnetic dipole moment of this piece is $m = MAt$. It can also be written as $m = IA$. $\therefore$ $I = Mt$ and the surface current density $K_b = I/t = M$. The subscript b denotes a bound current. This is because the current is due to bound electrons in atoms. The slab of Fig. 6.1b is made of a number of such pieces. For uniform magnetization, the internal currents cancel. So the resultant of this current distribution is a current around the

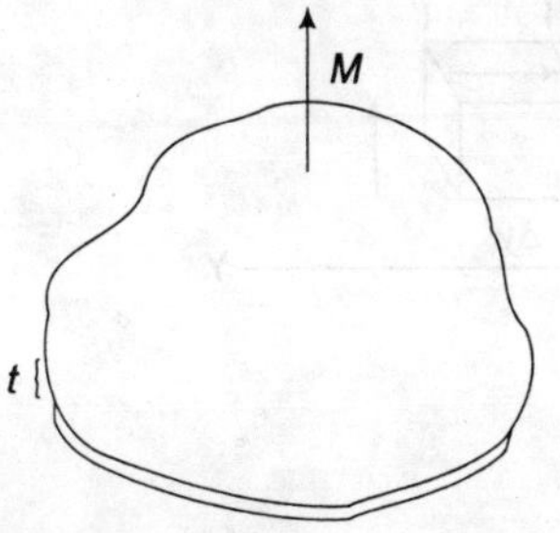

Fig. 6.1a

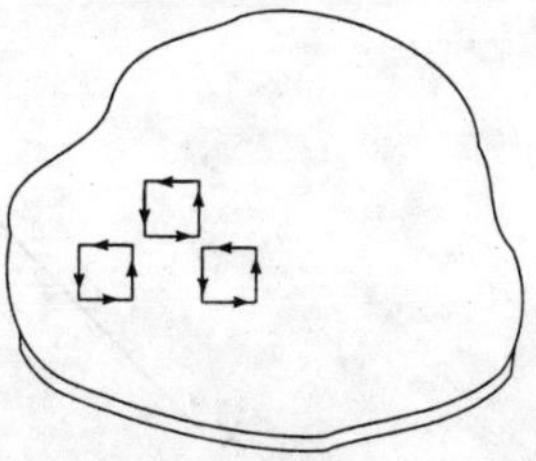

Fig. 6.1b

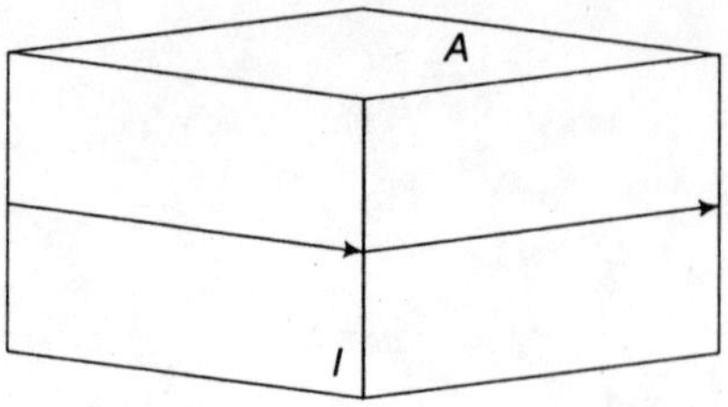

Fig. 6.2

periphery of the slab Fig. 6.1b. The normal $\hat{\mathbf{n}}$ to this periphery is outward. Therefore, the direction of the surface current is given by $\mathbf{K}_b = \mathbf{M} \times \hat{\mathbf{n}}$.

We will now consider non-uniform magnetization. In Fig. 6.3, we have two adjacent elementary volumes $\Delta x \Delta y \Delta z$ with current loops in the YZ plane, in a magnetic slab. The magnetization is along the x-axis. The magnetic moment of the element on the left is $M_x \Delta x\, \Delta y\, \Delta z = I \Delta y \Delta z$ and that on the right is $\left(M_x + \frac{\partial M_x}{\partial y}\Delta y\right)\Delta x\,\Delta y\,\Delta z = I'\Delta y\,\Delta z$

$$\therefore \quad I = M_x \Delta x$$

$$I' = \left(M_x + \frac{\partial M_x}{\partial y}\Delta y\right)\Delta x$$

∴ The net current in the direction of the positive z-axis is

$$I - I' = -\frac{\partial M_x}{\partial y}\Delta y\,\Delta x \tag{6.1.1}$$

We will also have another contribution to the current from volume elements magnetized in the direction of the y-axis (Fig. 6.4). Again we take

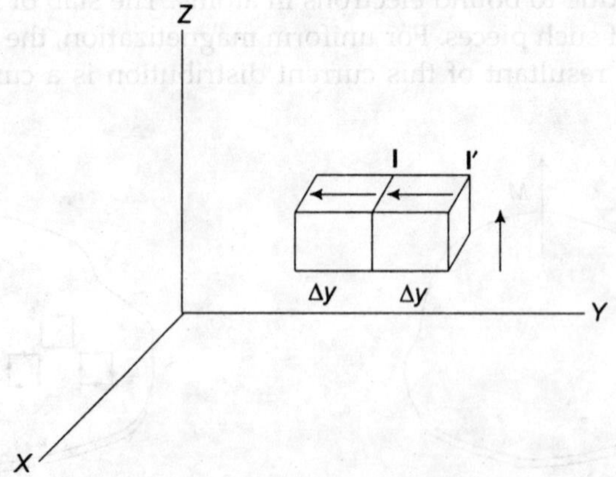

Fig. 6.3

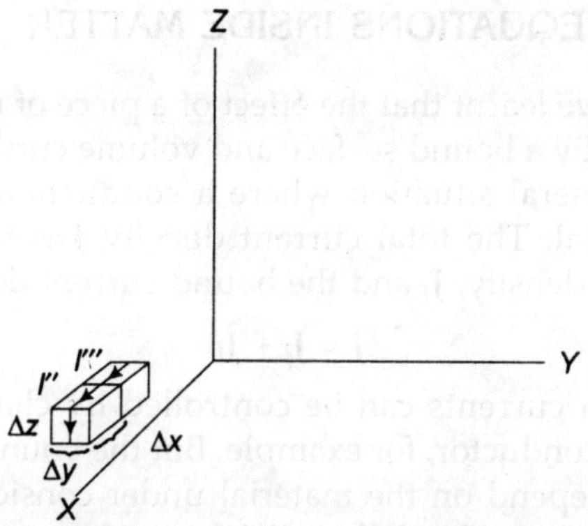

Fig. 6.4

two adjacent volume elements $\Delta x\,\Delta y\,\Delta z$ with current loops in the XZ plane. The magnetic moment of the volume element with loop current I''' is $M_y\,\Delta x\,\Delta y\,\Delta z = I'''\ \Delta x\,\Delta z$ and that with loop current I'' is $\left(M_y + \dfrac{\partial M_y}{\partial x}\Delta x\right)$ $\Delta x\,\Delta y\,\Delta z = I''\ \Delta x\,\Delta z$

$$\therefore \qquad I''' = M_y \Delta y$$

$$I'' = \left(M_y + \frac{\partial M_y}{\partial x}\Delta x\right)\Delta y$$

The net current in the direction of the positive z-axis is:

$$I'' - I''' = \frac{\partial M_y}{\partial x}\Delta x\,\Delta y \tag{6.1.2}$$

$\therefore$ The net current in the z-direction from (6.1.1) and (6.1.2) is

$$I_z = \left(\frac{\partial M_y}{\partial x} - \frac{\partial M_y}{\partial y}\right)\Delta x\,\Delta z$$

and the current density is

$$J_z = \frac{\partial M_y}{\partial x} - \frac{\partial M_x}{\partial y}$$

The right hand side is the z-component of the vector $\nabla \times \mathbf{M}$. So we may write the magnetization current density

$$\mathbf{J}_b = \nabla \times \mathbf{M} \tag{6.1.3}$$

6.2 THE FIELD EQUATIONS INSIDE MATTER

In the last section, we learnt that the effect of a piece of magnetized material can be represented by a bound surface and volume current density. Now let us consider the general situation where a conduction current also flows through the material. The total current density **J** is then the sum of the conduction current density, $\mathbf{J}_f$ and the bound current density $\mathbf{J}_b$:

$$\mathbf{J} = \mathbf{J}_f + \mathbf{J}_b$$

The conduction currents can be controlled by changing the potential difference across a conductor, for example. But the bound currents cannot be controlled – they depend on the material under consideration. Using eqn. (6.1.3), we can then write the differential form of Ampere's law as:

$$\nabla \times \mathbf{B} = \mu_0(\mathbf{J}_f + \nabla \times M)$$

or, $$\nabla \times \frac{\mathbf{B}}{\mu_0} - \nabla \times \mathbf{M} = \mathbf{J}_f$$

or, $$\nabla \times \left(\frac{\mathbf{B}}{\mu_0} - \mathbf{M}\right) = \mathbf{J}_f$$

The vector within brackets is called the auxiliary field or the field **H**.[⊕]

$$\mathbf{H} = \frac{\mathbf{B}}{\mu_0} - \mathbf{M} \tag{6.2.1}$$

$$\nabla \times \mathbf{H} = \mathbf{J}_f \tag{6.2.2}$$

Let us integrate (6.2.2) over a surface S, bound by a curve C, inside the material. Then we have

$$\int_S (\nabla \times \mathbf{H}).d\mathbf{a} = \int_S \mathbf{J}_f.d\mathbf{a}$$

The left hand side can be rewritten using Stokes' theorem and the right hand side is the free current I_f, i.e. the conduction current flowing in the material. So we get the integral form of (6.2.2):

$$\int \mathbf{H}.d\mathbf{l} = \mathbf{I}_f \tag{6.2.3}$$

6.3 BOUNDARY CONDITIONS ON THE FIELD VECTORS

Let us now suppose that we have two media 1 and 2 adjacent to each other. We shall now consider how the fields change across the interface. Let $\mathbf{B}_1$ and $\mathbf{B}_2$ be the magnetic fields in the two media. Consider a Gaussian pillbox of base area A, straddling the interface (Fig. 6.5). We shall use the integral form of (5.5.1) over the surfaces of the box

[⊕][In some books, **B** is referred to as the magnetic flux density and **H** as the magnetic field.]

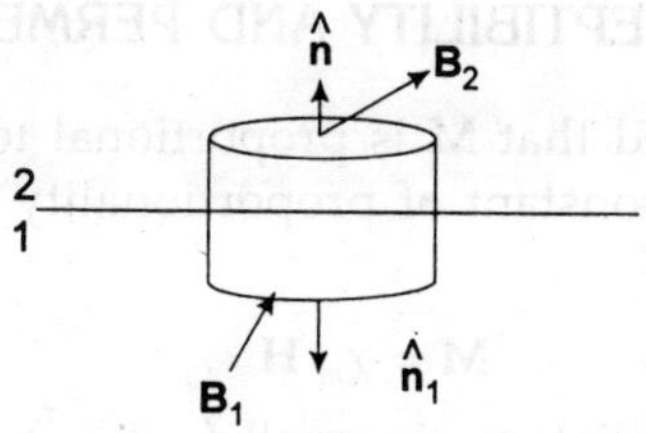

Fig. 6.5

$$\int_S \mathbf{B}.d\mathbf{a} = 0 \tag{6.3.1}$$

The curved surface of the box is infinitesimally small in thickness. So we can neglect the integral over it. The remaining two surfaces give:

$$\mathbf{B}_2.\hat{\mathbf{n}}_2 A + B_1.\hat{\mathbf{n}}_1 A = 0$$

Now $\hat{\mathbf{n}}_1$ and $\hat{\mathbf{n}}_2$ are unit vectors normal to the top and bottom surfaces of the box and $\hat{\mathbf{n}}_1 = -\hat{\mathbf{n}}_2$. So we have

$$B_{2n} - B_{1n} = 0 \tag{6.3.2}$$

where the subscript n denotes the normal component. Thus the normal component of **B** is contibnuous across the interface.

Consider now a rectangular loop (Amperian loop) of infinitesimal width and length 1 across the interface between two media 1 and 2 (Fig. 6.6). Let $\mathbf{H}_1$ and $\mathbf{H}_2$ be the auxiliary fields on the two sides. Let there be a free current density $\mathbf{K}_f$ normal to the plane of the diagram at the interface, pointing away from the reader. We will use (6.2.3) around the loop. Taking the vector **I** to point from left to right, we get,

$$\mathbf{H}_2.\mathbf{I} - \mathbf{H}_1.\mathbf{I} = \mathbf{K}_f.(\hat{\mathbf{n}}_2 \times \mathbf{I})$$

where $\hat{\mathbf{n}}_2$ is the unit vector from medium 1 to medium 2 across the interface. Interchanging the dot and the cross products on the right hand side, we get

$$(\mathbf{H}_2 - \mathbf{H}_1).\mathbf{I} = (\mathbf{K}_f \times \hat{\mathbf{n}}_2).\mathbf{I}$$

or, $$(\mathbf{H}_2 - \mathbf{H}_1)_t = \mathbf{K}_f. \times \hat{\mathbf{n}}_2 \tag{6.3.3}$$

where t denotes the tangential component. Thus, if there are no free surface currents, the tangential component of the auxiliary field is continuous across the interface.

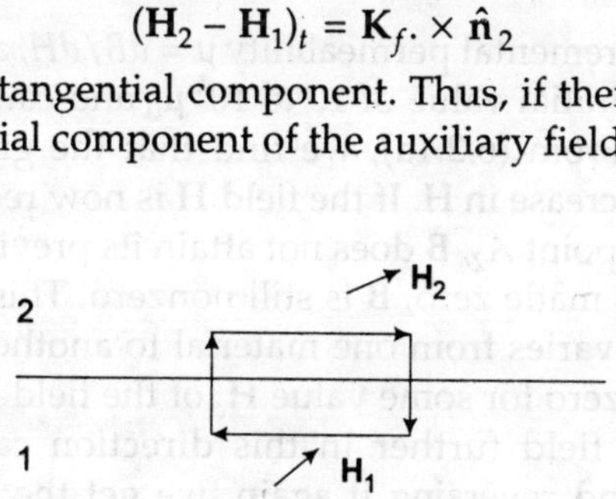

Fig. 6.6

6.4 MAGNETIC SUSCEPTIBILITY AND PERMEABILITY

In some materials, we find that **M** is proportional to **H**. These are linear, isotropic media and the constant of proportionality is called the *magnetic susceptibility* χ_m :

$$\mathbf{M} = \chi_m \mathbf{H} \tag{6.4.1}$$

For diamagnetic materials χ_m is small ($\sim 10^{-5}$) and negative and for paramagnetic materials, χ_m is small and positive. For linear materials, **B** is also proportional to **H**, the constant of proportionality being called the *permeability* μ:

$$\mathbf{B} = \mu \mathbf{H} \tag{6.4.2}$$

From equation (6.2.1)

$$\mathbf{B} = \mu_0(\mathbf{H} + \mathbf{M}) \tag{6.2.1a}$$

Combining this with (6.4.1) and comparing with (6.4.2), we get,

$$\mu = \mu_0(1 + \chi_m) \tag{6.4.3}$$

Ferromagnetic materials do not satisfy a simple relation like (6.4.2). If we apply a field **H** to a sample of this material and gradually increase it, **B** increase rapidly at first and then very little (Fig. 6.7). This charge in behaviour occurs when **M** attains its maximum value, called *saturation magnetization*.

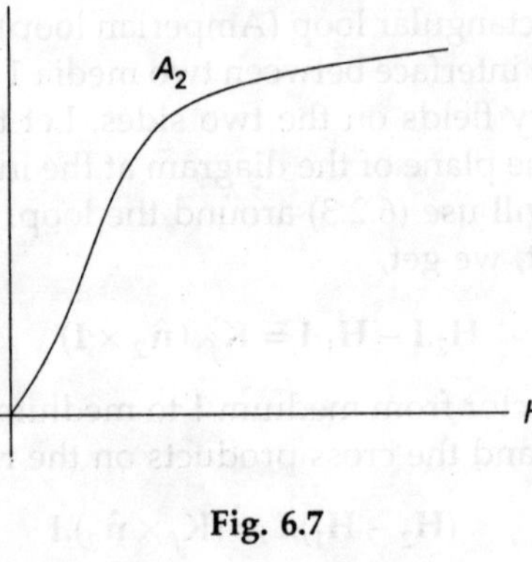

Fig. 6.7

We can define the incremental permeability $\mu = dB/dH$, assuming **B** and **H** to be parallel. It has an initial value of 10 to 10^4 μ_0 and can be as large as $10^6 \mu_0$ for some materials. From (6.2.1a), we find that the gradual increase in **B** thereafter is due to increase in **H**. If the field **H** is now reduced from its value corresponding to the point A_2, **B** does not attain its previous values (Fig. 6.8). Moreover, when **H** is made zero, **B** is still nonzero. This value of **B** is called the *remanence* $\mathbf{B}_r$ and varies from one material to another. If the field is now reversed, **B** becomes zero for some value $\mathbf{H}_c$ of the field **H**, called the *coercive force*. Increasing the field further in this direction causes **B** to saturate. Reducing the field and reversing it again, we get the portion $A_2 A_1$ of the

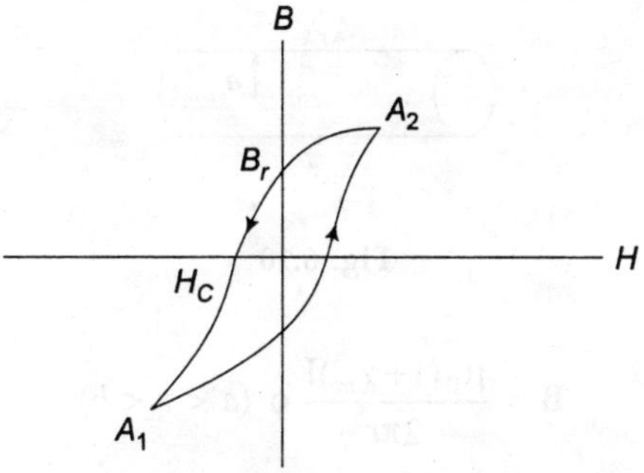

Fig. 6.8

curve. This closed loop is called a *hysterisis* loop. The word 'hysterisis' means 'to lag behind'. We see from this loop that the value of B depends on the history of the sample. It lags behind H. When a ferromagnetic material is taken through a cycle, the domains are either rotated or changed in size (Tutorial 1, Chapter 5). Work has to be done and the sample gets heated. This is particularly true if the material is taken through a cycle a number of times. An example is the core of a transformer with a coil carrying alternating current wound around it.

Example: A coaxial cable consists of two very long cylindrical tubes, separated by linear insulting material of magnetic susceptibility χ_m. A current I flows down the inner conductor and returns along the outer one (see Fig. 6.9); in each case the current distributes itself uniformly over the surface. find the magnetic field in the region between the tubes.

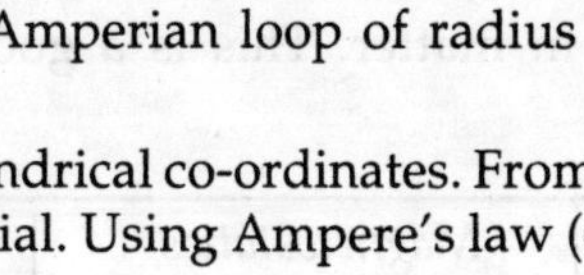

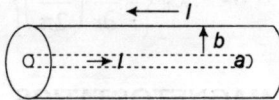

Fig. 6.9

Solution: Consider an Amperian loop of radius $r (a < r < b)$ between the tubes (Fig. 6.10).

We shall be using cylindrical co-ordinates. From symmetry, we expect the field **H** to be circumferential. Using Ampere's law (6.2.3) around such a loop, we get

$$H.2\pi r = I$$

$$\therefore \qquad \mathbf{H} = \frac{I}{2nr}\hat{\phi} \ (a < r < b)$$

The medium is linear with magnetic susceptibility χ_m. So we get

$$\mathbf{B} = \mu_0 (1 + \chi_m)\mathbf{H}$$

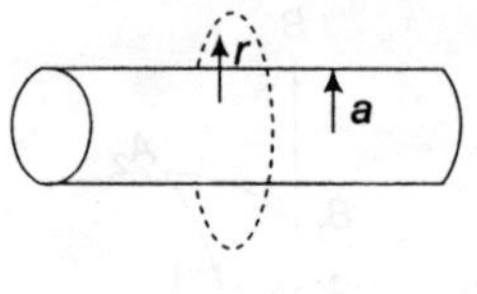

Fig. 6.10

$$\mathbf{B} = \frac{\mu_0(1+\chi_m)\mathbf{I}}{2\pi r}\,\hat{\phi}\ (a < r < b)$$

Let us find the magnetization currents $\mathbf{K}_b$ and $\mathbf{J}_b$. For that, we first need write down **M**. It is

$$\mathbf{M} = \chi_m\mathbf{H} = \frac{\chi_m \mathbf{I}}{2\pi r}\,\hat{\phi}$$

For the surface of the material adjacent to the inner tube, the outward normal is opposite in direction to the radial vector from the axis : $\hat{\mathbf{n}} = -\hat{\mathbf{r}}$

$\therefore$ At $r = a$, $\mathbf{K}_b = \mathbf{M}\times\hat{\mathbf{n}} = \dfrac{-\chi_m I}{2\pi a}\,\hat{\phi}\times\hat{\mathbf{r}} = \dfrac{\chi_m I}{2\pi a}\,\hat{\mathbf{k}}$

For the surface of the material adjacent to the outer tube, the outward normal is in the same direction as the radial vector from the axis : $\hat{\mathbf{n}} = -\hat{\mathbf{r}}$

$\therefore$ At $r = b$ $\mathbf{K}_b = \mathbf{M}\times\hat{\mathbf{n}} = \dfrac{\chi_m I}{2\pi b}\,\hat{\phi}\times\hat{\mathbf{r}} = \dfrac{-\chi_m I}{2\pi b}\,\hat{\mathbf{k}}$. The magnetization (volume) current density is

$$\mathbf{J}_b = \nabla\times\mathbf{M} = \chi_m\left\{\frac{1}{r}\frac{\partial}{\partial r}\left(\frac{I}{2\pi}\right)\right\}\hat{\mathbf{k}} = 0$$

6.5 COMPARISON OF MAGNETOSTATICS AND ELECTROSTATICS

In Chapter 4, we studied electric fields in matter and in this chapter, we have studied magnetic fields in matter. This is a good time to reflect on the properties of these fields.

Electrostatics	*Magnetostatics*	*Comment*
In vacuum $\nabla.\mathbf{E} = \dfrac{\rho}{\varepsilon_0}$ (1.8.11)	$\nabla.\mathbf{B} = 0$ (5.5.1)	There are no magnetic monopoles
$\nabla\times\mathbf{E} = 0$ (2.5.1)	$\nabla\times\mathbf{B} = \mu_0\mathbf{J}$ (5.5.4)	Electrostatic field is a conservative field
In matter $\nabla.\mathbf{D} = \rho$ (4.3.6)	$\nabla\times\mathbf{H} = \mathbf{J}_f$ (6.2.2)	

$E_{2t} = E_{1t}$ (4.2.2)	$\mathbf{H}_{2t} - \mathbf{H}_{1t} = \mathbf{K}_f \times \hat{\mathbf{n}}_2$ (6.3.3)	Tangential component of electric field is continuous across interface
$D_{2n} - D_{1n} = \sigma$ (4.4.1)	$B_{2n} = B_{1n}$ (6.3.2)	Normal component of magnetic field is continuous across interface

The table below summarizes the important results of this chapter in *SI* and Gaussian units:

Physical quantity/equation	*SI*	*Gaussian*
Magnetization current density (6.1.3)	$\mathbf{J}_b = \nabla \times \mathbf{M}$	$\mathbf{J}_b = c(\nabla \times \mathbf{M})$
Auxiliary field **H** (6.2.1)	$\mathbf{H} = \dfrac{\mathbf{B}}{\mu_0} - \mathbf{M}$	$\mathbf{H} = \mathbf{B} - 4\pi\mathbf{M}$
(6.2.2)	$\nabla \times \mathbf{H} = \mathbf{J}_f$	$\nabla \times \mathbf{H} = \dfrac{4\pi\mathbf{J}_f}{c}$

7

DC Circuits

7.1 THE CONTINUITY EQUATION

Let us consider a region of space where currents are flowing. For example, a conductor, or on electrolyte, or a diode. The current density is **J**. Then we learnt in the last chapter that the current I flowing across any surface S is

$$I = \int_S \mathbf{J}.d\mathbf{a} \tag{7.1.1}$$

A steady current distribution is one in which **J** is not a function of time. It is a function of co-ordinates only. A non-steady distribution is one in which **J** is a function of time as well as space. One of the fundamental principles of physics is the conservation of electric charge. Electric charge can neither be created nor destroyed. Let us first look at the consequence of this principle for non-steady currents. Let there be a volume V containing charge with density ρ. The total charge within this volume at any instant of time is $\int_V \rho dv$. Let S be the surface bounding this volume. If I is the current flowing across S out of this volume, then that must be due to a decrease in the amount of charge there. Since current is rate of flow of charge, we have

$$I = \frac{-d}{dt}\int_V \rho dv$$

where the negative sign appears because a decrease in charge corresponds to a flow of current out of V and vice versa. ∴ From (7.1.1),

$$\int_S \mathbf{J}.d\mathbf{a} = \frac{-d}{dt}\int_V \rho dv$$

or,

$$\int_S \mathbf{J}.d\mathbf{a} = -\int_v \frac{\partial \rho}{\partial t}\, dv$$

where the partial derivative is used because ρ is a function of co-ordinates as well and we are focussing our attention on a fixed volume element. The left hand side of this equation can be transformed using the divergence theorem. So we have

$$\int_V \nabla.\mathbf{J}dv = -\int_v \frac{\partial\rho}{\partial t}dv$$

or,

$$\int_V \left(\nabla.\mathbf{J} + \frac{\partial\rho}{\partial t}\right)dv = 0$$

Since this is true for any arbitrary volume, the integrand must be zero. Thus

$$\nabla.\mathbf{J} + \frac{\partial\rho}{\partial t} = 0 \tag{7.1.2}$$

This is *the equation of continuity*. For a steady current distribution, **J** and ρ are not functions of time. So there cannot be any decrease in the amount of charge in a given volume *V*. Any current flowing out of *V* must be compensated by a flow of current into *V*. So we must have

$$\int_S \mathbf{J}.d\mathbf{a} = 0$$

over any closed surface bounding a volume *V*. In other words

$$\int_V \nabla.\mathbf{J}dv = 0$$

∴

$$\vec{\nabla}.\mathbf{J} = 0 \tag{7.1.3}$$

7.2 KIRCHHOFF'S LAWS

We will now analyze electrical circuits. There are two rules due to Kirchhoff which help us to find currents and potential differences in any branch of a circuit.

Kirchhoff's current law: At any junction, the algebraic sum of the currents must be zero.

In the circuit of Fig. 7.1, a and b are junctions or nodes. If we use the convention that current flowing into the junction is negative and current flowing out is positive, this law tells us that $-I_3 + I_1 + I_2 = 0$ or,

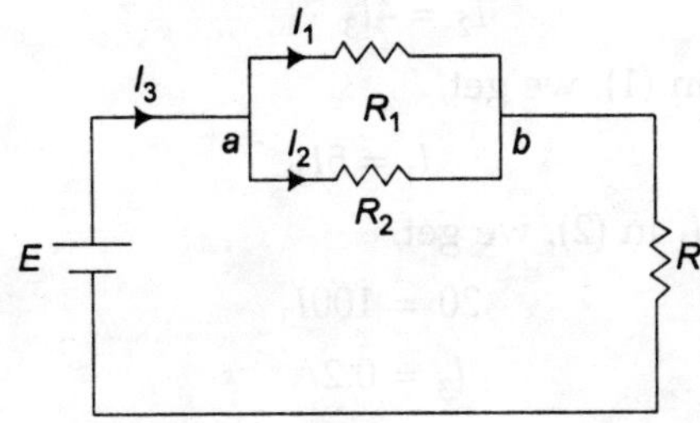

Fig. 7.1

$$I_3 = I_1 + I_2 \quad (7.2.1)$$

where I_1 is the current in the branch containing resistance R_1 and I_2 is the current in the branch containing resistance R_2. Kirchhoff's current law is just an expression of conservation of charge.

Kirchoff's voltage law: For any closed path, that is traversed in a single direction, the algebraic sum of voltages in zero.

In the circuit of Fig. 7.1, if we consider the closed path consisting of the battery of *emf E*, resistance R_2 and resistance R_3, then traversing the loop clockwise, we get, $E - I_2 R_2 - I_3 R_3 = 0$, or,

$$E = I_2R_2 + I_3R_3 \quad (7.2.2)$$

where a drop in potential is prefixed with a negative sign and a rise in potential with a positive sign.

Example 1: Find the current through th 5 Ω resistor in the circuit below:

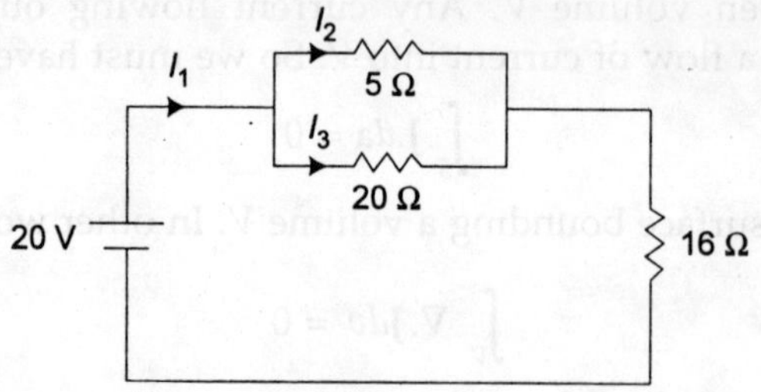

Fig. 7.2

Solution: From Kirchhoff's current law,

$$I_1 = I_2 + I_3 \quad (1)$$

Applying Kirchoff's voltage law around law around the loop containing the 20 Ω and 16 Ω resistors, we get,

$$20 = 20I_3 + 16I_1 \quad (2)$$

Applying Kirchhoff's voltage law around the loop containing the 5Ω and 20 Ω resistors. we get,

$$-5I_2 + 20I_3 = 0$$

or,
$$I_2 = 4I_3 \quad (3)$$

Substituting (3) in (1), we get,

$$I_1 = 5I_3 \quad (4)$$

Substituting for I_1 in (2), we get,

$$20 = 100I_3$$

or,
$$I_3 = 0.2\text{A}$$

∴ From (3), $I_2 = 0.8\text{A}$

The current $I_1 = 1\text{A}$

7.3 NETWORK THEOREMS

A network is a collection of electric elements such as resistors, coils capacitors and sources of energy, connected to form several interrelated circuits.[6] An energy source such as a battery is called an *active* element. Resistors, capacitors and inductors are *passive* elements. We shall now study two theorems about linear nerworks, which help us to simplify complicated circuits.

Thevenin's theorem: The current in any resistance[⊕] R, connected to two terminals of a network, is the same as if R were connected to a voltage source, whose voltage is the open-circuited voltage at the terminals in question and whose resistance is the resistance of the network looking back from the terminals, with all voltage sources replaced by resistances equal to the internal resistances of the voltage sources.

In Fig. 7.3, V_{oc} is the open-circuit voltage at the termanals referred to in the theorem and R_{Th} is the resistance of the network measured back from the terminals 1 and 2. It is sometimes called the Thevenin resistance.

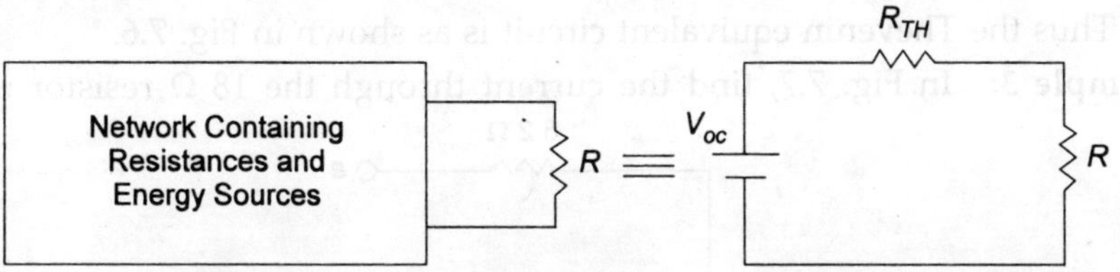

Fig. 7.3

Example 2: In Fig. 7.4, replace the network to the left of the terminals *ab* with its Thevenin equivalent.

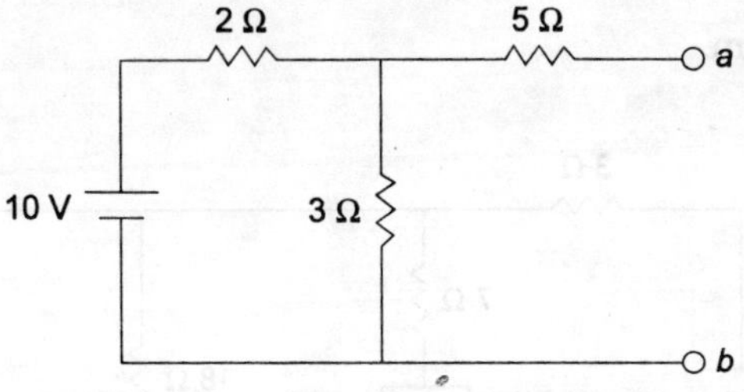

Fig. 7.4

[⊕][The general statement contains the work 'impedance', but we have omitted it since we have not yet talked about ac circuits]

Solution: With the terminals *a* and *b* open-circuited, the current through the 3 Ω resistance is

$$I = \frac{10}{5} = 2 \text{ A}$$

∴ The voltage across 3 Ω is 2 × 3 = 6 V. This is the open-circuit voltage V_{oc} between *a* and *b*. To find R_{Th}, we have to replace the battery by a short circuit because its internal resistance is assumed to be zero (Fig. 7.5).

$$R_{Th} = 5 + \frac{2 \times 3}{2+3} = 6.2 \ \Omega$$

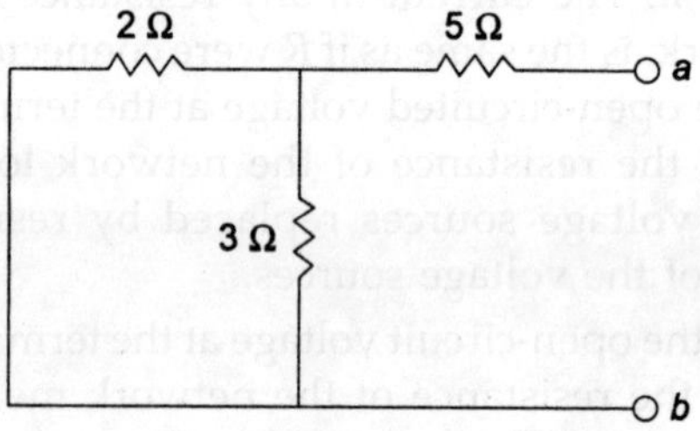

Fig. 7.5

Thus the Thevenin equivalent circuit is as shown in Fig. 7.6.

Example 3: In Fig. 7.7, find the current through the 18 Ω resistor using Thevenin's theorem.

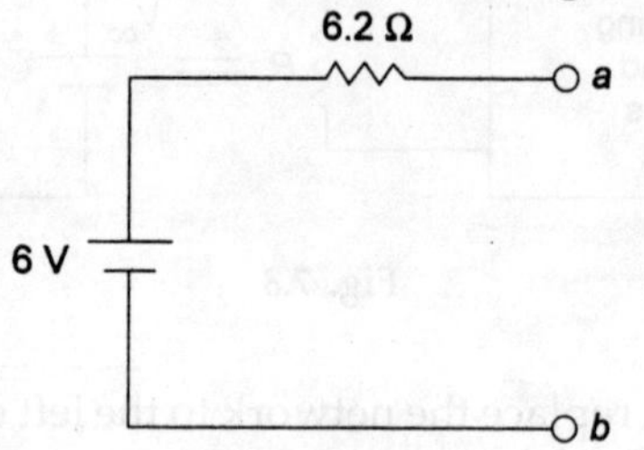

Fig. 7.6

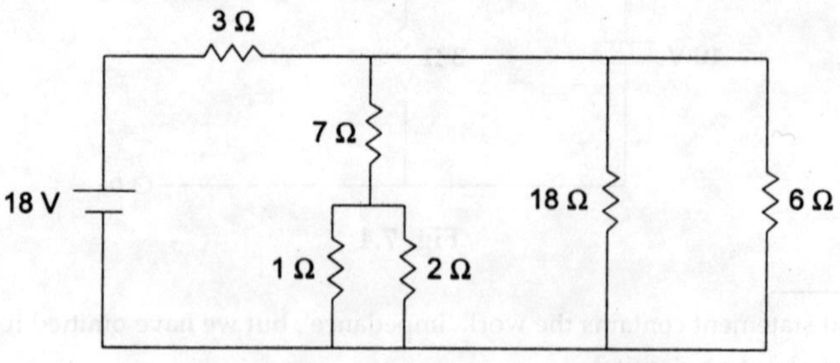

Fig. 7.7

Solution: To solve this problem, using Thevenin's theorem, remove the 18 Ω resistor, leaving the circuit open there, calling the terminals 1 and 2.

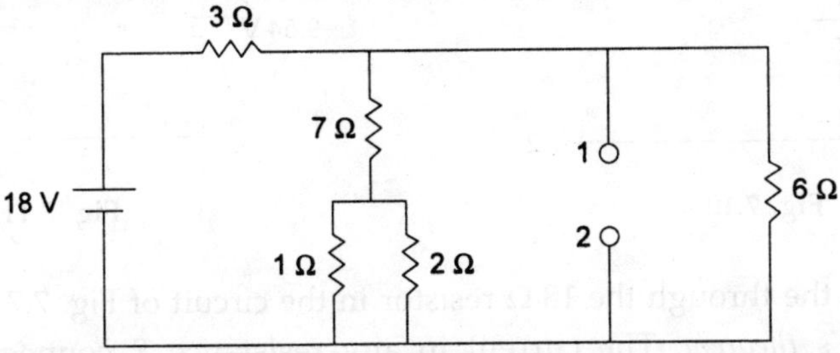

Fig. 7.8

Now let us simplify the circuit in stages. The 1 Ω and 2 Ω resistors give an equivalent resistance of 0.67 Ω. That in series with 7 Ω gives 7.67 Ω. This in parallel with 6 Ω gives 3.37 Ω. So the circuit simplifies to

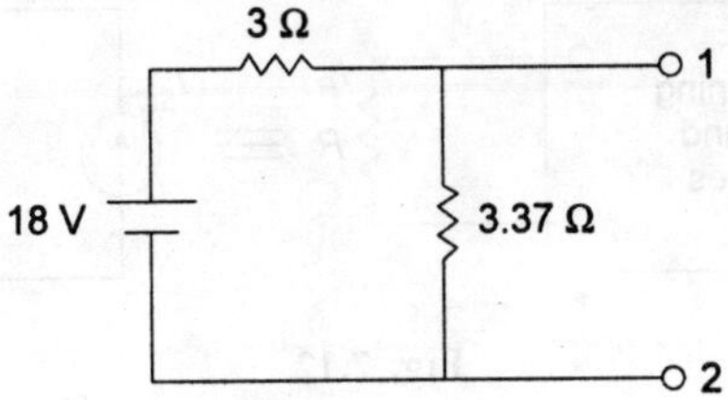

Fig. 7.9

Let us find the Thevenin equivalent circuit of Fig. 7.9. The current through the 3.37 Ω resistor is

$$\frac{18}{6.37} = 2.83 \text{ A}$$

∴ The p.d. across it and so across terminals 1 and 2 is $V_{oc} = 2.83 \times 3.37 =$ 9.54 V. The Thévenin resistance R_{Th} is obtained by replacing the battery with a short circuit. So

$$R_{Th} = \frac{3 \times 3.37}{6.37} = 1.59 \text{ } \Omega$$

The Thévenin equivalent circuit is shown in Fig. 7.10

Now place the 18 Ω resistor between 1 and 2. The current I in the circuit is

$$I = \frac{9.54}{19.59} = 0.49 \text{ A}$$

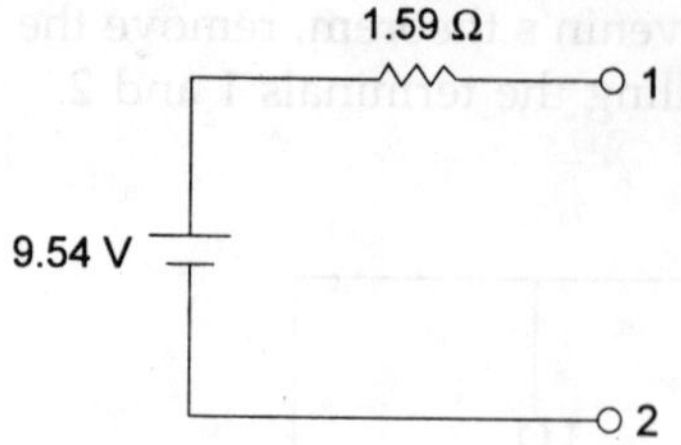

Fig. 7.10

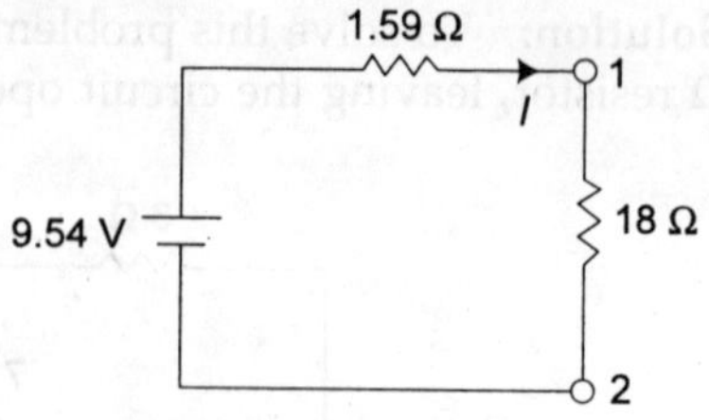

Fig. 7.11

This is the through the 18 Ω resistor in the circuit of Fig. 7.7

Norton's theorem: The current in any resistance R connected to two terminals of a network is the same as if R were connected to a constant current source whose current is equal to the current which flows through the two terminals when these terminals are short-circuited, the constant current source being in shunt with a resistance equal to the resistance of the network looking back from the terminals in question.

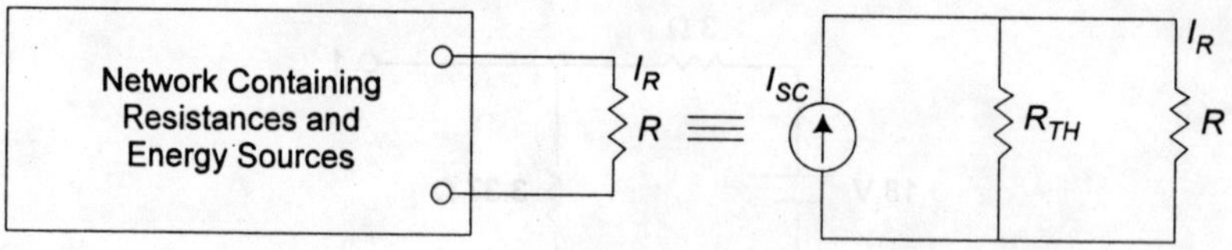

Fig. 7.12

In Fig. 7.12 $\mathbf{I}_{SC}$ is the current source. The shunt resistance is the same as the the Thevenin resistance R_{Th} of fig. 7.3.

Example 4: Find the Norton equivalent circuit for the network shown below:

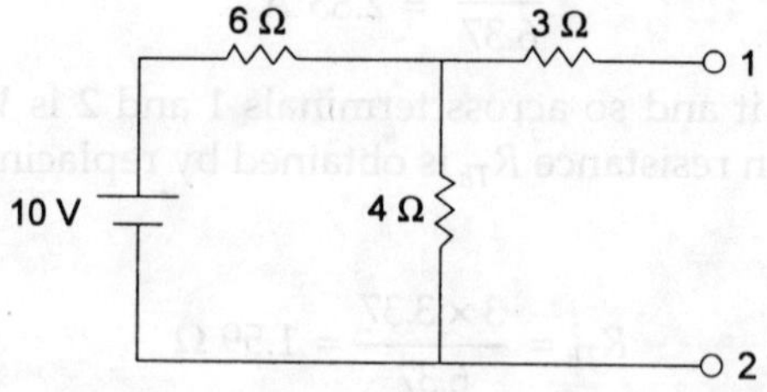

Fig. 7.13

Solution: We have to find the short-circuit current when terminals 1 and 2 are short-circuited. See Fig. 7.14. The equivalent resistance of the circuit is

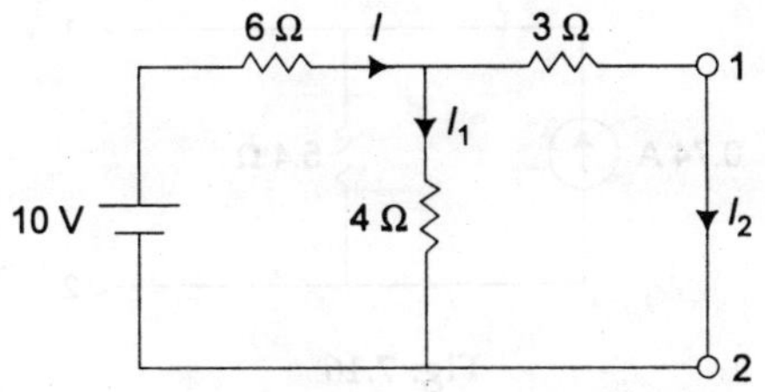

Fig. 7.14

$$6 + \frac{4 \times 3}{4 + 3} = 6 + \frac{12}{7} = 6 + 1.71 = 7.71\ \Omega$$

∴ The total current I in the circuit is

$$I = \frac{10}{7.71} = 1.30\ \text{A}$$

The current I_2 through the 3 Ω resistor is

$$I_2 = \frac{4}{7} \times 1.30 = 0.74\ \text{A}$$

To find the equivalent resistance looking back from the terminals 1 and 2 in Fig. 7.13, replace the battery by a short circuit (Fig. 7.15). The equivalent resistance is

$$3 + \frac{4 \times 6}{4 + 6} = 3 + 2.4 = 5.4\ \Omega$$

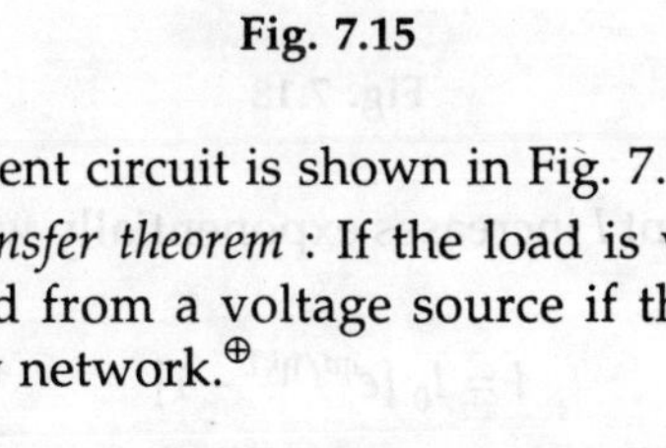

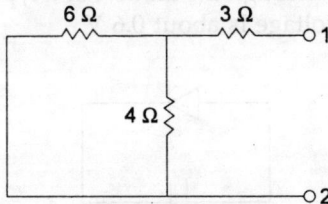

Fig. 7.15

∴ The Norton equivalent circuit is shown in Fig. 7.16

Maximum power transfer theorem : If the load is varied, then maximum power will be absorbed from a voltage source if the load is equal to the resistance of the supply network.[⊕]

[⊕][The general statement involves impedances and will be restated in the chapter on ac circuits]

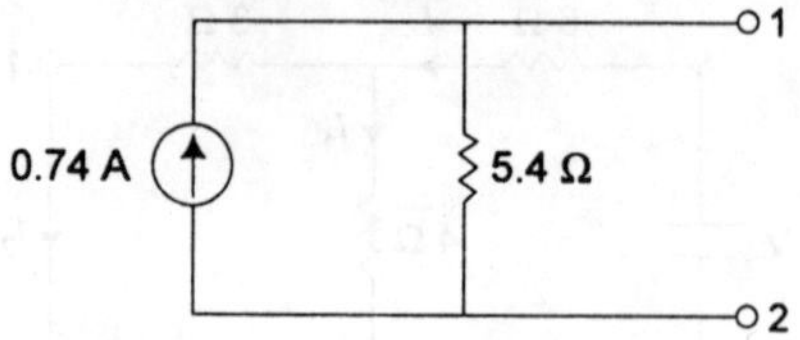

Fig. 7.16

7.4 NONOHMIC DEVICES

We shall now briefly consider two devices which do not obey Ohm's law. First, the *pn* junction diode. The circuit symbol of the device is shown in Fig. 7.17. The arrow points from the *p* to the *n* side. A *p*-type semiconductor has positively charged holes as majority carriers of current. An *n*-type semiconductor has negatively charged electrons as majority carriers. At the junction between the two types of semiconductors, there is a depletion region. In this region, there are no mobile charge carriers. There are positively charged donor ions in the *n*-side and negatively charged acceptor ions in the

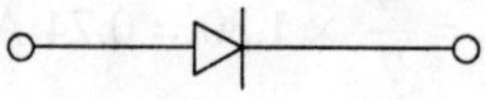

Fig. 7.17

p-side. They are in fixed sites. There is a potential barrier across the junction. If the positive terminal of a voltage source is connected to the *p*-side of the diode and the negative terminal to the *n*-side, the diode is said to be forward biased (Fig. 7.18). There is a threshold voltage, called the cutin voltage below which there is no current through the diode. For a typical silicon diode BY 127 available in India, this voltage is about 0.6 V.

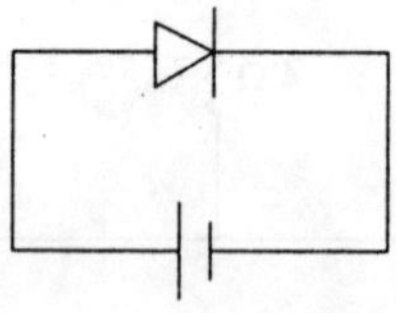

Fig. 7.18

Thereafter the current I increases exponentially with voltage V according to the relation

$$I = I_0\,[e^{qv/\eta kT} - 1] \tag{7.4.1}$$

where I_0 is the *reverse saturation current*, q is the magnitude of the electronic charge, k is Boltzmann's constant (1.38×10^{-23} J/K), T is the absolute temperature and η is a numerical factor which is 1 for a germanium diode and 2 a silicon diode. If the negative terminal of a voltage source is connected

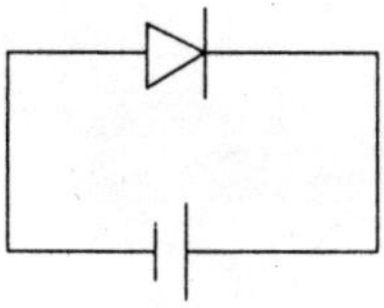

Fig. 7.19

to the p-side of the diode and the positive terminal to the n-side, the diode is said to be reverse biased (Fig. 7.19). The current which flows, I_0 is negligibly small, about a few microamps for a germanium diode and nanoamps for a silicon diode, at about 300 K. This current is due to minority carriers (i.e. electrons in the p-side and holes in the n-side) and so is a function of temperature.

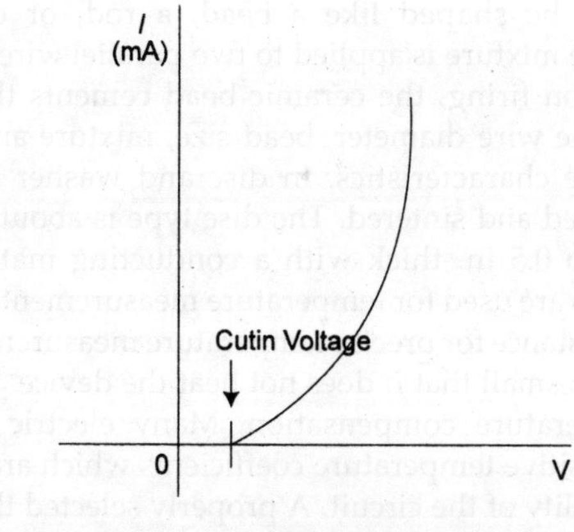

Fig. 7.20

The second nonohmic device which we shall consider is a thermistor. A thermistor is a device with a high negative temperature coefficient of resistance. In other words, the resistance decreases with increase of temperature. It is usually made of sintered mixtures of NiO, Mn_2O_3 or Co_2O_3. For example, a thermistor mix of manganese and nickel oxides has a temperature coefficient of – 4.4% per °C at 25° C. For a small current flowing, the device obeys Ohm's law. As power is dissipated, its temperature rises. Thus resistance decreases and voltage drop also decreases. With further increase in current, temperature increases and resistance decreases. It is assumed that thermal equilibrium is reached at each step. The voltage drop increases but by an amount less than it would have been had the resistance stayed constant. At some value of the current, the voltage reaches a maximum. At higher values of current, temperature rises but voltage drop decreases. In this region, the thermistor has negative resistance. A voltampere characteristic of a typical thermistor is shown in Fig. 7.21.

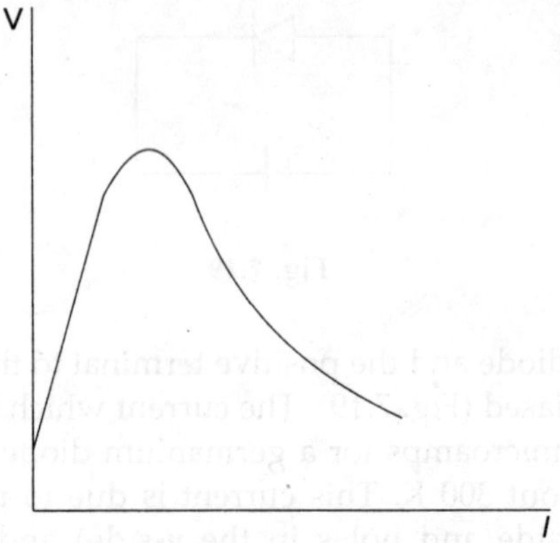

Fig. 7.21

Thermistors can be shaped like a bead, a rod, or disc.[6] In bead thermistors, the oxide mixture is applied to two parallel wires (0.001 to 0.005 in. in diameter). Upon firing, the ceramic bead cements the wires which become the leads. The wire diameter, bead size, mixture and wire spacing determine the device characteristics. In disc and washer thermistors, an oxide-binder is pressed and sintered. The disc type is about 0.2 to 0.6 in. in diameter and 0.04 to 0.5 in. thick with a conducting material and leads attached. Thermistors are used for temperature measurement. The thermistor must have a high resistance for precise temperature measurement. The power dissipated must be so small that it does not heat the device. Thermistors are also used for temperature compensation. Many electric and electronic components have positive temperature coefficients which are detrimental to the temperature stability of the circuit. A properly selected thermistor in the circuit containing such a component will provide temperature compensation.

7.5 TRANSIENTS IN DC CIRCUITS

We shall now come back to our study of *DC* circuits. We will consider what happens immediately after a source of *emf* is applied to a circuit or is switched off. The behaviour of the circuit during this period is called a transient. In other words, the transient occurs between the switching on and the attainment of a steady state by the circuit. The word 'transient' means remaining for a short time, temporary, transitory. First, we consider a series combination of an inductor L and a resistance R. The inductor is basically a coil which has a property of opposing a sudden change in current through it. A sudden change in current would result in a change in magnetic flux through the coil. Faraday's law of electromagnetic induction$^{\oplus}$ tells us that whenever the flux through a circuit is changed, an emf is induced in it. Lenz's law says that the direction of induced emf is such as to oppose the change producing it. So this emf is a back *emf*. It is given in magnitude by Ldi/dt,

$^{\oplus}$[See chapter 11]

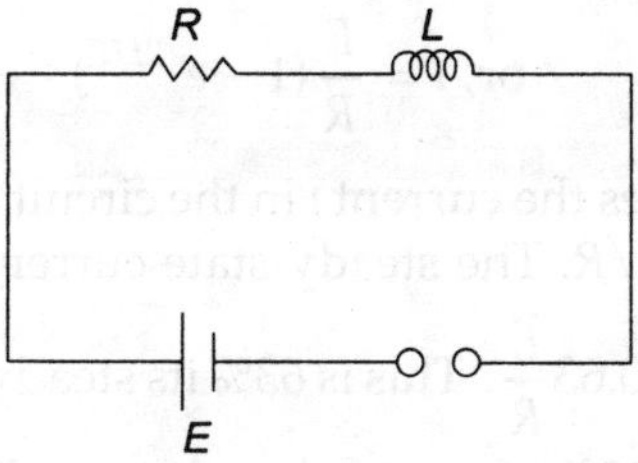

Fig. 7.22

where L is the self inductance of the coil and di/dt is the rate of change of current. Suppose the switch in Fig. 7.22 is initially open as shown.

Immediately after it is closed, the entire battery *emf* E appears across L. The current in the circuit is zero. This is because the inductor opposes a change in current through it (the current was zero initially). The current gradually starts increasing from zero. If i is the current at time t and is changing at the rate di/dt, then Kirchhoff's voltage law tells us that

$$E - \frac{Ldi}{dt} = Ri$$

where Ldi/dt is the magnitude of the back *emf* in the inductor.

$$\therefore \qquad E = Ri + \frac{Ldi}{dt}$$

or

$$\frac{di}{dt} + \frac{Ri}{L} = \frac{E}{L} \tag{7.5.1}$$

This is a linear, first order differential equation with constant coefficients. To solve it, multiply throughout by the integrating factor $e^{Rt/L}$. We get,

$$\frac{d}{dt}\left(e^{Rt/L}\, i\right) = \frac{Ee^{Rt/L}}{L}$$

or, integrating both sides with respect to t, we get,

$$e^{Rt/L} i = \frac{E}{L}\int e^{Rt/L} dt + a_1$$

where a_1 is a constant of integration.

$$ie^{Rt/L} = \frac{E}{R} e^{Rt/L} + a_1$$

At $t = 0$, $i = 0$. So we have $a_1 = -\frac{E}{R}$. Thus

$$ie^{Rt/L} = \frac{E}{R}\left(e^{Rt/L} - 1\right)$$

$$\text{or, } i = \frac{E}{R}(1 - e^{-Rt/L}) \tag{7.5.2}$$

Equation (7.5.2) gives the current i in the circuit at any instant t. As $t \to \infty$, $e^{-Rt/L} \to 0$ and so $i \to E/R$. The steady-state current in the circuit is E/R. At $t = L/R.\ i = \frac{E}{R}(1 - e^{-1}) = 0.63\frac{E}{R}$. This is 63% its steady value. The time required by the current to attain 63% of its steady value is called the *time constant* of the circuit. In this case, the time constant = L/R.

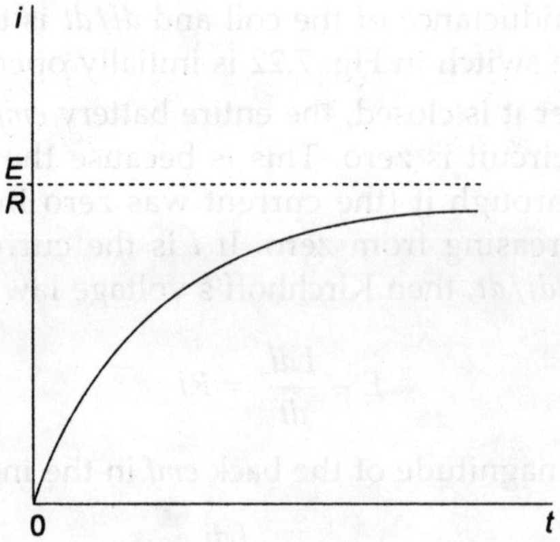

Fig. 7.23

Suppose now that a current I is flowing in the closed circuit of Fig. 7.22, when the key is suddenly opened. A spark will be seen across the switch as the key is opened. This is because the inductor prevents the current from suddenly changing from 1 to 0. A large induced voltage appears across the switch contacts. If, instead, we have a closed circuit with the battery removed, the *emf* equation of the circuit is

$$\frac{Ldi}{dt} + Ri = 0 \tag{7.5.3}$$

because the source of *emf* has been removed. Solving, we have,

$$\frac{di}{i} = \frac{-Rt}{L}$$

or, integrating,

$$1n\ i = \frac{-Rt}{L} + b_1$$

where b_1 is a constant of integration. We can also write the solution as

$$i = b_2 e^{-Rt/L}$$

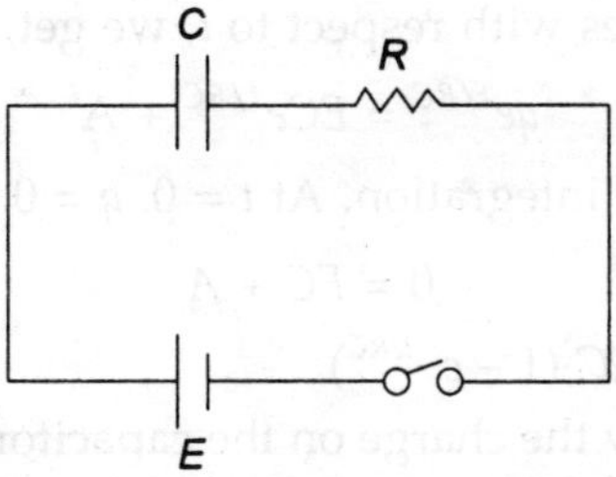

Fig. 7.24

where b_2 is another constant of integration. At $t = 0, i = I \therefore b_2 = I$. So the solution is

$$i = Ie^{-Rt/L} \tag{7.5.4}$$

The current decays exponentially to zero.

Next let us consider a series combination of a capacitor C and resistance R (Fig. 7.24).

Initially the key is open. Immediately after the key is closed, a current $I_0 = E/R$ starts flowing. The capacitor is uncharged to begin with. So the *p.d.* across it is zero initially. This current starts charging the capacitor. As charge accumulates on the capacitor, the *p.d.* across it increases. So the *p.d.* across the resistor decreases. Since the *p.d.* across the resistor at any instant t is Ri, where i is the current at that instant, the current decreases. This goes on until the capacitor is charged to a *p.d.* of E. The current in the circuit in then zero. Let q be the charge on the capacitor at any instant of time t. The *emf* equation of the circuit is

$$\frac{q}{C} + Ri = E$$

Since the current $i = dq/dt$, this becomes

$$\frac{q}{C} + \frac{Rdq}{dt} = E \tag{7.5.5}$$

or,

$$\frac{dq}{dt} + \frac{q}{RC} = \frac{E}{R}$$

To solve this equation, multiply throughout by $e^{t/RC}$

$$e^{t/RC}\frac{dq}{dt} + e^{t/RC}\frac{q}{RC} = \frac{E}{R}e^{t/RC}$$

The left hand side is $\frac{d}{dt}(qe^{t/RC})$. So we have

$$\frac{d}{dt}(qe^{t/RC}) = \frac{E}{R}e^{t/RC}$$

Integrating both sides with respect to t, we get,

$$qe^{t/RC} = ECe^{t/RC} + A$$

where A is a constant of integration. At $t = 0$, $q = 0$. Thus

$$0 = EC + A$$

∴ The solution is $q = EC\,(1 - e^{-t/RC})$ (7.5.6)

The time required by the charge on the capacitor to reach 63% of its final steady value EC is called the time constant of this circuit. The time constatnt is RC. The current in the circuit at any instant t is given by $i = dq/dt$.

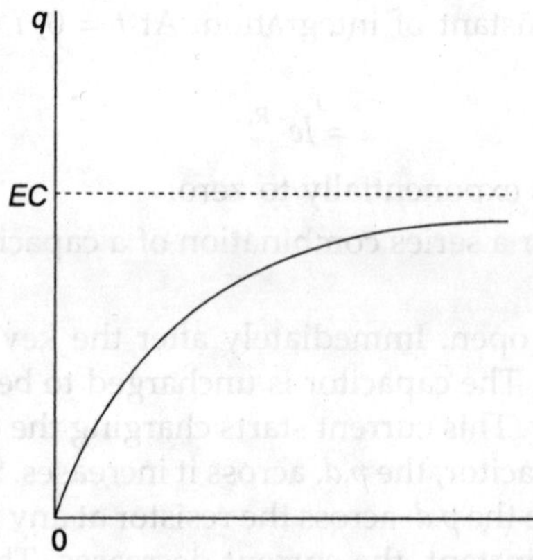

Fig. 7.25

$$\therefore \qquad i = \frac{Ee^{-t/RC}}{R} \qquad (7.5.7)$$

It decays exponentially to zero from an initial value E/R. Summing up, we find that voltage across a capacitor cannot change suddenly. It changes only gradually with a time constant RC.

Let us now study the discharge of the capacitor through a resistor. The source of *emf* has been removed. Suppose q_0 is the charge on the capacitor initially. The *emf* equation of the circuit is

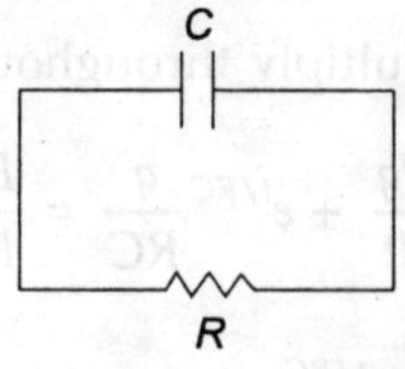

Fig. 7.26

$$Ri + \frac{q}{C} = 0$$

or, $$\frac{Rdq}{dt} + \frac{d}{C} = 0$$

The solution of this equation with the initial condition $q = q_0$ at $t = 0$ is

$$q = q_0 e^{-t/RC} \tag{7.5.9}$$

So the charge on the capacitor decreases exponentially from an initial value q_0.

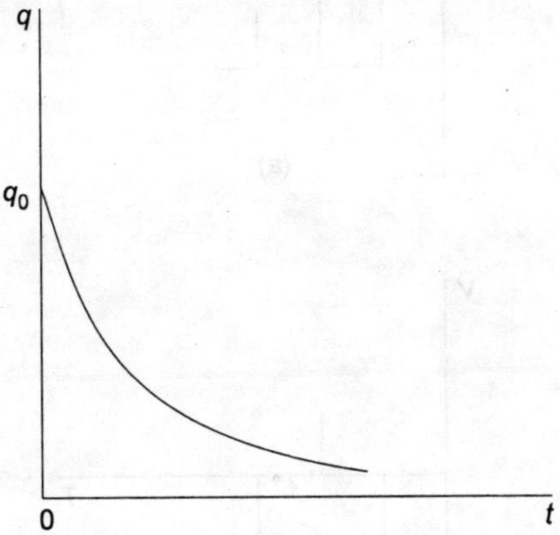

Fig. 7.27

We will look at two simple applications of RC circuits. In the circuit of Fig. 7.28, the voltage across the capacitor C is $V_{in} - V$. So the charge on the plates is $C\ (V_{in} - V)$ at any instant and the current is

$$i = \frac{Cd}{dt}(V_{in} - V)$$

Also, $i = \frac{V}{R}$. Thus $\frac{dV_{in}}{dt} - \frac{dV}{dt} = \frac{V}{RC}$. If we choose R and C small enough so that $\frac{dV}{dt} << \frac{dV_{in}}{dt}$, then

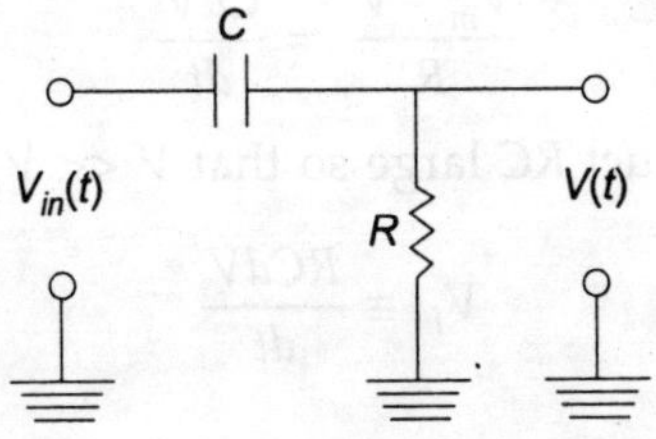

Fig. 7.28

$$V \cong RC \frac{dV_{in}}{dt}$$

Thus the output voltage is the time derivative of the input voltage. The output waveform for a square wave input waveform is shown below:

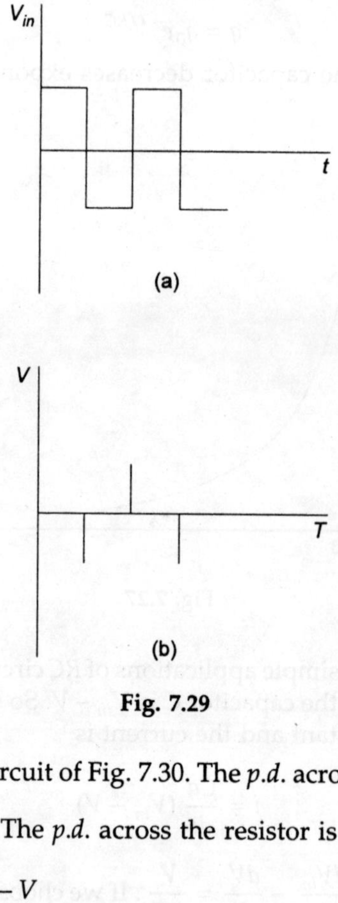

Fig. 7.29

Next consider the circuit of Fig. 7.30. The *p.d.* across the capacitor is *V*; so the current is $i = \frac{CdV}{dt}$. The *p.d.* across the resistor is $V_{in} - V$. So the current is also given by $i = \frac{V_{in} - V}{R}$

Thus we have

$$\frac{V_{in} - V}{R} = \frac{CdV}{dt}$$

If we keep the product *RC* large so that $V << V_{in}$,

$$V_{in} \cong \frac{RCdV}{dt}$$

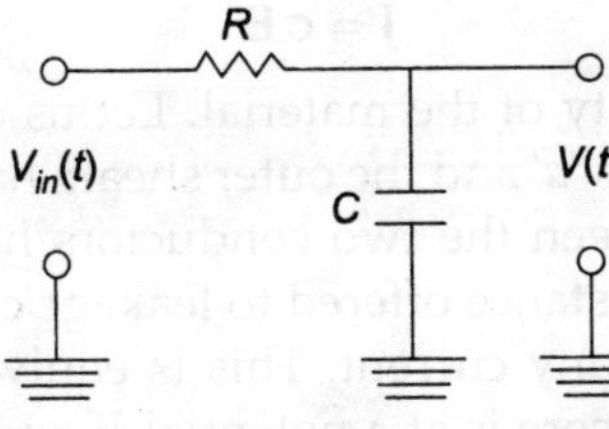

Fig. 7.30

$$\text{or, } V = \frac{1}{RC}\int V_{in}dt + \text{const.}$$

The output voltage is an integral of the input voltage.

Tutorial 1

Let us consider an example of a steady current distribution. A vacuum diode has two metal plates, a cathode and an anode. The cathode is coated with a material which emits electrons copiously when heated. In a plane-parallel diode, the two plates are parallel to each other (Fig. 7.31). The anode is maintained at a positive potential with respect to the cathode. The electrons are emitted with a low velocity **v** perpendicular to the cathode. They are then accelerated by the positive potential of the anode. Take the y-axis parallel to the plates and the x-axis perpendicular to them. The current consists of electrons going from the cathode to the anode. The charge density $\rho = -ne$

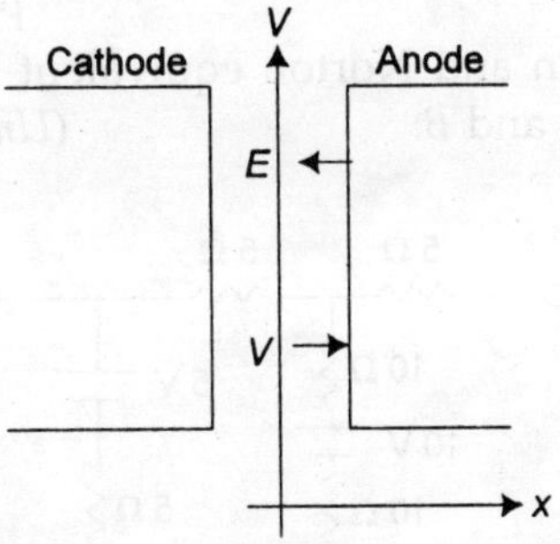

Fig. 7.31

where n is the number of electrons per unit volume. The current density $\mathbf{J} = \rho\mathbf{v}$. In the plane-parallel diode, **J** has no y or z componenets. Using the continuity equation, show that if conditions are stready, **J** is independent of x.

Tutorial 2

We can write Ohm's law in terms of the current density **J** and electric field **E** as

$$\mathbf{J} = \sigma \mathbf{E} \tag{1}$$

where σ is the conductivity of the material. Let us consider a coaxial cable. The inner core has a radius 'a' and the outer sheath has a radius 'b' ($b > a$). The medium (dielectric) between the two conductors has a conductivity σ. We wish to determine the resistance offered to leakage currents. We assume that the cable is not carrying any current. This is equivalent to saying that the entire length of the inner core is at a potential V and the entire length of the sheath is at zero potential. Neglect the megnetic field produced by leakage currents. Recall the electric field between the two conductors from problem 5, chapter 2. The field is radial. Using this field and eqn.(1), calculate the resistance of a length 1 of this coaxial cable.

Problems

1. In a Wheastone's bridge arrangement, a galvanometer of resistance 10 Ω is connected to the junction of R_1 and R_2 and of R_3 and R_4. R_1 = 30 Ω R_2 = 20 Ω, R_3 = 40 Ω, R_4 = 30 Ω. The battery has *emf* 1.5 V and internal resistance 1 Ω. Calculate the current flowing through the galvanometer. *(University of Calcutta, 1998)*

   [Ans. 0.03A]

2. A *dc* generator (*emf* = 20 V) delivers a maximum power of 10 W to an external load resistance. Calculate the internal resistance of the source and its short circuit current. How much power will it dissipate when its terminals are shorted? *(University of Calcutta, 1999)*

   [Ans. Intenal resistance = 10 Ω, short circuit current = 2 A. power dissipated = 40 W]

3. Find the Thévenin and Norton equivalent circuit for the following circuit between A and B: *(University of Calcutta. 2000)*

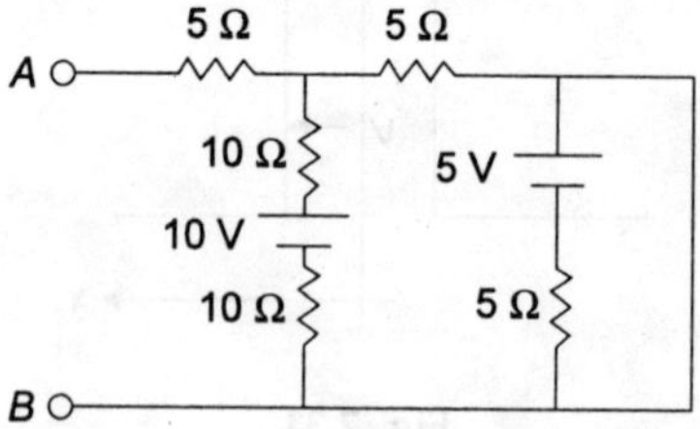

[Ans. Thévenin equivalent circuit:

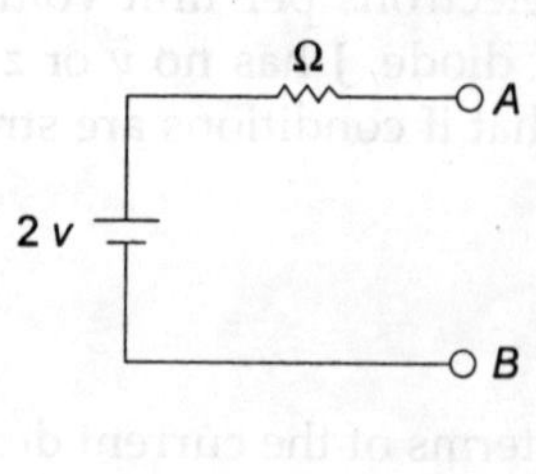

Norton equivalent circuit:

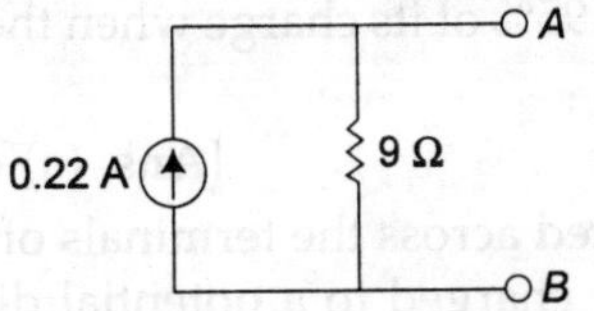

]

4. Determine the Thévenin and Norton equivalent circuits of the given network between the terminals *A* and *B*.

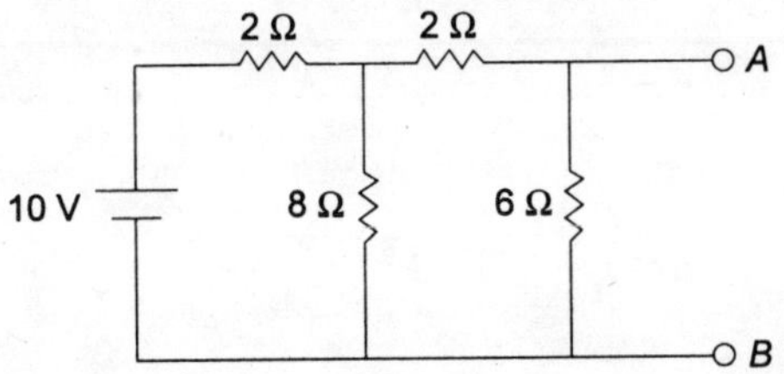

[**Ans.** Thevenin equivalent circuit:

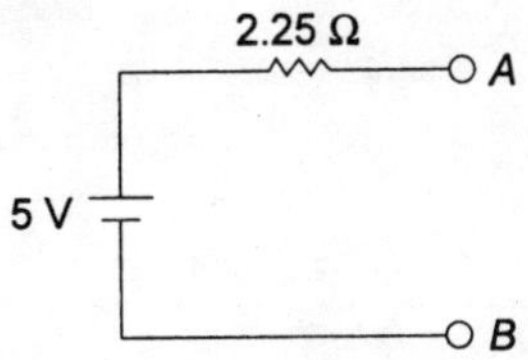

Norton equivalent circuit:

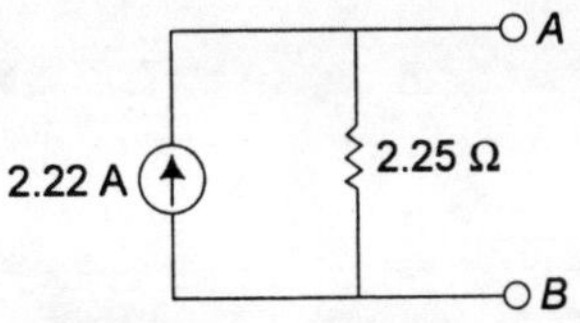

]

5. A direct voltage of 100 V is applied to a coil of resistance 10 Ω and inductance 10 H. What is the value of the current 0.1 sec. after switching on the time taken for the current to reach one half of its final value?

[**Ans.** current = 0.95 A, time = 0.69 sec.]

6. A coil of resistance 20 Ω and inductance 0.5 H is switched on to a direct current, 200 V supply. Calculate the rate of change of current (a) at the instant of closing the switch and (b) when $t = L/R$. Find also (c) the final steady value of the current.

[**Ans.** (a) 400 A/s. (b) 147 A/s. (c) 10A.]

7. A 5 μ*F* capacitor is connected to a constant voltage source through a resistance of 2 MΩ. Calculate the time taken for the capacitor to lose (a) 50% (b) 63.2% (c) 95% of its charge when the voltage source is short-circuited.

[**Ans.** (a) 6.93 s (b) 10 s (c) 29.9 s]

8. A resistor is connected across the terminals of a 20 μ*F* capacitor which has been previously charged to a potential difference of 500 V. If the potential difference falls to 300 V in 0.5 min., calculate the resistance in megohms.

[Ans. 2.94 MΩ]

8

AC Circuits

8.1 DEFINITIONS

An alternating current or voltage in one which reverses its direction periodically. The figures below show some examples. Fig. 8.1 (a) shows a sine wave, (b) a triangle wave and (c) a square wave. In this book, we shall almost exclusively be studying sinusoidal waves, i.e. sine or cosine waveforms. Let us define certain quantities first. The number of oscillations per second is called the *frequency* of the alternating current. The time T required to complete one oscillation is called its *period*. $\therefore$ $T = 1/f$. The *angular*

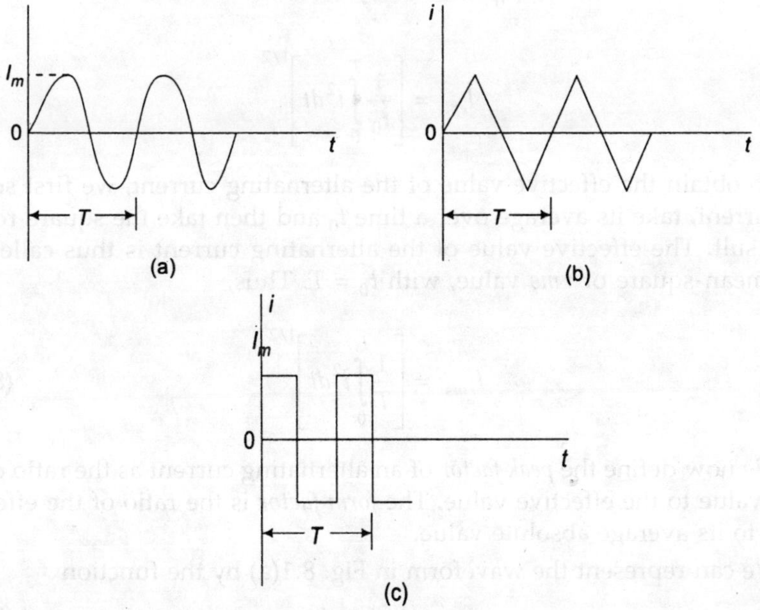

Fig. 8.1

frequency ω is given by $\omega = 2\pi f$ rad/s. The maximum value of the alternating current is called its *peak value* I_m. The average or *dc* value of the alternating current over a complete cycle is defined by

$$I_{dc} = \frac{1}{T}\int_0^T i\,dt \tag{8.1.1}$$

For currents like those shown in Fig. 8.1(a), (b) and (c), this average is clearly zero. If we use a dc ammeter to measure an alternating current, a zero reading will result. So to measure the current, we use its heating effect. The effective value of an alternating current is defined to be that steady current which produces the same heating effect in a resistor over the same time interval as the alternating current. Let t_0 be the time interval and the resistance be R. The heat produced by the alternating current in R in this interval is

$$R\int_0^{t_0} i^2 dt$$

This is the same as the heat produced by the effective value of the current $I_{eff.}$, in the same time interval. Thus

$$R\int_0^{t_0} i^2 dt = RI_{eff}^2 t_0$$

$$I_{eff.} = \left[\frac{1}{t_0}\int_0^T i^2 dt\right]^{1/2}$$

To obtain the effective value of the alternating current, we first square the current, take its average over a time t_0 and then take the square root of the result. The effective value of the alternating current is thus called the root-mean-square or *rms* value, with $t_0 = T$. Thus,

$$I_{rms} = \left[\frac{1}{T}\int_0^{t_0} i^2 dt\right]^{1/2} \tag{8.1.2}$$

We now define the *peak factor* of an alternating current as the ratio of the peak value to the effective value. The *form factor* is the ratio of the effective value to its average absolute value.

We can represent the waveform in Fig. 8.1(a) by the function

$$i = I_m \sin\omega t \tag{8.1.3}$$

Let us use this representation to calculate the average absolute value and the *rms* value. The average absolute value is the same as the average value over half a cycle. It is

$$\frac{2}{T}\int_0^{T/2} idt = \frac{2}{T}\int_0^{T/2} I_m \sin\omega t dt$$

$$= \frac{4}{\omega T} = \frac{2}{\pi} \text{ (using } \omega T = 2\pi)$$

The *rms* value is $$I_{rms} = \left[\frac{1}{T}\int_0^T I_m^2 \sin^2 \omega t dt\right]^{1/2}$$

$$= \left[\frac{I_m^2}{2}\right]^{1/2} = \frac{I_m}{\sqrt{2}} = 0.71 I_m$$

$\therefore$ The peak factor of the sinusoidal wave is $\sqrt{2}$. The form factor is $\frac{0.71}{0.64} = 1.11$

We can also represent alternating currents and voltages by rotating vectors or phasors. Let us briefly consider complex numbers. Let $\hat{\mathbf{n}}$ be a unit vector in some direction. Then $a\hat{\mathbf{n}}$ is a vector in the same direction and $-\hat{\mathbf{n}}$ is a vector in the opposite direction. Since $j^2 = -1$, we can think of multiplication by *j* twice as rotation of vector $\hat{\mathbf{n}}$ by 180°. So multiplication by *j* once is equivalent to rotation of vector $\hat{\mathbf{n}}$ by 90°. Thus $j\hat{\mathbf{n}}$ is a vector of unit length, making an angle 90° with $\hat{\mathbf{n}}$. If a and *b* are scalars, then $(a + bj)\hat{\mathbf{n}}$ is a vector making an angle $\tan^{-1}\left(\frac{b}{a}\right)$ with respect to $\hat{\mathbf{n}}$. It has length $\sqrt{(a^2 + b^2)}$ We can also express this as $re^{j\theta}\hat{\mathbf{n}}$. Since $e^{j\theta} = \cos\theta + j\sin\theta$ with

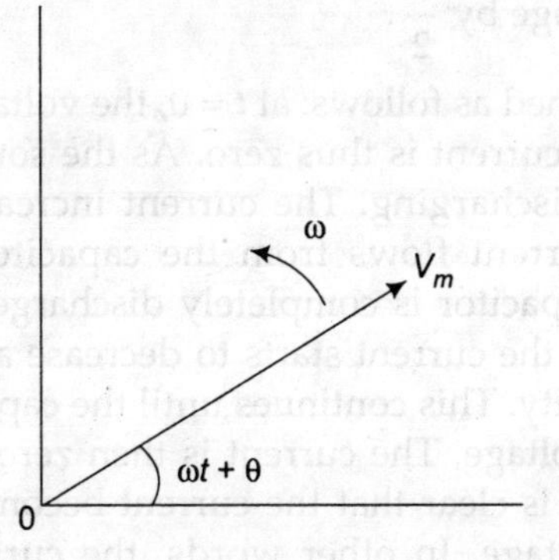

Fig. 8.2

$r = \sqrt{(a^2 + b^2)}$ and $\tan\theta = \frac{b}{a}$. An alternating current or voltage is represented by a rotating vector or phasor, rotating at an angular speed of ω rad/s. Thus an alternating voltage can be represented by $V_m e^{j(\omega t+0)}\hat{\mathbf{n}}$. V_m is the peak value. $V_m e^{j(\omega t+0)}\, \hat{\mathbf{n}} = \{V_m \cos(\omega t + \theta) + jV_m \sin(\omega t + \theta)\}\hat{\mathbf{n}}$. Taking the dot product of this with $\hat{\mathbf{n}}$, we have $V_m e^{j(\omega t+0)}\hat{\mathbf{n}}.\hat{\mathbf{n}} - V_m \cos(\omega t + \theta)$ (since $j\hat{\mathbf{n}}.\hat{\mathbf{n}} = 0$), the physical voltage.

8.2 SIMPLE CIRCUITS

Let us now see the response of capacitors and inductors to an *ac* voltage source. First we shall considèr a capacitor. Let $v = V_m \cos \omega t$ be the applied voltage.

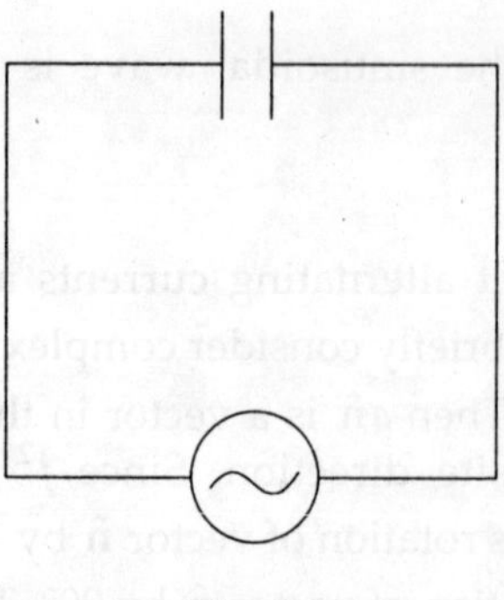

Fig. 8.3

The charge on the capacitor at any instant t is then $q = CV_m \cos \omega t$. The current in the circuit is $i = \frac{dq}{dt} = -\omega CV_m \sin \omega t = \omega CV_m \cos\left(\omega t + \frac{\pi}{2}\right)$. The current leads the voltage by $\frac{\pi}{2}$.

This can be explained as follows: at $t = 0$, the voltage across the capacitor is maximum. V_m The current is thus zero. As the source voltage decreases, the capacitor starts discharging. The current increases, the negative sign implying that the current flows from the capacitor to the source. This continues until the capacitor is completely discharged. The current is then maximum. Thereafter the current starts to decrease as the source voltage is now of opposite polarity. This continues until the capacitor is charged to the maximum negative voltage. The current is then zero. From the two curves (Fig. 8.4a and 8.4b) it is clear that the current becomes zero a quarter of a period before the voltage. In other words, the current leads the voltage across the capacitor by a phase difference of $\pi/2$. The ratio of the peak value

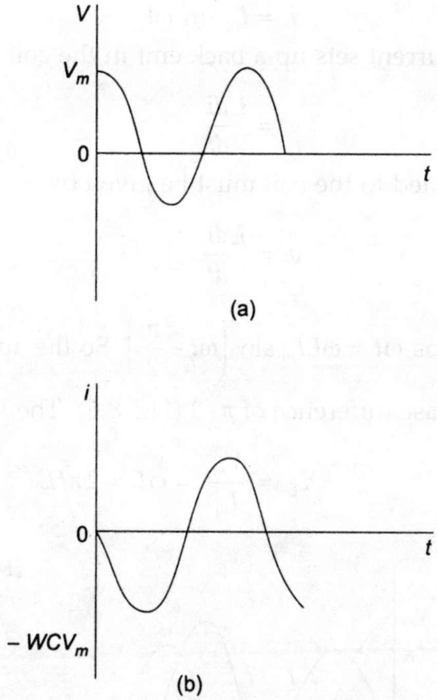

Fig. 8.4

of the voltage to the peak value of the current is called the *capacitive reactance* of the capacitor. It is denoted by X_C and has a unit of ohms. Thus

$$X_C = \frac{V_m}{I_m}$$

The peak value of the current is $I_m = \omega C V_m$. ∴ The reactance of the capacitor is $X_C = 1/\omega C = 1/2\pi f C$ where f is the frequency of the supply.

Next let us suppose that an alternating voltage is applied to a coil of self inductance L. Let the current in the circuit be given by

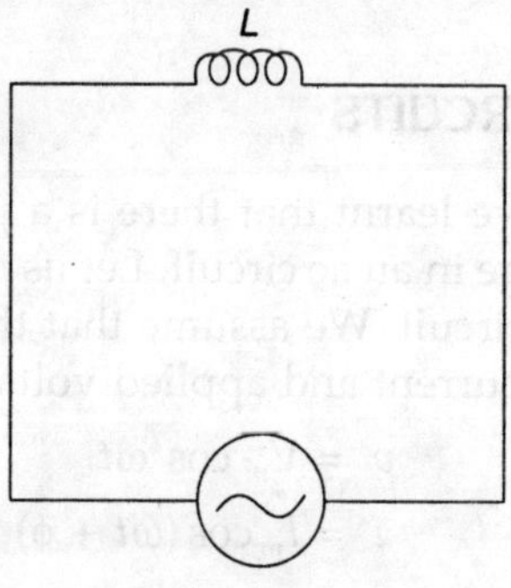

Fig. 8.5

$$i = I_m \sin \omega t \tag{8.2.1}$$

This changing current sets up a back emf in the coil of magnitude

$$e = \frac{L\,di}{dt}$$

The voltage applied to the coil must be given by

$$v = \frac{L\,di}{dt}$$

and so is $v = \omega L I_m \cos \omega t = \omega L I_m \sin\left(\omega t + \frac{\pi}{2}\right)$. So the applied voltage leads the currents by a phase difference of $\pi/2$ (Fig. 8.6). The *inductive reactance* is

$$X_L = \frac{V_m}{I_m} = \omega L = 2\pi f L$$

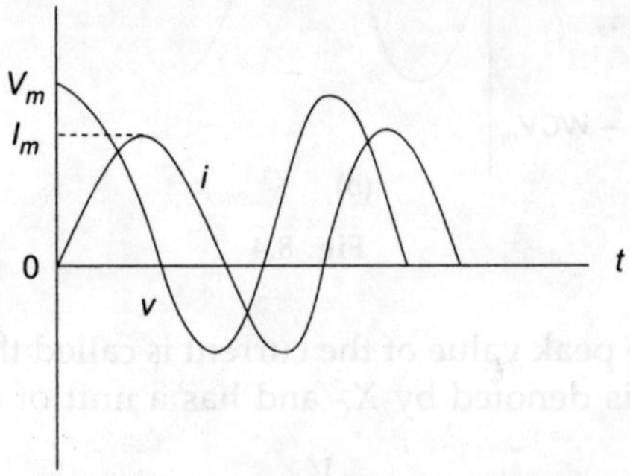

Fig. 8.6

In a resistor R, the current and voltage are in phase with each other. On applying an ac voltage of $v = V_m \cos \omega t$ to the resistor, we get a current $I = \frac{V_m}{R} \cos \omega t$ flowing through it.

8.3 POWER IN AC CIRCUITS

In the last section, we have learnt that there is a phase difference between current and applied voltage in an ac circuit. Let us now calculate the average power consumed by the circuit. We assume that there is an arbitrary phase difference ϕ between the current and applied voltage. Let

$$v = V_m \cos \omega t$$

$$i = I_m \cos(\omega t + \phi)$$

The instantaneous power P consumed by the circuit is

$$P = vi$$

and the average power is

$$P_{av} = \frac{1}{T}\int_0^T vi\,dt = \frac{V_m I_m}{T}\int_0^T \cos\omega t \cos(\omega t + \phi)\,dt$$

$$\int_0^T \cos\omega t \cos(\omega t + \phi)\,dt = \frac{1}{2} T \cos\phi$$

$$\therefore \qquad P_{av} = \frac{1}{2} V_m I_m \cos\phi \qquad (8.3.1)$$

Since the *rms* values of v and i are $V_m/\sqrt{2}$ and $I_m/\sqrt{2}$ respectively, we can write (8.3.1) as

$$P_{av.} = V_{rms} I_{rms} \cos\phi \qquad (8.3.2)$$

The product $V_{rms} I_{rms}$ is called the *apparent power* and $\cos\phi$ is called the *power factor* of the circuit. Thus

Average power = Apparent power x power factor

The unit of apparent power is voltamperes or *VA*. For a resistance, $\phi = 0$ and so $P_{av.} = V_{rms} I_{rms}$. For a capacitor or an inductor, $\phi = \pm\pi/2$ and so $P_{av.} = 0$. Thus a pure inductor or a pure capacitor does not consume any power on the average. They are called *wattless* elements. In general, the component $I_{rms}\cos\phi$ is called the power component of the current because when multiplied by V_{rms} it gives the average power. The component $I_{rms}\sin\phi$ is called the wattless component of the current because it does not contribute to the average power. The product $V_{rms} I_{rms} \sin\phi$ is called the quadrature power and its unit is voltamperes reactive or var.

8.4 THE SERIES RL CIRCUIT

We will now consider what happens when an AC voltage is applied to a series combination of R and L. We will assume that the transients have died away and steady state conditions prevail. The emf equation of the circuit is

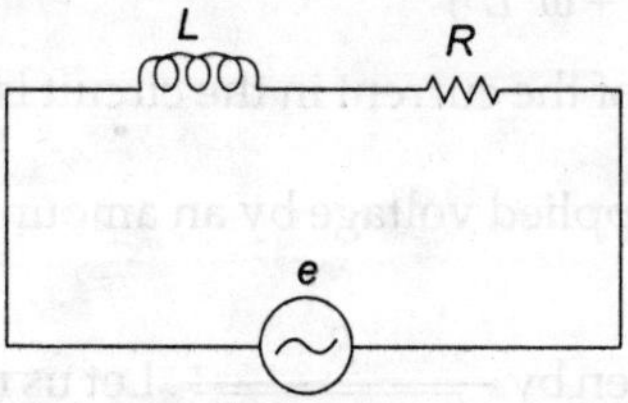

Fig. 8.7

$$e = Ri + \frac{L\,di}{dt} \tag{8.4.1}$$

We shall solve it using two methods: sinusoids and phasors. Let us take

$$e = E_m \cos \omega t$$

and
$$i = I_m \cos(\omega t + \phi) \tag{8.4.2}$$

Substituting is (8.4.1) we get:

$$E_m \cos \omega t = RI_m \cos(\omega t + \phi) - \omega L I_m \sin(\omega t + \phi)$$

Collecting coefficients of cos ωt and sin ωt. We get,

$$(E_m - RI_m \cos\phi + \omega L I_m \sin\phi)\cos\omega t + (RI_m \sin\phi + \omega L I_m \cos\phi)\sin\omega t = 0$$

This is of the form

$$A \cos \omega t + B \sin \omega t = 0$$

with A and B constants. This can be true for all time t only if both A and B are zero. So we have

$$R \sin\phi + \omega L \cos\phi = 0$$

or,
$$\tan\phi = \frac{-\omega L}{R} \tag{8.4.3}$$

and
$$E_m - RI_m \cos\phi + \omega L I_m \sin\phi = 0 \tag{8.4.4}$$

From (8.4.3), we have,

$$\sin\phi = \frac{-\omega L}{\sqrt{(R^2 + \omega^2 L^2)}}, \cos\phi = \frac{R}{\sqrt{(R^2 + \omega^2 L^2)}}$$

Putting this in (8.4.4), we get

$$I_m = \frac{E_m}{\sqrt{(R^2 + \omega^2 L^2)}} \tag{8.4.5}$$

Thus the instantaneous current in the circuit is

$$i = \frac{E_m}{\sqrt{(R^2 + \omega^2 L^2)}} \cos\left(\omega t - \tan^{-1}\frac{\omega L}{R}\right) \tag{8.4.6}$$

Thus the peak value of the current in the circuit is given by (8.4.5) and the current lags behind the applied voltage by an amount $\tan^{-1}\left(\frac{\omega L}{R}\right)$. The power factor of this circuit is given by $\frac{R}{\sqrt{(R^2 + \omega^2 L^2)}}$. Let us now use phasors to solve the same problem. We take

$$\mathbf{E} = E_m e^{jwt}\hat{\mathbf{n}} \tag{8.4.7}$$

and $$\mathbf{I} = Ie^{j\omega t}\hat{\mathbf{n}}$$

where E_m is real and I may be complex.

$$\frac{d\mathbf{I}}{dt} = j\omega Ie^{j\omega t}\hat{\mathbf{n}} = j\omega\mathbf{I}$$

The instantaneous voltage and current are given by

$$e = \mathbf{E}.\hat{\mathbf{n}}$$

$$i = \mathbf{I}.\hat{\mathbf{n}}$$

Substitute this in (8.4.1). We have,

$$\mathbf{E}.\hat{\mathbf{n}} = R\mathbf{I}.\hat{\mathbf{n}} + j\omega L\mathbf{I}.\hat{\mathbf{n}}$$

or $$\mathbf{I} = \frac{\mathbf{E}}{R + j\omega L} \tag{8.4.8}$$

We can write this as $\mathbf{I} = \mathbf{E}/Z$ with $Z = R + j\omega L$. Z is called the *complex impedance* of the circuit. Thus

$$\mathbf{I} = \frac{\mathbf{E}}{\sqrt{(R^2 + \omega^2 L^2)}} e^{-j\theta} \text{ where } \tan\theta = \frac{\omega L}{R}$$

Using (8.4.7) $$\mathbf{I} = \frac{E_m e^{j(\omega t - \theta)}\hat{\mathbf{n}}}{\sqrt{(R^2 + \omega^2 L^2)}}$$

∴ The instantaneous current is $i = \mathbf{I}.\hat{\mathbf{n}}$

∴ $$i = \frac{E_m}{\sqrt{(R^2 + w^2 L^2)}} \cos\left(\omega t - \tan^{-1}\frac{\omega L}{R}\right)$$

We can represent this result by a phasor diagram.

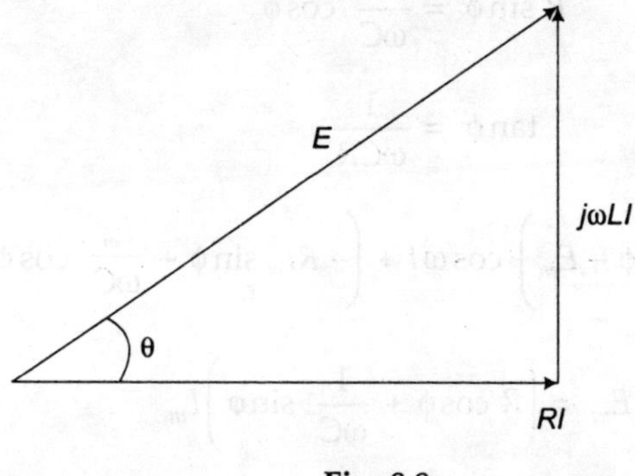

Fig. 8.8

8.5 THE SERIES RC CIRCUIT

Next let us consider a series combination of R and C (Fig. 8.9). The *emf* equation for this circuit is

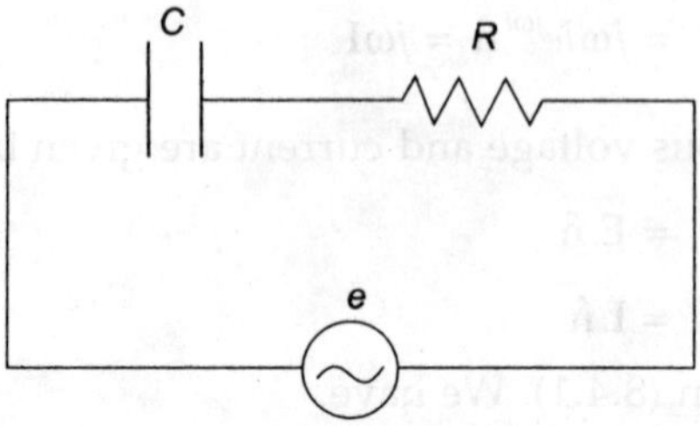

Fig. 8.9

$$e = \frac{q}{C} + Ri \qquad (8.5.1)$$

We will assume, as before

$$e = E_m \cos\omega t$$
$$i = I_m \cos(\omega t + \phi)$$

The charge q on the capacitor, at any instant t, is given by

$$q = \int i dt = \frac{I_m}{\omega}\sin(\omega t + \phi)$$

Putting this in the *emf* equation, we get,

$$E_m \cos\omega t = RI_m \cos(\omega t + \phi) + \frac{I_m}{\omega C}\sin(\omega t + \phi)$$

Collecting coefficients of cos ωt and sin ωt, we get:

Equating the coefficients to zero, we get,

$$R\sin\phi = \frac{1}{\omega C}\cos\phi$$

or,
$$\tan\phi = \frac{1}{\omega CR} \qquad (8.5.2)$$

$$\left(RI_m\cos\phi + \frac{I_m}{\omega C}\sin\phi - E_m\right)\cos\omega t + \left(-RI_m\sin\phi + \frac{I_m}{\omega C}\cos\phi\right)\sin\omega t = 0$$

and,
$$E_m = \left(R\cos\phi + \frac{1}{\omega C}\sin\phi\right)I_m \qquad (8.5.3)$$

with $\cos\phi = \dfrac{R}{\sqrt{\left(R^2 + \dfrac{1}{\omega^2 C^2}\right)}}$ and $\sin\phi = \dfrac{1/\omega C}{\sqrt{\left(R^2 + \dfrac{1}{\omega^2 C^2}\right)}}$

we have,
$$E_m = I_m \sqrt{\left(R^2 + \frac{1}{\omega^2 C^2}\right)} \tag{8.5.4}$$

Thus, the instantaneous current in the circuit is given by

$$I = \frac{E_m}{\sqrt{\left(R^2 + \frac{1}{\omega^2 C^2}\right)}} \cos\left(\omega t + \tan^{-1}\frac{1}{\omega CR}\right) \tag{8.5.5}$$

The current in the circuit leads the applied voltage by phase angle of $\tan^{-1}\left(\dfrac{1}{\omega CR}\right)$. The power factor of this circuit is $\dfrac{R}{\sqrt{\left(R^2 + \dfrac{1}{\omega^2 C^2}\right)}}$. Let us now use phasors to solve the same problem. We take

$$\mathbf{E} = E_m e^{j\omega t}\hat{\mathbf{n}}$$

$$\mathbf{I} = I e^{j\omega t}\hat{\mathbf{n}}$$

where E_m is real and I may be complex.

$$e = \mathbf{E}.\hat{\mathbf{n}},\ i = \mathbf{I}.\hat{\mathbf{n}}$$

$$\int \mathbf{I}dt = \frac{1e^{j\omega t}\hat{\mathbf{n}}}{jw} = \frac{1}{j\omega}$$

Substituting in the *emf* equation, we get,

$$\mathbf{E}.\hat{\mathbf{n}} = R\mathbf{I}.\hat{\mathbf{n}} + \frac{1\,\mathbf{I}.\hat{\mathbf{n}}}{j\omega C}$$

or,
$$\mathbf{I} = \frac{E}{R + \frac{1}{j\omega C}}$$

The impedance is
$$Z = R + \frac{1}{j\omega C} = R - \frac{j}{\omega C}$$

Thus $$\mathbf{I} = \frac{\mathbf{E}}{\sqrt{\left(R^2 + \frac{1}{\omega^2 C^2}\right)}} e^{j\theta} \text{ with } \tan\theta = \frac{1}{\omega CR}$$

or, $$\mathbf{I} = \frac{\mathbf{E}_m}{\sqrt{\left(R^2 + \frac{1}{\omega^2 C^2}\right)}} e^{j(\omega t + \theta)} \hat{\mathbf{n}}$$

The instantaneous current is

$$i = \mathbf{I}.\hat{\mathbf{n}}$$

$$= \frac{\mathbf{E}_m}{\sqrt{\left(R^2 + \frac{1}{\omega^2 C^2}\right)}} \cos\left(\omega t + \tan^{-1}\frac{1}{\omega RC}\right)$$

The phasor diagram for the circuit is shown below:

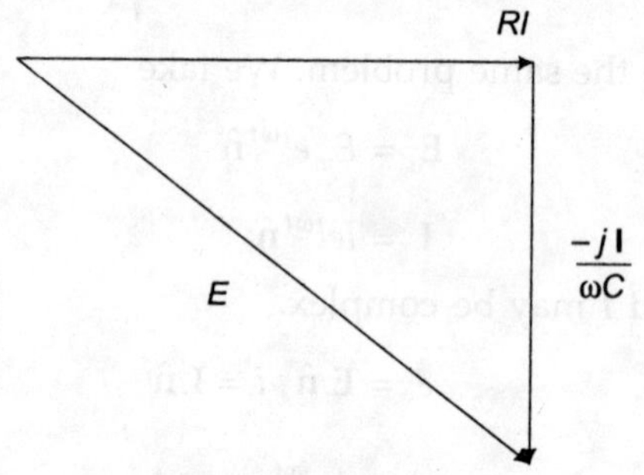

Fig. 8.10

Example 1: A resistor of 100 Ω is connected in series with a 50 μF capacitor to a supply at 200 V, 50 Hz. Find (a) the impedance, (b) the current, (c) the power factor, (d) the phase angle, (e) the voltage across the resistor and across the capacitor.

Solution: The supply voltage quoted is the *rms* value unless otherwise stated, in all problems. Since $I_{rms} = I_m/\sqrt{2}$ and $E_{rms} = E_m/\sqrt{2}$, we still have

$$I_{rms} = \frac{E_{rms}}{\sqrt{\left(R^2 + \frac{1}{\omega^2 C^2}\right)}}$$

$$\omega = 2\pi \times 50 = 314 \text{ rad/s.} \quad \frac{1}{\omega C} = 63.7\ \Omega$$

$$Z = 100 - j\,63.7,\ |Z| = 118.6\ \Omega$$

$$I_{rms} = \frac{200}{118.6} = 1.69\ \text{A}$$

Power factor $\cos\phi = \dfrac{R}{\sqrt{\left(R^2 + \dfrac{1}{\omega^2 C^2}\right)}} = \dfrac{R}{|Z|} = 0.843$

Phase angle $\phi = 32°32'$

RMS voltage across resistor $V_R = I_{rms}R = 169$ V

RMS voltage across capacitor $V_C = \dfrac{I_{rms}}{\omega C} = 1.69 \times 63.7 = 108$ V

Note $E_{rms}^2 = V_R^2 + V_C^2$

Example 2: An inductive coil takes 10 A and dissipates 1000 W when connected to a supply at 250 V, 25 Hz. Calculate (a) the impedance (b) the effective resistance (c) the inductance (d) the angle of lag.

Solution: The inductive coil can be represented as an inductor L in series with a resistance R (the effective resistance of the coil). $I_{rms} = 10$ A, $E_{rms} = 250$ V.

$\therefore$ Impedance $|Z| = \dfrac{E_{rms}}{I_{rms}} = \dfrac{250}{10} = 25\ \Omega$

Power dissipated $= E_{rms} I_{rms} \dfrac{R}{|Z|}$

$$1000 = 250 \times 10 \times \frac{R}{25} \qquad \therefore R = 10\ \Omega$$

$$\omega L = 2\pi \times 25L = 50\pi L$$

$$|Z|^2 = R^2 + \omega^2 L^2 \qquad \therefore 625 = 100 + (50\pi L)^2$$

$\therefore$ $L = 0.146$ H

$$\cos\theta = \frac{R}{|Z|} = \frac{10}{25} = 0.4 \qquad \therefore \theta = 66°25'$$

8.6 THE SERIES RESONANT CIRCUIT

Let us now consider a circuit in which all the three elements R, L and C are present and are connected in series (Fig. 8.11). The emf equation of the circuit is

$$e = Ri + \frac{1}{C}\int i\,dt + L\frac{di}{dt} \tag{8.6.1}$$

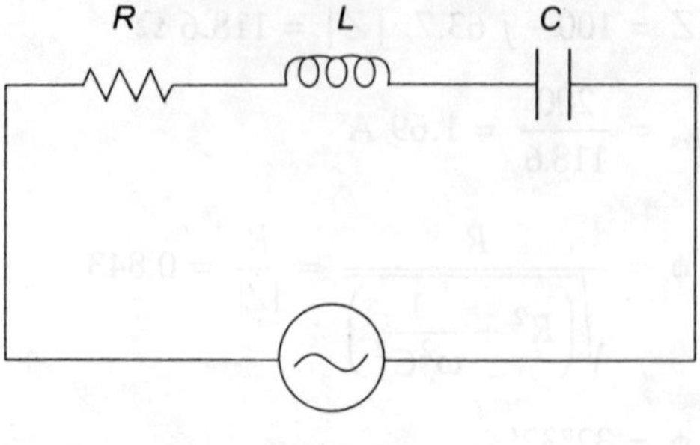

Fig. 8.11

We will merely state the result here. The solution is left for the tutorial.

$$\mathbf{I} = \frac{E}{R + j\left(\omega L - \dfrac{1}{\omega C}\right)} \tag{8.6.2}$$

$$\mathbf{I} = \frac{E_m e^{j\omega t}}{|Z| e^{i\theta}} \hat{\mathbf{n}} \text{ where } \tan\theta = \left(\omega L - \frac{1}{\omega C}\right) / R$$

$$\therefore \qquad \mathbf{I} = \frac{E_m e^{i(\omega t - 0)}}{\sqrt{R^2 + \left(wL - \dfrac{1}{\omega C}\right)^2}} \hat{\mathbf{n}}$$

The peak value of the current is

$$I_m = \frac{E_m}{\sqrt{R^2 + \left(wL - \dfrac{1}{\omega C}\right)^2}} \tag{8.6.3}$$

If we vary the frequency of the ac supply, keeping E_m fixed, I_m reaches its maximum value when $\omega_0 L = \dfrac{1}{\omega_0 C}$ or,

$$\omega_0 = \frac{1}{\sqrt{LC}} \tag{8.6.4}$$

It decreases if ω is either decreased or increased from this value. At this frequency, $\theta = 0$. Thus, the current and voltage are in phase. The circuit is then said to be in resonance with the applied voltage. This is why the circuit of Fig. 8.11 is called a series resonant circuit. In this case, the current is maximum at resonance. This circuit is also called an *acceptor* circuit. The circuit accepts the current, so to speak. How sharply peaked is the curve of Fig.8.12? To get a quantitative measure of that, we define the Q factor as

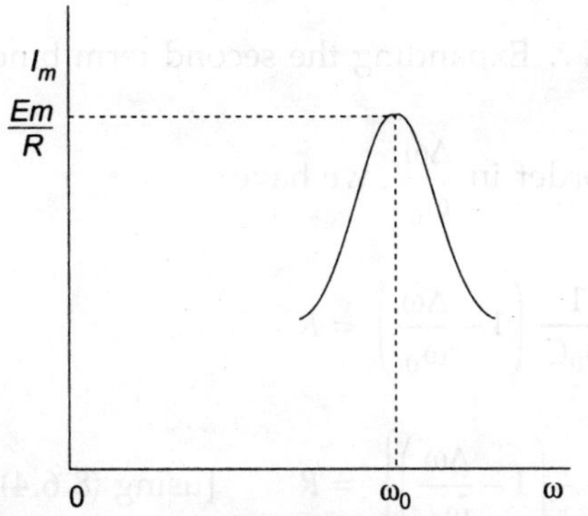

Fig. 8.12

$$Q = \frac{f_0}{\Delta f} \tag{8.6.5}$$

where f_0 (= $\omega_0/2\pi$) is the resonant frequency and Δf is called the *bandwidth*. It is the difference

$$\Delta f = f_2 - f_1$$

where f_2 and f_1 are the upper and lower half-power frequencies. At these frequencies, the power dissipated by the circuit is half the power it dissipates at resonance. The peak value of the current at resonance is $I_m = E_m/R$. Thus, at frequencies f_2 and f_1, $I_m = E_m/R\sqrt{2}$. This is because power is proportional to I^2 and so the current becomes $1/\sqrt{2}$ times its value at resonance $\therefore$ from (8.6.3), we have,

$$2R^2 = R^2 + \left(\omega L - \frac{1}{\omega C}\right)^2$$

$$R^2 = \left(\omega L - \frac{1}{\omega C}\right)^2$$

$$\omega L - \frac{1}{\omega C} = \pm R$$

Let us put $\omega = \omega_0 + \Delta\omega$ and take the positive sign in the above equation.

$$(\omega_0 + \Delta\omega)L - \frac{1}{(w_0 + \Delta\omega)C} = R$$

$$\omega_0 L\left(1 + \frac{\Delta\omega}{\omega_0}\right) - \frac{1}{\omega_0 C\left(1 + \frac{\Delta\omega}{\omega_0}\right)} = R$$

Now $\frac{\Delta\omega}{\omega_0} << 1$. ∴ Expanding the second term binomially and keeping only terms of first order in $\frac{\Delta\omega}{\omega_0}$, we have

$$\omega_0 L\left(1+\frac{\Delta\omega}{\omega_0}\right) - \frac{1}{\omega_0 C}\left(1-\frac{\Delta\omega}{\omega_0}\right) = R$$

or, $$\omega_0 L\left\{1+\frac{\Delta\omega}{\omega_0} - \left(1-\frac{\Delta\omega}{\omega_0}\right)\right\} = R$$ [using (8.6.4)] or, $\frac{2\,\Delta\omega}{\omega_0}.\omega_0 L = R$

The bandwidth is now $\frac{1}{2\pi}\{\omega_0 + \Delta\omega - (\omega_0 - \Delta\omega)\}$ because f_2 and f_1 are $\frac{\omega_0 + \Delta\omega}{2\pi}$ and $\frac{\omega_0 - \Delta\omega}{2\pi}$. ∴ It is $\frac{1.2\,\Delta\omega}{2\pi}$.

∴ $$Q = \frac{\omega_0}{2\,\Delta\omega} = \frac{\omega_0 L}{R} \tag{8.6.6}$$

Since for a series resonant circuit, $\omega_0 = 1/\sqrt{LC}$,

$$Q = \frac{1}{R}\sqrt{\frac{L}{C}}$$

8.7 THE PARALLEL RESONANT CIRCUIT

Consider now the circuit of Fig. 8.13

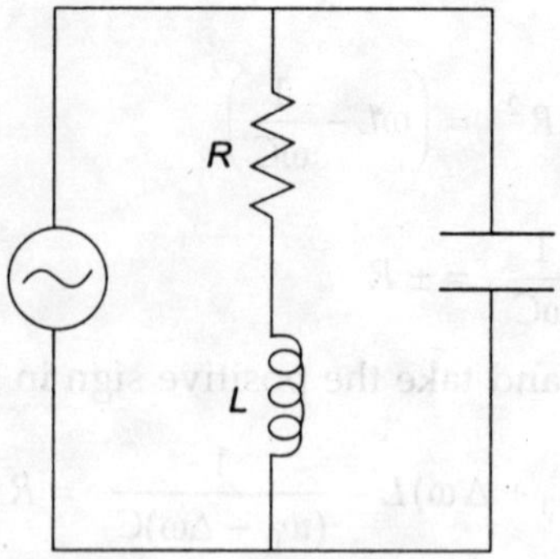

Fig. 8.13

This circuit is called a parallel resonant circuit or a tank circuit. The R usually refers to the losses in the inductance. Let us define the *admittance* Y as

$$Y = \frac{1}{Z} \tag{8.7.1}$$

It is useful in analyzing impedances in parallel. For purely resistive circuit, we know that the equivalent conductance of a parallel combination of resistors is the sum of their conductances. Similarly, the equivalent admittance of the above circuit is

$$Y = \frac{1}{R + j\omega L} + j\omega C$$

$$= \frac{R - jwL}{R^2 + \omega^2 L^2} + j\omega C$$

$$\therefore \qquad Y = \frac{R}{R^2 + \omega^2 L^2} + j\omega\left(C - \frac{L}{R^2 + \omega^2 L^2}\right)$$

The instantaneous current is

$$i = Ye$$

Resonance occurs when the current is in phase with the voltage. If the frequency of the ac source is varied, there will be a frequency f_0 at which the imaginary part of the admittance is zero. The admittance is then simply a conductance

$$C = \frac{L}{R^2 + \omega_0^2 L^2}$$

or,

$$\omega_0 = \left(\frac{1}{LC} - \frac{R^2}{L^2}\right)^{1/2} \tag{8.7.2}$$

This is the resonant (angular) frequency. At this frequency, the admittance is minimum because it has only a nonzero real part. Thus the current has a minimum amplitude at resonance (Fig. 8.14). This is why this circuit is also called a *rejector circuit*. The circuit rejects the current, so to speak. In this figure, I_m is the peak value of the current. At resonance, the admittance is

$$Y_0 = \frac{R}{R^2 + w_0{}^2 L^2} = \frac{RC}{L}$$

The impedance, which is simply a resistance, is

$$Z_0 = \frac{L}{CR} \tag{8.7.3}$$

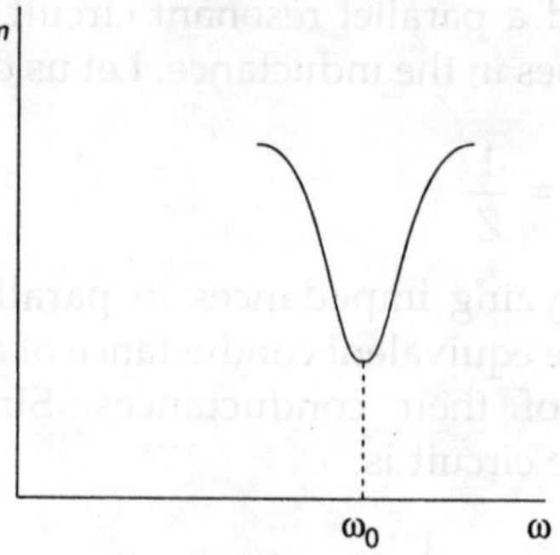

Fig. 8.14

If the resistance in the circuit is small, we may write

$$\omega_0 \cong \frac{1}{\sqrt{LC}} \tag{8.7.4}$$

The complex impedance of the circuit is

$$Z = \frac{R + j\omega L}{(1 - \omega^2 LC) + j\omega CR}$$

The impedance is maximum at resonance, the value being given by (8.7.3). For the case of small R, we can calculate the Q factor of the circuit quite simply. At the half power frequencies, we must have

$$|Z|^2 = \frac{Z_0{}^2}{2}$$

or,

$$\frac{R^2 + \omega^2 L^2}{\left(1 - \omega^2 + CR\right)^2 + \omega^2 C^2 R^2} = \frac{L^2}{2C^2R^2}$$

This is true if in the denominator,

$$1 - \omega^2 LC = \omega CR \tag{8.7.5}$$

Let $\omega = \omega_0 \pm \Delta\omega$, where $\Delta\omega$ is small, be the two frequencies, so that the bandwidth is $2\,\Delta\omega$. Let $\omega = \omega_0 + \Delta\omega$ in (8.7.5) and put $\omega \cong \omega_0$ in the right hand side. Then, we have,

$$1 - (\omega_0 + \Delta\omega)^2 LC \cong \omega_0 CR$$

or,

$$1 - \omega_0^2 \left(1 + \frac{\Delta\omega}{\omega_0}\right)^2 LC \cong \omega_0 CR$$

or, using $\omega_0{}^2 LC \cong 1$ and expanding binomially, we get

$$1 - \left(1 + \frac{2\Delta\omega}{\omega_0}\right) \cong \omega_0 CR$$

or $$\frac{2\Delta\omega}{\omega_0} CR \cong \omega_0 CR$$

$\therefore$ $$Q = \frac{\omega_0}{2\Delta\omega} \cong \frac{1}{\omega_0 CR} = \frac{\omega_0 L}{R}$$

At resonance, we may write

$$Z_0 = \frac{\omega_0 L}{R} \cdot \frac{1}{\omega_0 C} = \omega_0 \mathrm{L.Q}$$

The impedance of the capacitive branch is $1/\omega_0 C$ (in magnitude) at resonance and that of the inductive branch is $\approx \omega_0 L$ (in magnitude, neglecting R). $\therefore$ At resonance, the impedance of the circuit is approximately Q times the impedance of either branch.

In the chapter of dc circuits, we had mentioned that the maximum power transfer theorem would be restated after introducing the concept of impedance. The statement is: The maximum power will be absorbed by one network from another joined to it at two terminals, when the impedance of the receiving network is varied, if the impedances looking into the two networks at the junction are complex conjugates of each other.

Skin effect: When a steady current (i.e. dc) flows through a conductor, the current density is uniform throughout its cross section. But when an alternating current flows through it, the current density is not uniform. There is a concentration of current in the outer layers. At radio frequencies (a few MHz to a few hundred MHz) the current flows almost entirely along the surface of the conductor. This is known as the skin effect. Conductors that are required to carry high frequency alternating currents are made up of a number of strands of fine wire, insulated from each other, in order to have a large surface for any given area of cross section, since the central parts of thick wires would not carry any appreciable part of the current and would thus be useless.

8.8 A.C. BRIDGES

We will now study the measurement of inductances using A.C. Bridges. A general A.C. bridge may be represented as follows:

D is a detector—usually a pair of headphones which give a minimum sound when the bridge is balanced. The condition of balance is just a generalization of the condition of balance for a Wheatstone bridge:

$$\frac{Z_1}{Z_2} = \frac{Z_3}{Z_4} \tag{8.8.1}$$

Since the Z's are complex quantities, (8.8.1) implies a balance of the component of current in phase with the A.C. voltage source as well as a

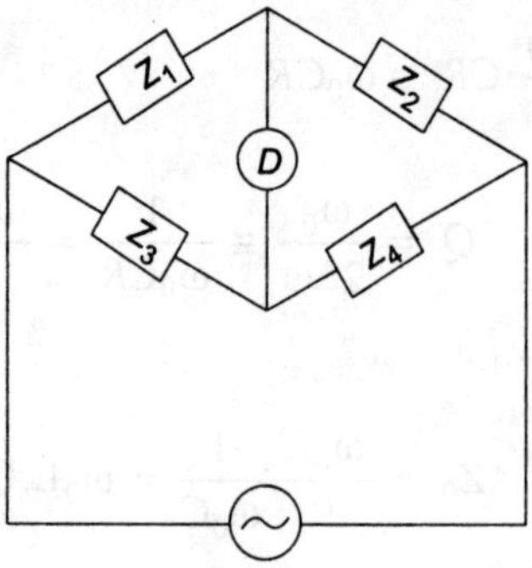

Fig. 8.15

balance of the 90°-out-of-phase component. Thus, one of the conditions involves resistors and can be obtained with a dc source and a moving-coil galvanometer.

Let us first consider Owen's bridge (Fig. 8.16) which is used to measure inductance in terms of a standard fixed capacitor. Comparing Fig. (8.15) and (8.16), we find,

$$Z_1 = R_1 + \frac{1}{j\omega C_1},\ Z_2 = \frac{1}{j\omega C_2},\ Z_3 = R_3 + j\omega L,\ Z_4 = R_4$$

Using condition (8.8.1), we have,

$$\left(R_1 + \frac{1}{j\omega C_1}\right) j\omega C_2 = \frac{R_3 + j\omega L}{R_4}$$

or,
$$j\omega C_2 R_1 + \frac{C_2}{C_1} = \frac{R_3}{R_4} + \frac{j\omega L}{R_4}$$

Equating real and imaginary parts, we get,

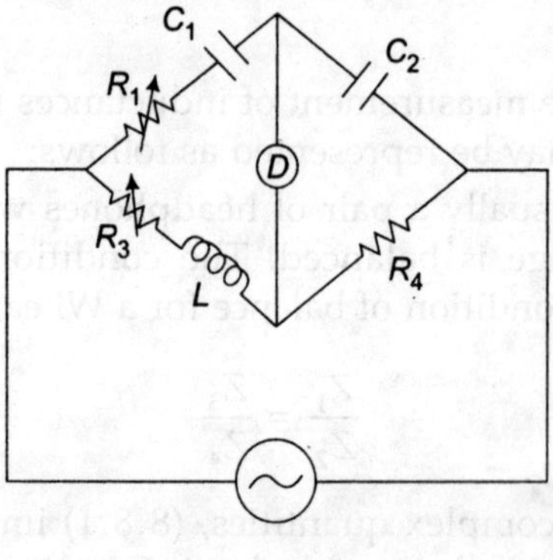

Fig. 8.16

$$\frac{C_2}{C_1} = \frac{R_3}{R_4} \tag{8.8.2}$$

and
$$L = C_2 R_1 R_4 \tag{8.8.3}$$

The condition (8.8.2) is obtained by varying R_3 and condition (8.8.3) by varying R_1. According to Owen, residual sound in the headphones, due to inductance of leads and terminals, can be eliminated. This can be done by first obtaining a balance with L in the circuit and then obtaining a balance with a short-circuit across L. If R_1 and R_1' are the respective values of R_1 for these balances, then

$$L = C_2 R_4 (R_1 - R_1') \tag{8.8.4}$$

Next let us consider Anderson's bridge (Fig. 8.17). This circuit does not appear to be of the form of Fig. 8.15. To convert it into that form, we have to use the star-delta transformation. We shall not derive this transformation in this book. We will merely state the result. We are going to transform the set (r, R_2, R_4) to the set (R_5, R_6, R_7) indicated by the dotted lines (Fig. 8.18).

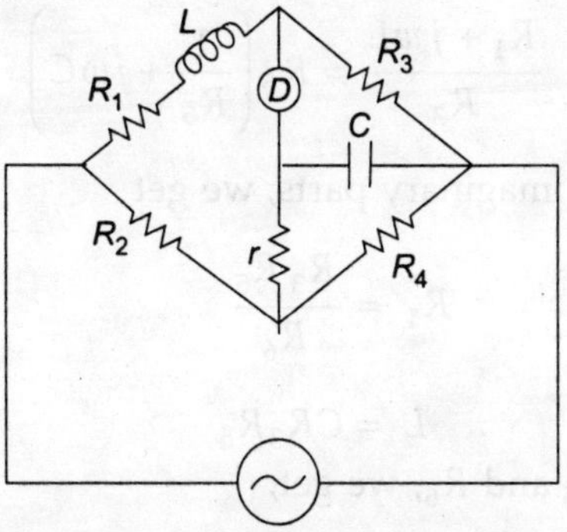

Fig. 8.17

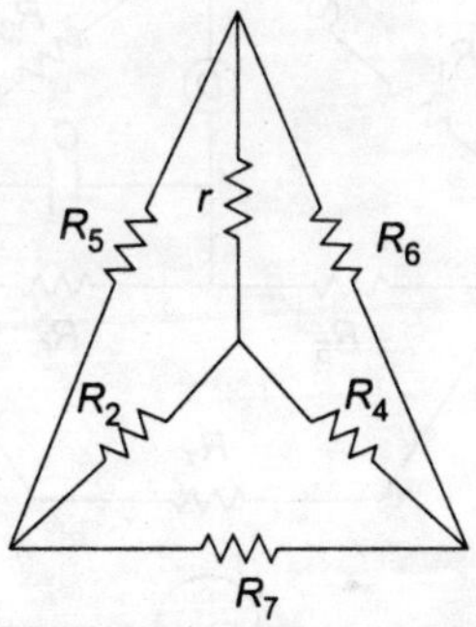

Fig. 8.18

$$R_5 = \frac{R_2 r + R_4 r + R_2 R_4}{R_4} \tag{8.8.5a}$$

$$R_6 = \frac{R_2 r + R_4 r + R_2 R_4}{R_2} \tag{8.8.5b}$$

$$R_7 = \frac{R_2 r + R_4 r + R_2 R_4}{r} \tag{8.8.5c}$$

The circuit of Fig. 8.17 now becomes the circuit of Fig. 8.19

The resistance R_7 shunts the source and so does not affect the balance. The circuit is now of the form (8.15), with $Z_1 = R_1 + jwL$, $Z_2 = R_3$, $Z_3 = R_5$ and $\frac{1}{Z_4}$ $= \frac{1}{R_6} + j\omega C$. Using the condition (8.8.1), we thus have

$$\frac{R_1 + jwL}{R_3} = R_5\left(\frac{1}{R_6} + j\omega C\right)$$

Equating the real and imaginary parts, we get

$$R_1 = \frac{R_3 R_5}{R_6}$$

and $$L = CR_3R_5$$

Or, substituting for R_5 and R_6, we get,

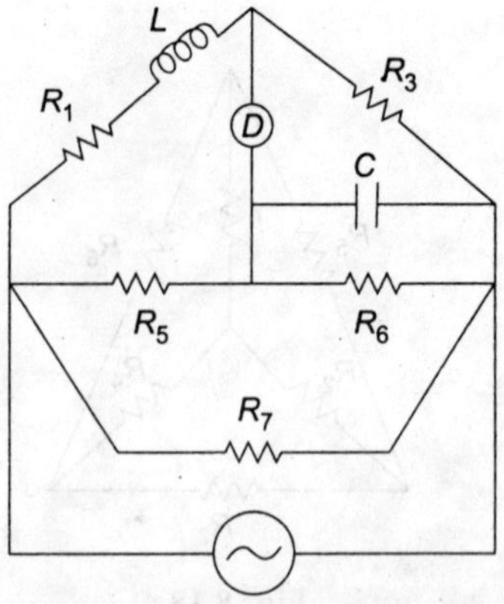

Fig. 8.19

$$R_1 = \frac{R_3 R_2}{R_4}$$

$$L = \frac{C R_3}{R_4} (R_2 r + R_4 r + R_2 R_4)$$

Tutorial

Consider the series RLC circuit. Solve the equation for the current using phasors. Draw the phasor diagram for frequencies less than resonance, at resonance and greater than resonance.

Problems

1. A 100 W, 200 V electric lamp has to be connected to AC mains with rms voltage 400 V and frequency 50 Hz. What capacity must be added in series with the lamp to obtain the normal glow for which the lamp has been designed? Calculate the power factor of the circuit.

   (University of Calcutta, 2002)

   [**Ans.** C = 4.6 μF, Power factor = 0.5]
2. A mercury vapour lamp rated at 2.5 A and 110 V is operated by a 220 V, 50 Hz main supply. Explain with numerical calculations how the lamp can run at its rated condition economically. (University of Calcutta, 2000)

   [*Hint:* we can either connect a resistance or an inductance in series with the lamp so that the *rms* voltage across the lamp is 110 V. Using inductance is more economical because the average power consumed by it is zero, whereas a resistance would consume power. The inductance used in this way is called a choking coil.]

   [**Ans.** L = 0.243 H]
3. A coil of resistance 10 Ω and inductance 0.1 H is connected in series with a 150 μF capacitor across a 200 V, 50 Hz supply. Calculate (a) the current, (b) the voltage across the coil and the capacitor.

   [**Ans.** (a) 14 A (b) V_L = 462 V, V_C = 297 V]
4. An alternating voltage 80 + j60 V is applied to a circuit and the current flowing is −4 + j10A. Find (a) the impedance of the circuit, (b) the power consumed and (c) the phase angle.

   [**Ans.** (a) $|Z|$ = 9.26 Ω (b) Power = 280.8 W (c) θ = 74° 58′, leading]
5. A coil of resistance 2 Ω and inductance 0.01 H is connected in series with a capacitor across 200 V mains. What must be the capacitance in order that maximum current occurs at a frequency of 50 Hz ? Find also the current and the voltage across the capacitor.

   [Ans. C = 1013 μF, I_{max} = 100A, V_C = 314.2 V]
6. A coil of resistance 20 Ω has an inductance of 0.2 H and is connected in parallel with a 100 μF capacitor. Calculate the frequency at which the circuit will act as a non-inductive resistance of R ohms. Find also the value of R. [Ans. f = 31.8 Hz, R = 100 Ω]

9

Generators and Rectifiers

9.1 THREE PHASE AC POWER GENERATION

The mechanism for generating an alternating voltage is Faraday's law of electromagnetic induction. In an ac generator or dynamo, a coil is rotated at a constant speed in a uniform magnetic field. Fig. 9.1 shows schematic diagram of a simple dynamo with only one turn of the coil. The ends of the coil are connected to flat brass rings *R*, which are supported on the shaft (on which the coil rotates) by discs of insulating material (not shown). Contact with the rings is made by small blocks of carbon *B*, supported by springs and

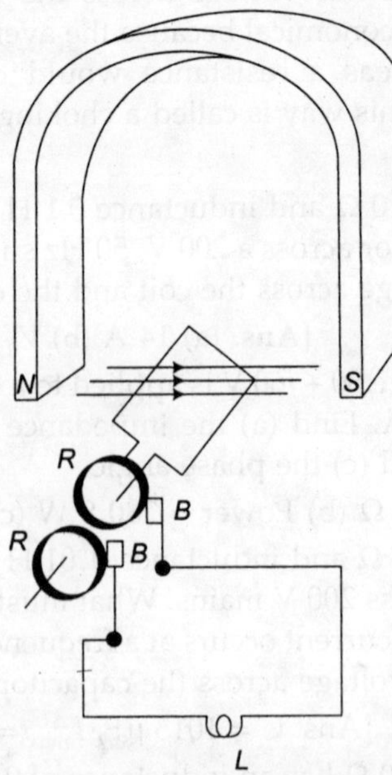

Fig. 9.1

shown connected to a lamp L. As the coil rotates, the emf induced in it is given by

$$E = 2\pi f A B \sin 2\pi f t \quad (9.1.1a)$$

where f is the frequency of rotation, A is the area of the coil and B is the magnetic field. If there are N turns, the emf is N times as great

$$E = 2\pi f N A B \sin 2\pi f t \quad (9.1.1b)$$

If practice, the magnetic field of a typical generator is provided by an electromagnet, called a *field magnet*. The coils of the electromagnet are fed with a steady current. The rotating coil is wound on an iron core and is called the armature. The generator is also called an alternator. The field magnet is called the stator because it is stationary. The armature is called the rotor because it is rotated.

There are certain economic and operating advantages of generating ac voltages using three phase systems. In this system, three coils are equally distributed about the circumference of the rotor (Fig. 9.2). The coils are displaced from one another by 120 degrees in space. Slip rings are not shown.

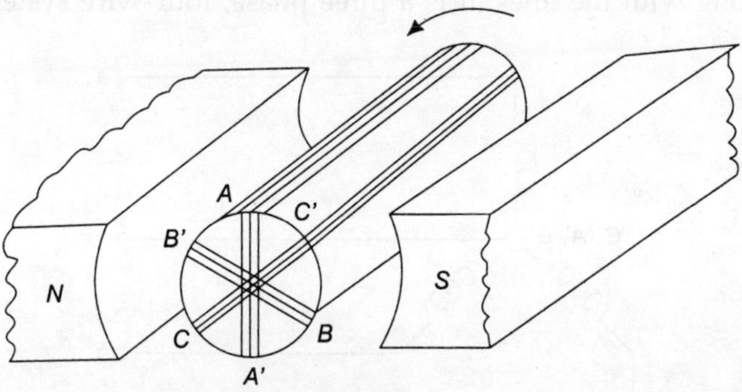

Fig. 9.2

As the rotor rotates in the anticlockwise direction, first the set A comes between the pole faces. Thereafter the set B comes between the poles. Finally, it is the turn of C. The voltages generated in the coils A, B and C are shown in Fig. 9.3. Voltage in coil B lags behind that in coil A by 120 electrical degrees

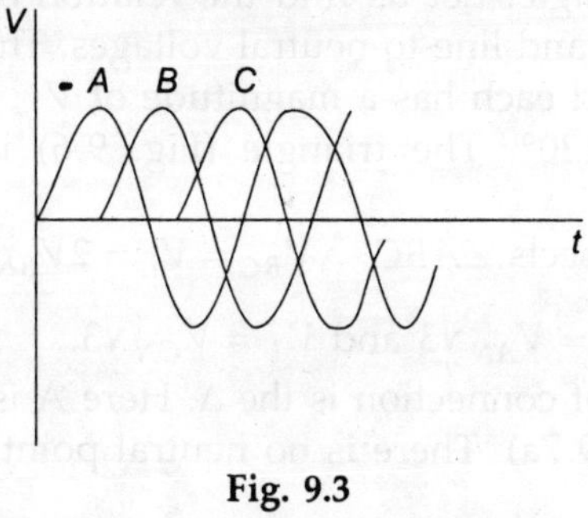

Fig. 9.3

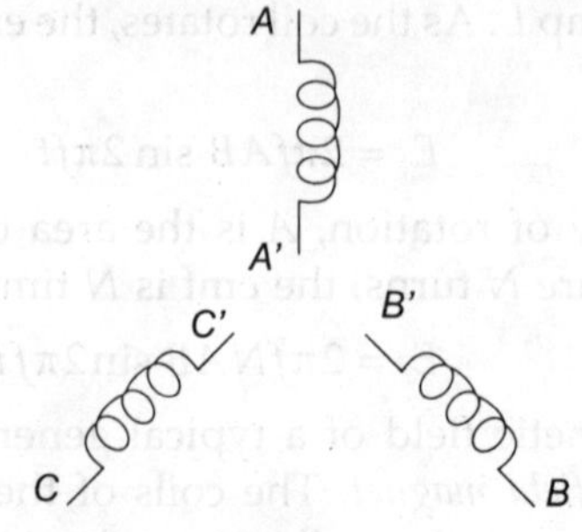

Fig. 9.4

and voltage in coil C lags behind that in coil A by 240 electrical degrees. The coils can be represented as shown in Fig. 9.4. The ends of the coils can be connected in two ways: Y and Δ. They are shown schematically in Figs. 9.5 and 9.7. In the Y-connection (Fig. 9.5), the ends A', B' and C' are joined at a common point, designated the neutral N.[(8)] The ends AB, C are brought out to become the lines a, b, c of the three phase system. If the neutral point is carried along with the lines, it is a three phase, four-wire system. A three

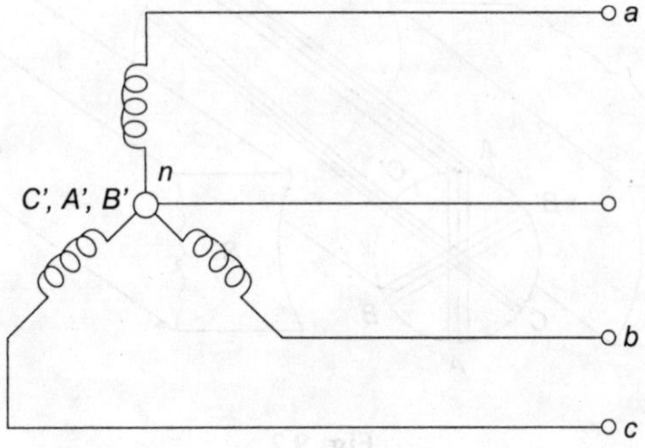

Fig. 9.5

phase system (or, in general, a polyphase system) may be balanced or unbalanced. A balanced system is one in which the magnitudes of all phase voltages are equal and the voltages of any two adjacent phases are displaced by the same phase angle. Let us find the relation between rms (or peak) values of line-to-line and line-to-neutral voltages. The line-to-line voltages are V_{AB}, V_{BC}, V_{CA} and each has a magnitude of V_L, say and adjacent ones differ in phase by 120°. The triangle (Fig. 9.6) is an equilateral one. Furtheremore, BN bisects $\angle ABC$. $\therefore V_{BC} = V_L = 2V_{BN} \cos 30° = 2V_{BN} \dfrac{\sqrt{3}}{2} = V_{BN}\sqrt{3}$. Similarly, $V_L = V_{AN}\sqrt{3}$ and $V_L = V_{CN}\sqrt{3}$.

The other mode of connection is the Δ. Here A is connected to B', B to C' and C to A' (Fig. 9.7a). There is no neutral point in this case. Here the

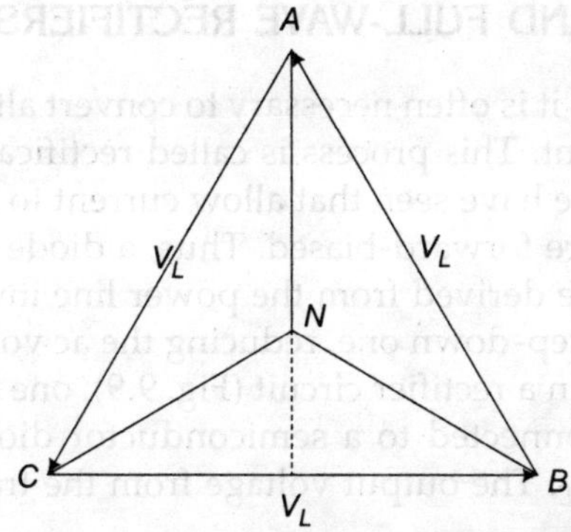

Fig. 9.6

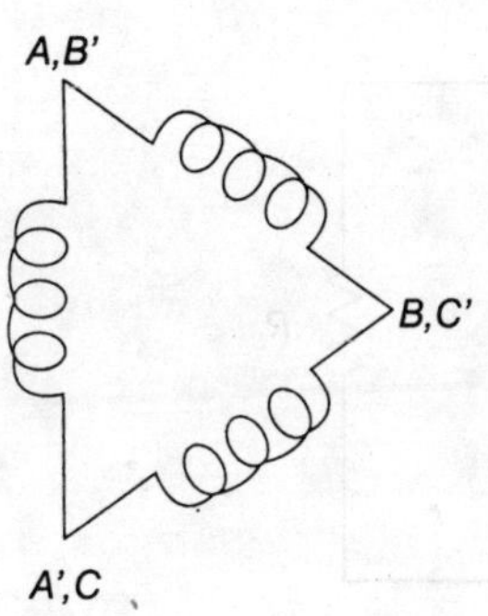

Fig. 9.7a

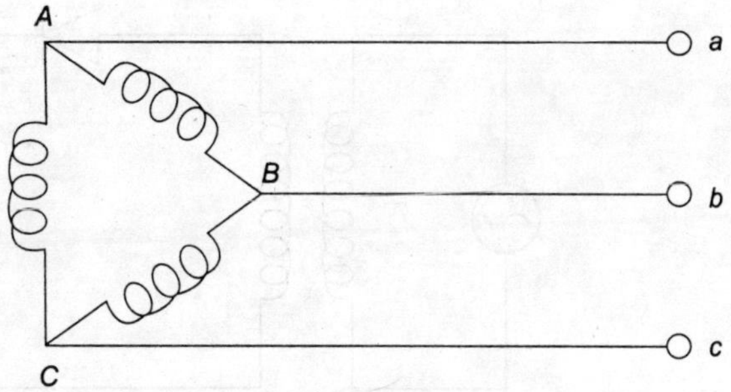

Fig. 9.7b

currents in adjacent phases are displaced by 120 electrical degrees and are equal in magnitude. say I_P. $\therefore I_{AB} = I_{BC} = I_{CA} = I_P$. Again, we have an equilateral triangle. The line current

$$I_{Bb} = 2I_p \cos 30° = I_P\sqrt{3}. \text{ or, } I_L = \sqrt{3}\, I_P.$$

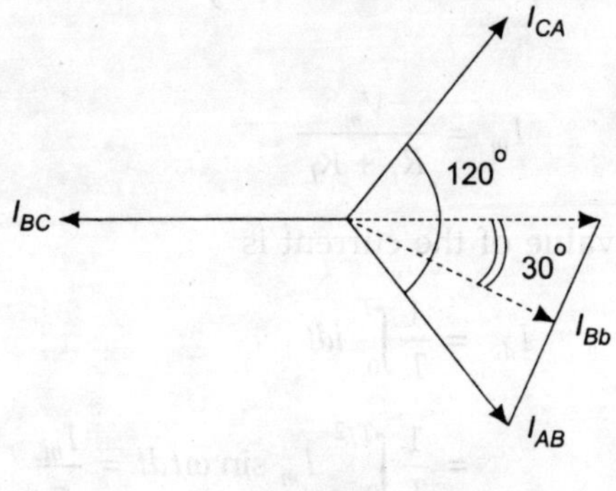

Fig. 9.8

9.2 HALF-WAVE AND FULL-WAVE RECTIFIERS

For many applications it is often necessary to convert alternating current into a unidirectional current. This process is called rectification. In our study of diodes in chapter 7, we have seen that allow current to flow in one direction only. i.e. when they are forward-biased. Thus, a diode can act as a rectifier. Typically, an ac source derived from the power line involves a transformer. The transformer is a step-down one, reducing the ac voltage from the power line to a lower value. In a rectifier circuit (Fig. 9.9), one end of the secondary of a transformer is connected to a semiconductor diode, which in turn is connected to a load R_L. The output voltage from the transformer secondary is of the form

$$v_1 = V_m \sin \omega t$$

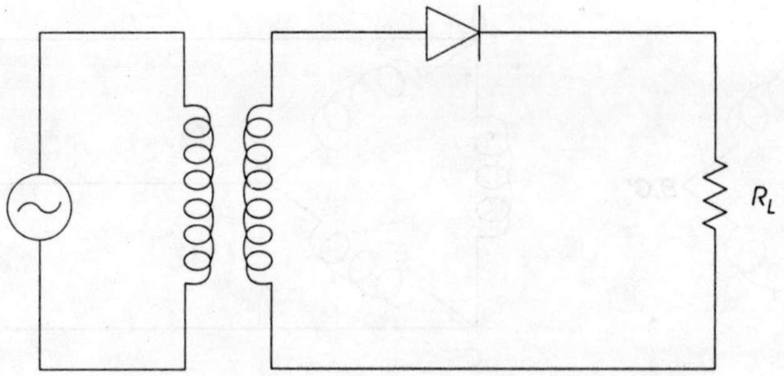

Fig. 9.9

V_m is usually large compared with the cutin voltage V_γ of the diode. So we may assume $V_\gamma \cong 0$. The diode conducts when it is forward biased. So the current flows only during the positive half-cycle of v_i (Fig. 9.10b). This is why this circuit is called a *half-wave rectifier*. The voltage across R_L will thus be nonzero only during this half. If we represent the diode by a resistance R_f during this half. If we represent the diode by a resistance R_f during this part of the cycle, then

$$I_m = \frac{V_m}{R_f + R_L} \tag{9.2.1}$$

The average, or dc value of the current is

$$I_{dc} = \frac{1}{T}\int_0^T i dt$$

$$= \frac{1}{T}\int_0^{T/2} I_m \sin \omega t dt = \frac{I_m}{\pi}$$

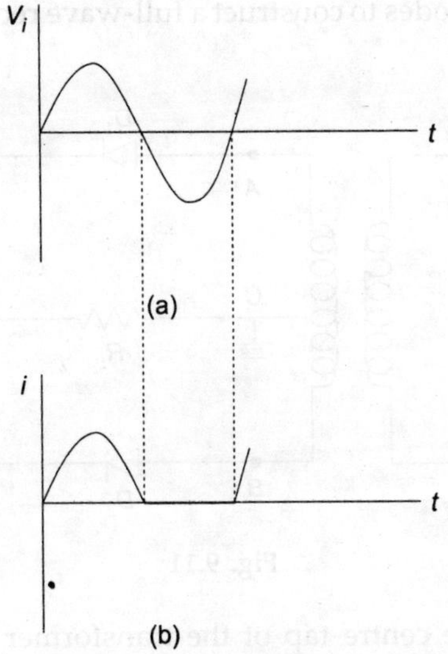

Fig. 9.10

∴ The average value of the load voltage is

$$V_{dc} = I_{dc} R_L = \frac{I_m R_L}{\pi} \tag{9.2.2}$$

Substituting for I_m from (9.2.1) into (9.2.2), we get

$$V_{dc} = \frac{V_m R_L}{(R_f + R_L)\pi} = \frac{V_m (R_f + R_L - R_f)}{\pi(R_f + R_L)}$$

$$= \frac{V_m}{\pi} - \frac{V_m R_f}{\pi(R_f + R_L)}$$

$$\therefore \quad V_{dc} = \frac{V_m}{\pi} - I_{dc} R_f \tag{9.2.3}$$

If we plot a graph of V_{dc} vs. I_{dc}, we get a straght line with a negative slope $-R_f$. The intercept on the V_{dc} axis is V_m/π. This is the average voltage when $I_{dc} = 0$. It is called the no-load voltage, corresponding to the output voltage for infinite load. It is obtained by open-circuiting the load, i.e. disconnecting the load from one of the terminals of the circuit to which it is connected. The variation of dc output voltage as a function of dc load current is called regulation. The percentage regulation for a particular load current is defined as

$$\text{\% regulation} = \frac{V_{\text{no load}} - V_{\text{load}}}{V_{\text{load}}} \times 100\%$$

We can use two diodes to construct a full-wave rectifier circuit (Fig. 9.11).

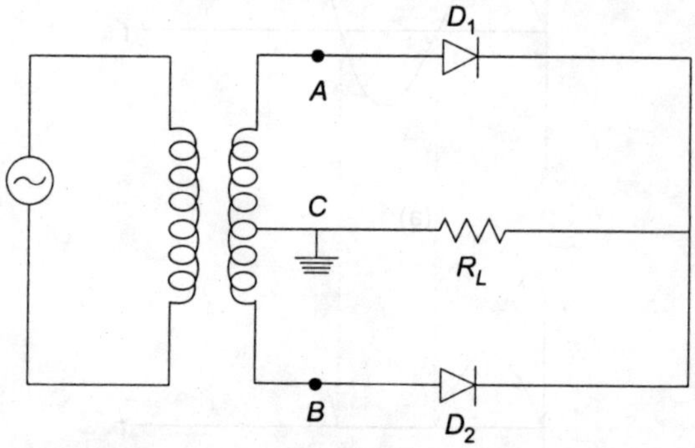

Fig. 9.11

The point C is the centre-tap of the transformer secondary. When the point A is positive with respect to C, diode D_1 conducts. During this period, B is negative with respect to C and so D_2 is off. This is one half-cycle of the input ac. During the next half-cycle, B is positive with respect to C and so D_2 conducts. At the same time, D_1 is off. Thus, during the entire cycle, current flows through R_L in the same direction. The output waveform across R_L looks like that shown in Fig. 9.12b. The average value of the current is now

$$I_{dc} = \frac{2}{T}\int_0^{T/2} I_m \sin \omega t dt = \frac{2I_m}{\pi}$$

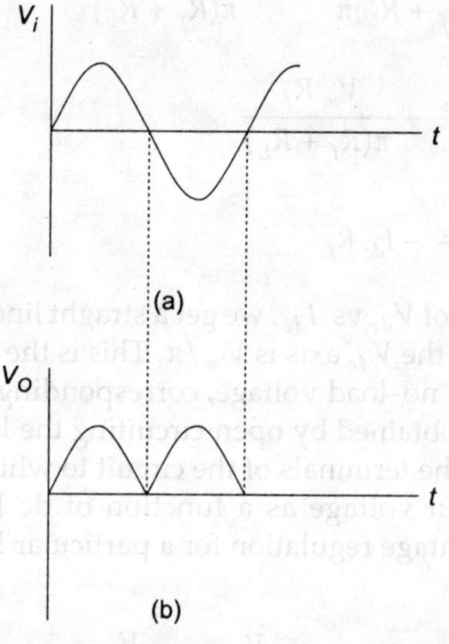

Fig. 9.12

$$\therefore \quad V_{dc} = \frac{2I_m R_L}{\pi} \tag{9.2.4}$$

Substituting for I_m from (9.2.1) into (9.2.4), we get

$$V_{dc} = \frac{2V_m R_L}{\pi(R_f + R_L)}$$

$$= \frac{2V_m(R_f + R_L - R_f)}{\pi(R_f + R_L)}$$

$$= \frac{2V_m}{\pi} - \frac{2V_m R_f}{(R_f + R_L)\pi}$$

$$\therefore \quad V_{dc} = \frac{2V_m}{\pi} - I_{dc} R_f \tag{9.2.5}$$

So if we plot a graph of V_{dc} vs. I_{dc}, we get a straight line with a negative slope $-R_f$ and no-load voltage $2V_m/\pi$. The no-load voltage for a full-wave rectifier is twice that for a half-wave rectifier.

There is another rectifier circuit which is often used for constructing regulated power supplies. This is the bridge rectifier circuit. It uses four diodes (Fig. 9.13)

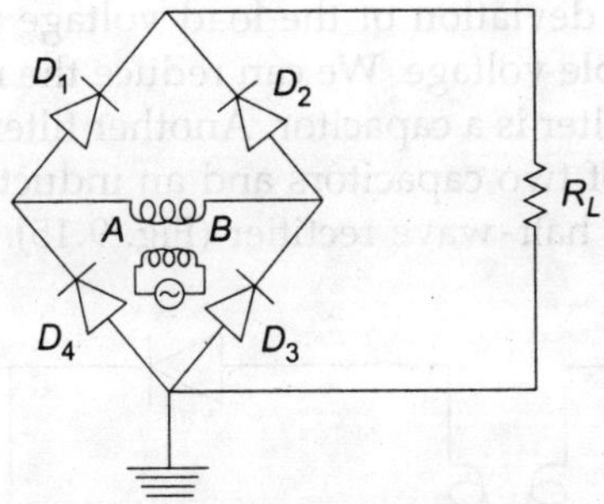

Fig. 9.13

When end A of the transformer secondary is positive with respect to B, diodes D_1 and D_3 are forward biased and conduct. D_2 and D_4 are reverse biased and so are off. When B is positive with respect to A, diodes D_2 and D_4 are forward biased and conduct. D_1 and D_3 are off. So the current flows through R_L in the same direction throughout one cycle.

An important feature of rectifier circuits is the maximum reverse voltage across the diode when it is reverse biased. This voltage is called the *peak inverse voltage*. For a half-wave rectifier, it is V_m. For a full-wave rectifier, let us suppose that B is negative with respect to C. If we neglect the voltage across the diode D_1, the entire voltage V_m appears across R_L. So if we apply Kirchhoff's voltage law around the circuit containing diode D_2 and R_L, we find that the voltage v across the diode satisfies the equation

$$2V_m + v = 0$$

or,

$$v = -2V_m,$$

For the bridge rectifier, a centre-tapped transformer is not required and the peak inverse voltage is V_m.

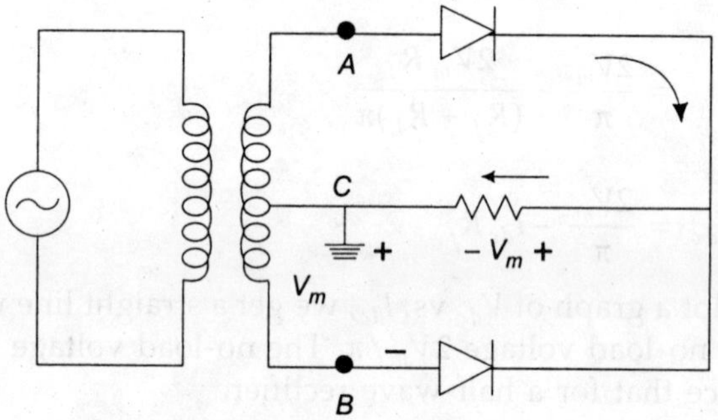

Fig. 9.14

We find that although the current from rectifiers is unidirectional, it varies with time. The deviation of the load voltage from an average or dc value is called the ripple voltage. We can reduce the ripple voltage by using a filter. The simplest filter is a capacitor. Another filter which is often used is a π-filter, consisting of two capacitors and an inductor. Let us consider the capacitor filter, with a half-wave rectifier (Fig. 9.15).

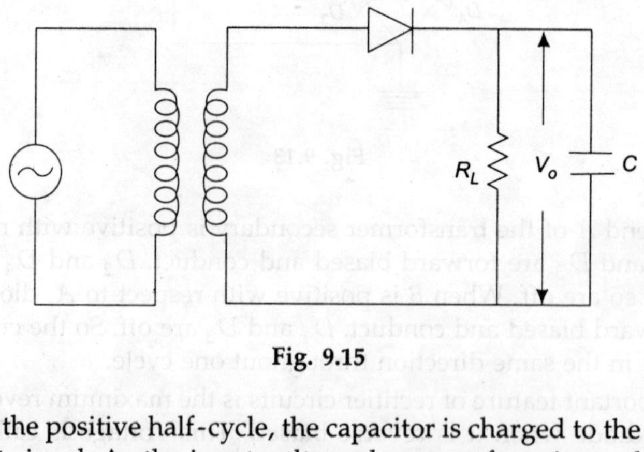

Fig. 9.15

During the positive half-cycle, the capacitor is charged to the peak value of the input signal. As the input voltage decreases from its peak value, the diode now becomes reverse biased (since the n-side, connected to C is at V_m

but the p-side is less then V_m) and stops conducting. The capacitor gradually discharges through R_L with a time constant CR_L. This continues until the input voltage starts increasing from zero and exceeds the capacitor voltage (Fig. 9.16). The dotted line represents the load voltage in absence of the filter. The solid line represents the load voltage with the filter. Clearly, the ripple voltage is reduced when the filter is used. A full-wave rectifier with a

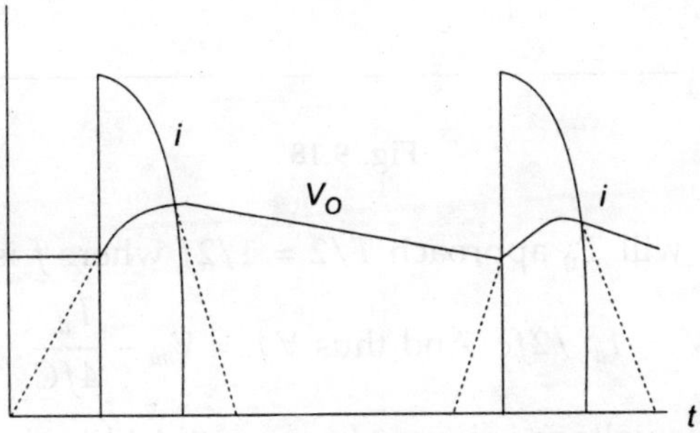

Fig. 9.16

capacitor filter is shown in Fig. 9.17. The load voltage without and with the filter is shown by dotted and solid lines respectively, in Fig. 9.18. We will now approximately calculate[9] the average or dc voltage in Fig. 9.18. If $\omega CR_1 >> 1$, the capacitor starts discharging, i.e. the diode stops conducting when $\omega t \approx \pi/2$. If V_r is the total capacitor discharge voltage, the average load voltage is

$$V_{dc} = V_m - \frac{V_r}{2}$$

Furthermore, if C is large, the capacitor discharge is almost linear over the time shown in Fig. 9.18. If T_0 is the total nonconducting time, the capacitor voltage changes by an amount $I_{dc}T_0/C$. The better the filtering

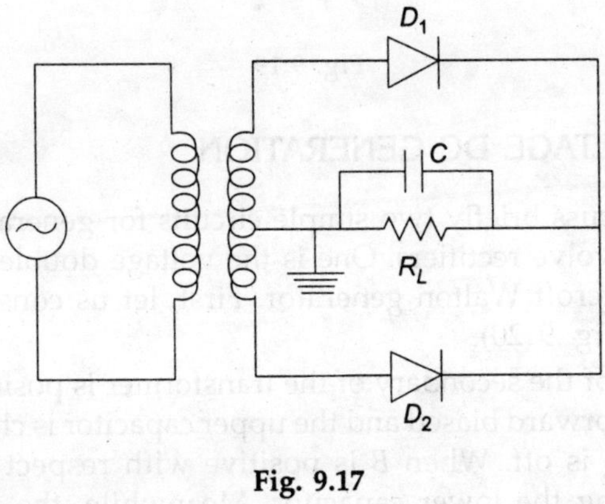

Fig. 9.17

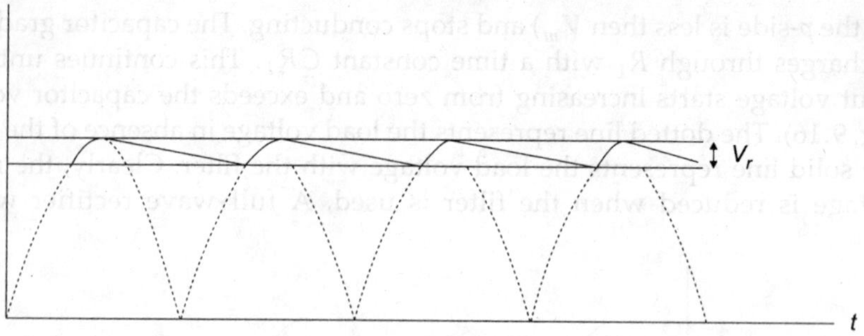

Fig. 9.18

action, the closer will T_0 approach $T/2 = 1/2f$ where f is the power line frequency. Then $V_r = I_{dc}/2fc$. And thus $V_{dc} = V_m - \frac{I_{dc}}{4fC}$.

Thus the ripple voltage is inversely proportional to the capacitance. To keep the ripple low, C must be high. For this purpose, large capacitances of the order of tens of microfarads are used. These capacitors are electrolytic capacitors (see note at the end of this chapter). Another filter which is also used is the π-filter (Fig. 9.19), consisting of two capacitors and an inductor.

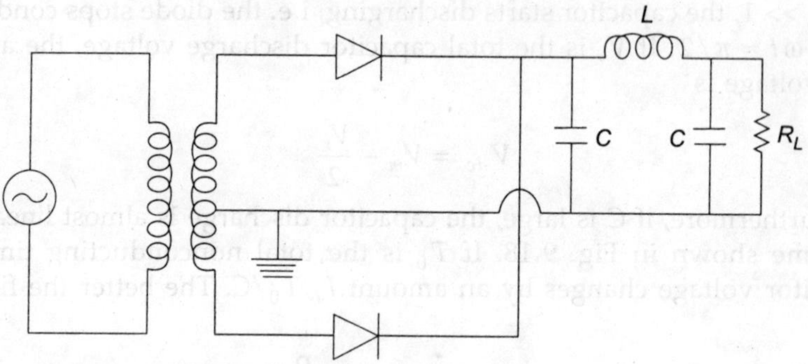

Fig. 9.19

9.3 HIGH VOLTAGE DC GENERATION

We will now discuss briefly two simple circuits for generation of high DC voltages. Both involve rectifiers. One is the voltage doubler circuit and the other is the Cockcroft Walton generator. First, let us consider the voltage doubler circuit (Fig. 9.20).

When end A of the secondary of the transformer is positive with respect to B, diode D_1 is forward biased and the upper capacitor is charged gradually to V_m. Diode D_2 is off. When B is positive with respect to A, diode D_2 conducts, charging the lower capacitor. Meanwhile, the upper capacitor

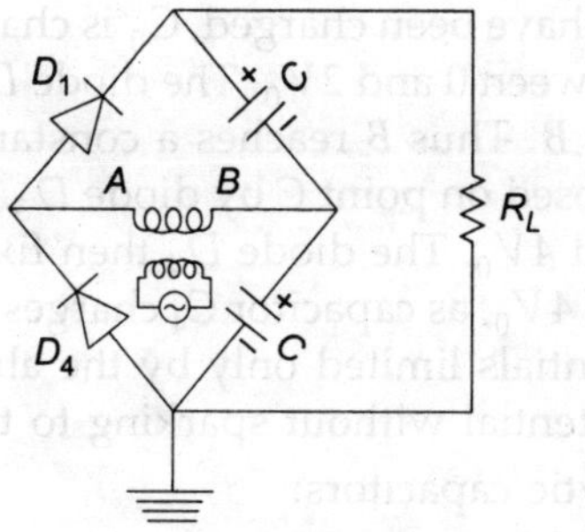

Fig. 9.20

discharges through R_L. Thus, when the transformer secondary reaches its peak, the voltage across both capacitors is nearly V_m with the polarities shown and so the voltage across the load is almost $2V_m$.

The second method of generating high DC voltages is the Cockcroft Walton generator[10] (Fig. 9.21). It can reach potentials of about 800 kV. Let the peak secondary transformer voltage be V_0 where $V_0 \sim 100\,\text{kV}$. The capacitors are charged through a suffciently large load R so that the time constants RC are large compared with the time $1/\omega$ that characterizes the variation of the transformer voltage. If we consider steady-state conditions

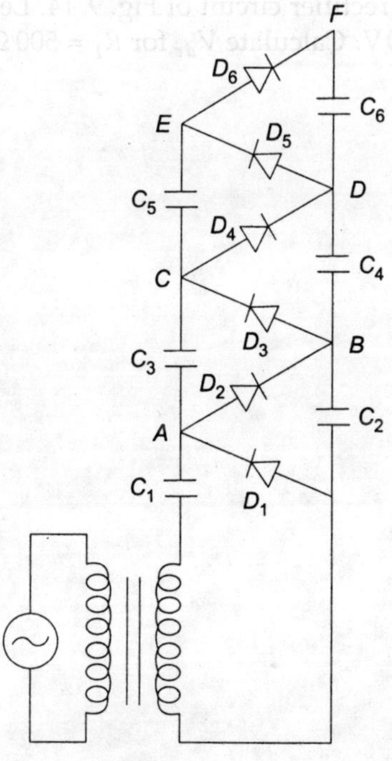

Fig. 9.21

when all the capacitors have been charged, C_1 is charge to V_0. ∴ The voltage at point A oscillates between 0 and $2V_0$. The diode D_2 conducts when A is at a higher potential than B. Thus B reaches a constant potential of $2V_0$. This potential of $2V_0$ is imposed on point C by diode D_3. Thus the ac voltage at C varies between $2V_0$ and $4V_0$. The diode D_4 then fixes the potential at point D to a constant value of $4V_0$, as capacitor C_4 charges to a voltage of $2V_0$. This goes on to higher potentials limited only by the ability of the high voltage terminal to hold its potential without sparking to the surroundings.

A note on electrolytic capacitors:

An electrolytic capacitor is constructed as follows: two sheets of aluminium foil, separated by muslin soaked in a special solution of ammonium borate is taken. These are rolled up and sealed in an insulating container. Wires attached to the foil strips are connected to a battery. Electrolysis takes place and a thin film of aluminium oxide forms of the foil connected to the positive terminal of the battery. This film is a very good insulator. A capacitor is thus formed with the oxide film as the dielectric. Since the film is very thin, large capacitances upto a few hundred microfarads can be made by this method. The two capacitor terminals are marked positive and negative. When the electrolytic capacitor is used in an electric circuit, this polarity must be maintained.

Tutorial

Consider the full-wave rectifier circuit of Fig. 9.14. Let the peak transformer secondary voltage be 20 V. Calculate V_{dc} for $R_L = 500\,\Omega$, $R_f = 5\,\Omega$ and also the %regulation.

10

Motion of Charged Particles in Electric and Magnetic Fields

10.1 THE CATHODE-RAY OSCILLOGRAPH

When a particle of charge q enters a region of space where there is an electric field **E** and a magnetic field **B**, it experiences a force

$$\mathbf{F} = q(\mathbf{E} + \mathbf{v} \times \mathbf{B})$$

where v is the velocity of the particle. This simple fact is used in the cathode-ray oscillograph and in accelerators for charged particles. We shall first study the former. The cathode-ray oscillograph is used for studying time-varying voltages. A plot of voltage (along the vertical axis) versus the time (along the horizontal axis) is observed on the screen. This is done by manipulating a beam of electrons onto a fluorescent screen. The oscillograph consists of five systems: (1) The cathode-ray tube and associated controls in focus, intensity or brightness, (2) the vertical or signal amplifier and its associated devices, (3) horizontal axis time-base circuits, called the sweep generator, (4) auxiliary facilities, (5) power supplies. We shall only be concerned with (1). The cathode-ray tube consists of the vacuum envelope, the electron gun and the phosphor screen. The electron gun is a system of electrodes which produces, controls, focusses and deflects an electron beam. The electron beam is produced by thermionic emission from the cathode C (Fig. 10.1). The cathode is surrounded by a control grid G. This grid is maintained at a negative potential with respect to the cathode. By varying the potential on the grid, the intensity of the spot on the screen can be controlled. Following the grid is a focusing system F. Electrons are emitted at various angles relative to the axis of the tube. The aim of this system of electrodes is to bring them back towards the tube axis. The anode A is the next electrode, being maintained at ground potential. The electrodes C, G, F are at negative potentials $-V_2$, $-V_3$ and $-V_1$. A is followed by the deflecting system P_y and P_x. A vertical electric field applied to the plates P_y deflects the beam vertically and a horizontal electric field applied to the plates P_x deflects the beam horizontally. This is electrostatic deflection. Many cathode-ray tubes e.g. in

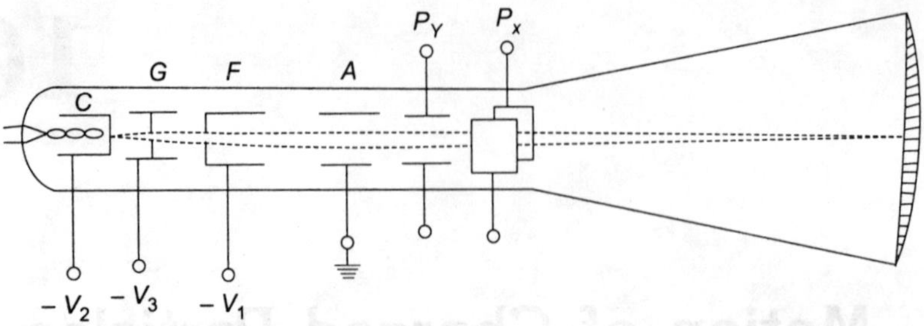

Fig. 10.1

TV and radar also use magnetic deflection systems. In this case, a magnetic field is produced perpendicular to the axis of the tube by means of coils outside the tube. The screen is coated with a material called a phosphor, which emits visible light on electron bombardment.

Let us first consider electrostatic deflection, assuming that the electron beam has been focussed and is moving along the axis of the tube. Let m = mass of electron q = magnitude of its change, V_2 = potential difference between cathode and last anode, E = electric field between the vertical deflection plates, L_1 = axial length of the deflecting plates and L_2 = distance from the

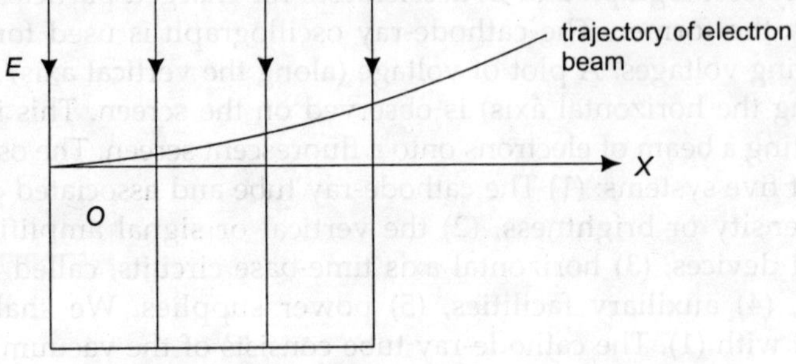

Fig. 10.2

centre of the deflecting plates to the screen. We will take the x-axis to be along the axis of the tube and the y-axis vertically upward. The kinetic energy of an electron as it enters the vertical deflecting plates can be equated to the loss in electrostatic potential energy of the electron as it goes from the cathode to the last anode. Thus

$$qV_2 = \frac{mv_x^2}{2} \tag{10.1.1}$$

After the electron enters the region shown in Fig. 10.2, it experiences a vertical acceleration 'a' given by

$$a = \frac{qE}{m} \tag{10.1.2}$$

Since initially the speed of the electron in the y-direction was zero, the speed after a time t in this direction is given by

$$v_y = at = \frac{qEt}{m} \tag{10.1.3}$$

and its displacement along the positive y-axis is

$$y = \frac{at^2}{2} = \frac{qEt^2}{2m} \tag{10.1.4}$$

The distance the electron has moved along the x-axis is

$$x = v_x t = \left(\frac{2q V_2}{m}\right)^{1/2} t \tag{10.1.5}$$

since there is no force on the electron in the x direction. Thus the path of the electron is a parabola, the equation being $x^2 = 4V_2 y/E$. A property of the parabola is that the tangent at any point (x, y) intersects the x-axis at the point $x/2$. So if the tangent to the emergent beam is produced back it meets the axis at the midpoint of the plates. The angle α which the emergent beam makes with the axis is given by

$$\tan\alpha = \frac{v_y}{v_x} = \frac{qEt}{m}\left(\frac{m}{2qV_2}\right)^{1/2}$$

$$= Et\left(\frac{q}{2mV_2}\right)^{1/2} \tag{10.1.6}$$

where t is time taken for the electron to traverse the length L_1 of the plates.

$$t = \frac{L_1}{v_x} = L_1\left(\frac{m}{2qV_2}\right)^{1/2}$$

Thus $$\tan\alpha = E\left(\frac{q}{2mV_2}\right)^{1/2} L_1\left(\frac{m}{2qV_2}\right)^{1/2}$$

$$= \frac{EL_1}{2V_2} \tag{10.1.7}$$

The vertical deflection of the beam from the axis when it arrives at the screen is

$$y = L_2 \tan\alpha = \frac{EL_1 L_2}{2V_2} \tag{10.1.8}$$

If d is the vertical distance between the vertical deflection plates and V_0 is the potential difference between them, then $E = V_0/d$. So

$$y = \frac{L_1 L_2 V_0}{2d V_2} \tag{10.1.9}$$

The effectiveness of a pair of deflecting electrodes is often described in terms of the deflection factor, i.e. the number of volts necessary to deflect the beam a unit distance at the phosphor screen, for a given accelerating potential.

We will next consider magnetic deflection. Let the electron beam be moving along the x-axis, as before, and a magnetic field exists perpendicular to the plane of the diagram, in the region between the plates. Under the action of the field, an electron describes an arc of a circle of radius R such that

$$\frac{mv_x^2}{R} = qv_x B$$

where B is the magnetic field along the positive z-axis .

$$R = \frac{mv_x}{qB}$$

Note that the speed of the electron is unchanged, as there is no force along the x-axis. If the length of the arc is s, $s = R\alpha$ (Fig. 10.3).

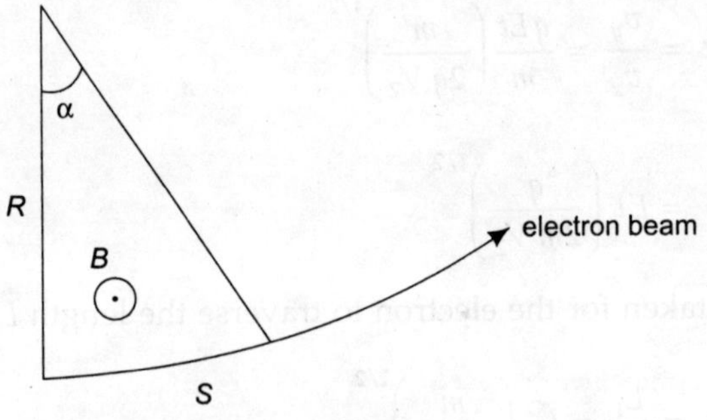

Fig. 10.3

$$\therefore \qquad \alpha = \frac{s}{R} = \frac{sqB}{mv_x}$$

$$= \frac{sqB}{m}\left(\frac{m}{2qV_2}\right)^{1/2}$$

$s \cong L_1$. But the tangent to the beam will not intersect the axis exactly at the midpoint of the field. So, strictly speaking, L_2 should not be measured from the midpoint. For small α, however, the error is negligible and the deflection $y' \cong L_2\alpha$.

$$\therefore \quad y' \cong L_1 B L_2 \left(\frac{q}{2m V_2} \right)^{1/2} \tag{10.1.10}$$

10.2 POSITIVE RAY ANALYSIS OF THOMSON

An early application of the motion of charged paricles in electric and magnetic fields was made in the last decade of the nineteenth century and the first two decades of the twentieth century. In a discharge tube, at pressures of 0.001 mm to 0.05 mm Hg, a well defined dark space is formed near the cathode. In the cathode dark space, electrons travel at high velocities (~10^9 cm/s.). The electrons knock off electrons from the atoms of the gas in the discharge tube. The positive ions moved towards the cathode with atoms of the gas in the discharge tube. The positive ions moved towards the cathode with much lower velocities and passed through perforations in the cathode towards the glass tube. These positive ions were called positive rays by Thomson. In order to study them further, Thomson subjected them to simultaneous electric and magnetic fields parallel to each other. Let us suppose that the ions are moving with speed v_z in the z direction. A magnetic field B is applied in the y direction. An electric field E is also applied in the y direction. Consider the magnetic field. Under its influence, a particle of charge q will be deflected towards the x-axis. Let v_x be the speed acquired by it, t be the time it spends in the field and 1 be the length of the magnet pole pieces (about 3 cm). Then, since change in momentum in the x direction = impulse in the x direction, we have,

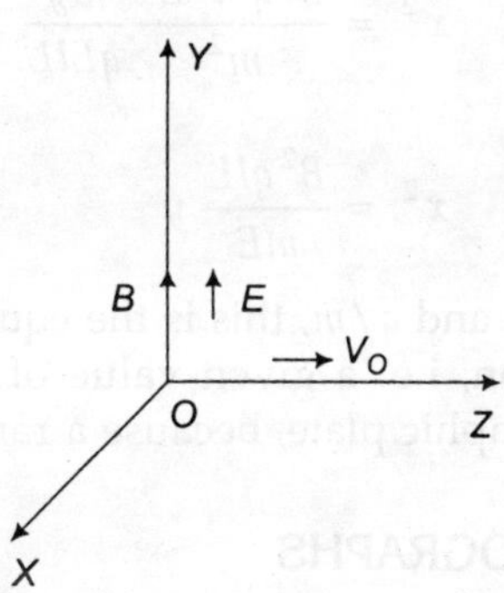

Fig. 10.4

$$m v_x = B q v_z t_1$$

There is no force on the particle in the z direction. ∴ The z component of its velocity remains unchanged. ∴ $t_1 = l/v_z$. So

$$m v_x = B q l \tag{10.2.1}$$

The electric field E exerts a force in the y direction. So we have,

$$mv_y = qEt_1$$

or,
$$mv_y = \frac{qEl}{v_z} \tag{10.2.2}$$

A photographic plate is placed so that its centre is at a distance $L = 40$ cm from the centre of the pole piece. Thus, the deflection in the x direction is

$$x = v_x t_2 = \frac{BqlL}{mv_z} \tag{10.2.3}$$

where t_2 is the time require by the particle to reach the photographic plate. The deflection in the y direction is

$$y = v_y t_2 = \frac{L}{V_z} \cdot \frac{qEl}{mv_z}$$

$$y = \frac{qElL}{mv_z^{\ 2}} \tag{10.2.4}$$

Squaring (10.2.3), we get

$$x^2 = \frac{B^2 q^2 l^2 L^2}{m^2 V_z^2}$$

or, using (10.2.4),
$$x^2 = \frac{B^2 q^2 l^2 L^2}{m^2} \cdot \frac{my}{qElL}$$

$$x^2 = \frac{B^2 qlL}{mE} y \tag{10.2.5}$$

For constant B, E, l, L and q/m, this is the equation of a parabola. For a given type of positive ion, i.e. a given value of q/m, a parabola will be described on the photographic plate, because a range of values of v_z exists.

10.3 MASS SPECTROGRAPHS

There were two disadvantages of Thomson's positive ray analysis. Firstly, two ions with same charge but masses close together could not be distinguished. If we define a quantity called the *resolving power* as m/dm where m is the mean mass and $m - dm/2$ and $m + dm/2$ are the masses of two ions which can just be distinguished as separate, then this was small in Thomson's apparatus. Secondly, an exposure of an hour or two was required for the photographs.

To improve on these features, the first mass spectrograph was designed by F.W. Aston in 1919[11]. A mass spectrograph is a device for accurate determination of atomic masses. The separate ions are focussed on to a photographic plate. One difference from Thomson's apparatus was that the electric and magnetic fields were not applied in the same region of space. Nor were they parallel to each other. Let us look at a schematic diagram of Aston's mass spectrograph (Fig. 10.5). S_1 and S_2 are two slits which form the chathode. The positive ions emerging from them are deflected by an electric field E applied between the plates P_1 and P_2. After traversing a distance 'b' in field-free region, they enter a region containing a magnetic field B perpendicular to the plane of the diagram, towards the reader. The ions describe an arc of a circle. Thereafter they move in a field-free region for a distance 'a' and impinge on a photographic plate.

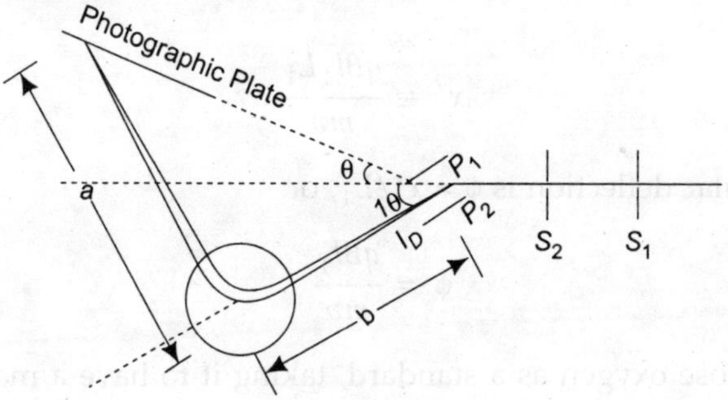

Fig. 10.5

Consider the effect of the electric field alone. We found in the last section that the deflection of an ion of charge q and mass m in an electric field E is

$$y = \frac{qElL}{mv_z^2}$$

in the direction of the field. Here l is the length of the region permeated by the electric field and L is approximately the distance traveled by the ion in field-free space. ∴ The angular deflection is

$$\theta = \frac{y}{L} = \frac{qEl}{mv_z^2} \tag{10.3.1}$$

Let us do away with the subscript z and simple write v. ∴ $\theta = qEl/mv^2$. After going a distance L, the ion enters a magnetic field directed at right angles to its velocity. We now choose our axes as $X'\,Y'\,Z'$ with v along the Y' axis and B along the Z' axis. The ion experience a forces qvB along the X' axis and thus an acceleration qvB/m. Let l_1 be the length of the (small) circular region of space over which the magnetic field exists. ∴ The time taken by the ion to go a distance l_1 in the y' direction is $t_1 = l_1/v$. ∴ The speed acquired by the ion in the x' direction in this time is $v_x' = qvBt_1/m = qBl_1/m$. If the beam now moves a distance L_1 in field-free space, the deflection in the x' direction is

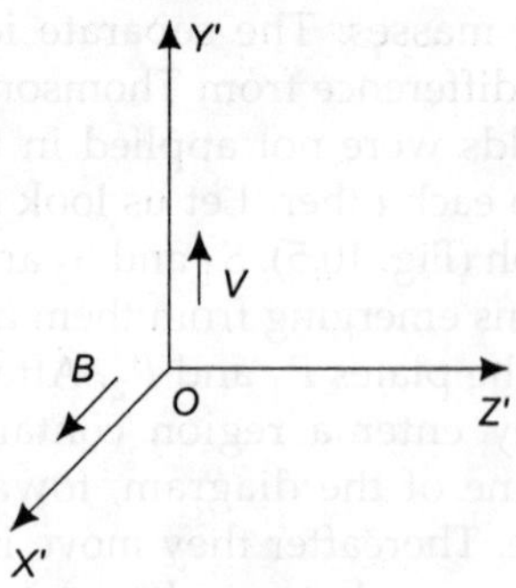

Fig. 10.6

$$x' \cong \frac{qBl_1L_1}{mv}$$

and the angular deflection is $\phi = x'/L_1$, or

$$\phi = \frac{qBl_1}{mv} \tag{10.3.2}$$

Aston chose oxygen as a standard, taking it to have a mass of 16.0000. Masses of other ions were obtained by the method of bracketing. Consider eqn. (10.3.1). The electric field E is proportional to the potential difference V between the plates. The potential difference is changed to a value V_1 (and so therefore does the electric field change) in such a way that ions of a different mass m_1 (but same charge q) and speed v_1 are deflected through the same angle θ. Then

$$\frac{qVl}{mv^2} = \frac{qV_1l}{m_1v_1^2}$$

or,
$$\frac{V}{V_1} = \frac{m_1V_1^2}{mv^2} \tag{10.3.3}$$

If the magnetic field is kept constant, then the deflection ϕ in (10.3.2) is the same for ions having the same ratio q/mv. Thus ions of the same charge q which are focussed to the same position on the photographic plate will have the same linear momentum

$$mv = m_1v_1 \tag{10.3.4}$$

Combining this with (10.3.3), we get

$$\frac{V}{V_1} = \frac{m}{m_1} \tag{10.3.5}$$

In the method of bracketing, three exposures were made on the same photographic plate. These were with (1) a potential difference V across the plates, (2) a potential difference $nV + \Delta V$, where n is any real number and ΔV is small, and (3) a potential difference $nV - \Delta V$. Thus three lines will be obtained on the photographic plate. Two lines due to a positive ion of mass m_1 will appear symmetrically bracketed on either side of mass m if

$$\frac{m}{m_1} = n$$

Aston obtained a resolving power of 120. Thereafter he incorporated a number of improvements like reducing the width of the slits, using superior vacuum pumps and providing a concentrated magnetic field over a longer path with the help of 'sickle-shaped' flat iron pole pieces. In this way, he increased the resolving power to 600 in 1925 in his second mass spectrograph.

The first mass spectrometer was constructed by A.J. Dempster in 1918. In a mass spectrometer the ions, after passing through a detector slit, are recorded electronically. In Dempster's instrument, positive ions were first accelerated by a potential difference V, so that

$$qV = \frac{1}{2}mv^2 \tag{10.3.6}$$

Care was taken to ensure that the initial energy of the ions was negligible compared to that obtained in an electric field. A beam of monoenergetic ions of speed v was produced due to (10.3.6). Thereafter, this beam was allowed to enter a uniform magnetic field perpendicular to the direction of the beam. The ions were made to move in a semicircle of radius R given by the relation

$$\frac{mv^2}{R} = qvB$$

$$\therefore \qquad m = \frac{BqR}{v} = BqR\sqrt{\frac{m}{2qV}}$$

$$\sqrt{m} = BR\sqrt{\frac{q}{2V}}$$

$$\therefore \qquad \frac{m}{q} = \frac{B^2R^2}{2V} \tag{10.3.7}$$

For given values of B, V, the radii of the paths proportional to $\sqrt{m/q}$. Thus the ions were sorted out according to their masses.

Dempster discovered that the magnetic field provided a focussing action on the ion beam. If a central ray of the beam is normal to the boundary both where it enters the field and where it leaves it, the beam will be refocussed at

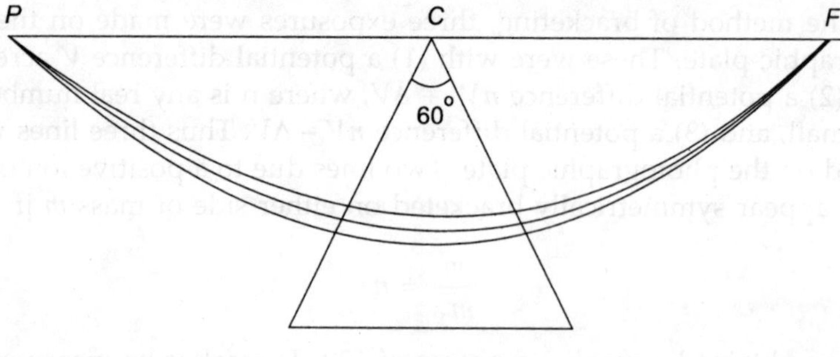

Fig. 10.7

a point or short line F (Fig. 10.7) in the field-free space beyond the magnetic field, provided that the centre of curvature C of the trajectory of the ion in the field lies on the straight line joining the source P and the focus F. It is assumed that the semi-angle of divergence of the beam is small and that the magnetic field changes discountinuously from zero to its full constant value at the boundary. The figure shows a sector magnetic field of angle 60°.

In 1933, K.T. Bainbridge designed another mass spectrograph. Here the ions were made to traverse electric and magnetic fields at right angles to each other. Such a system is called a system of crossed fields. If E and B are the magnitudes of the electric and magnetic fields (Fig. 10.8), then the ions whose speed v satisfies the relation

$$qvB = qE$$

i.e.
$$v = \frac{E}{B} \tag{10.3.8}$$

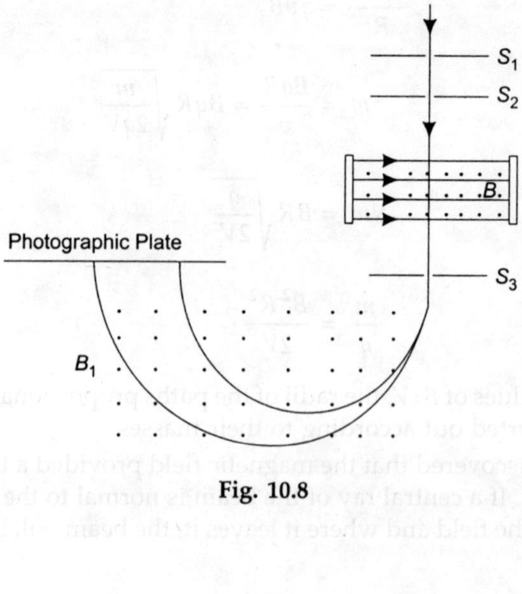

Fig. 10.8

will go undeviated and will pass through the slit S_3. This is because for ions of speed given by (10.3.8), the deflecting force due to the electric field is equal and opposite to that due to the magnetic field. The above arrangement is thus called a *velocity selector*. The ions which pass through the slit S_3 are now allowed to enter a magnetic field B_1. In this field, the ions describe a semicircle of radius R given by

$$R = \frac{mv}{qB_1} \tag{10.3.9}$$

Since v, q and B_1 are constant (for ions of the same charge), $R \propto m$. So we have a linear mass scale. The ions are incident on a photographic plate.

Later developments in mass spectrometry include producing ions under very precise conditions and using sector magnetic fields. The first ion source was made by Nier in 1940.

Example: In a certain region of space, there are a uniform electric field E and a uniform magnetic field B both directed along the z-axis. A particle of charge q and mass m is injected at time $t = 0$ with a velocity v_0 along the x-axis. Find the velocity of the particle at time t.

Solution:

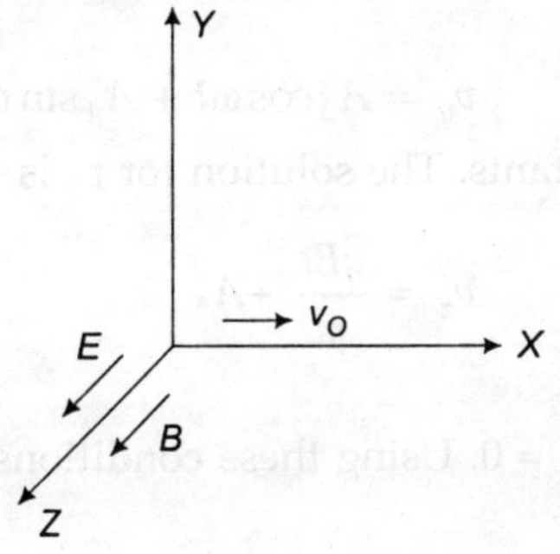

Fig. 10.9

Let the velocity of the particle at any instant t be $\mathbf{v} = v_x\hat{\mathbf{i}} + v_y\hat{\mathbf{j}} + v_z\hat{\mathbf{k}}$. The electric and magnetic fields are $\mathbf{E} = E\hat{\mathbf{k}}$ and $\mathbf{B} = B\hat{\mathbf{k}}$. The Lorentz force on the particle is thus

$$\mathbf{F} = qv_yB\hat{\mathbf{i}} - qv_xB\hat{\mathbf{j}} + qE\hat{\mathbf{k}}$$

The equations of motion along the three axes are therefore

$$\frac{dv_x}{dt} = \frac{qv_yB}{m} \tag{1}$$

$$\frac{dv_y}{dt} = \frac{-qv_xB}{m} \tag{2}$$

$$\frac{dv_z}{dt} = \frac{qE}{m} \tag{3}$$

Differentiating (1) with respect to t, we have,

$\frac{d^2 v_x}{dt^2} = \frac{qB}{M}\frac{dv_y}{dt}$. Using (2), we get $\frac{d^2 v_x}{dt^2} + \left(\frac{qB}{m}\right)^2 v_x = 0$. Let $\omega = \frac{qB}{M}$. ∴ The equation becomes

$$\frac{d^2 v_x}{dt^2} + \omega^2 v_x = 0 \tag{4}$$

The solution is

$$v_x = A_1 \cos\omega t + A_2 \sin\omega t \tag{5}$$

The equation for v_y is similarly,

$$\frac{d^2 v_y}{dt^2} + \omega^2 v_y = 0 \tag{6}$$

and the solution is

$$v_y = A_3 \cos\omega t + A_4 \sin\omega t \tag{7}$$

Here the A's are constants. The solution for v_z is

$$v_z = \frac{qEt}{m} + A_5 \tag{8}$$

At $t = 0$, $v_x = v_0$, $\frac{dv_x}{dt} = 0$. Using these conditions in (5), we get,

$$v_x = v_0 \cos\omega t \tag{9}$$

At $t = 0$, $v_y = 0$, $\frac{dv_y}{dt} = -\omega v_0$.

∴ $$v_y = -v_0 \sin\omega t \tag{10}$$

At $t = 0$, $v_z = 0$

∴ $$v_z = \frac{qEt}{m} \tag{11}$$

∴ The motion is circular in the XY plane and linear along the z axis. So the resultant motion is helical.

Tutorial 1

In this tutorial, we shall discuss a particle accelerator: the cyclotron. The cyclotron can accelerate charged particles like protons, deuterons, etc. It consists of an ion source I, two D shaped metal chambers called the dees, attached to an oscillator. A magnetic field exists perpendicular to the plane of the diagram. The dees are placed in a vacuum chamber. Suppose a positive ion is emitted by the ion source I. The oscillator voltage of the upper dee is negative with respect to the lower, so that the ion is accelerated upward. Once inside the dees, it is shielded from the effect of the electric field. The magnetic field, which is perpendicular to the plane of the dees, makes the particle move in a semicircular arc of radius

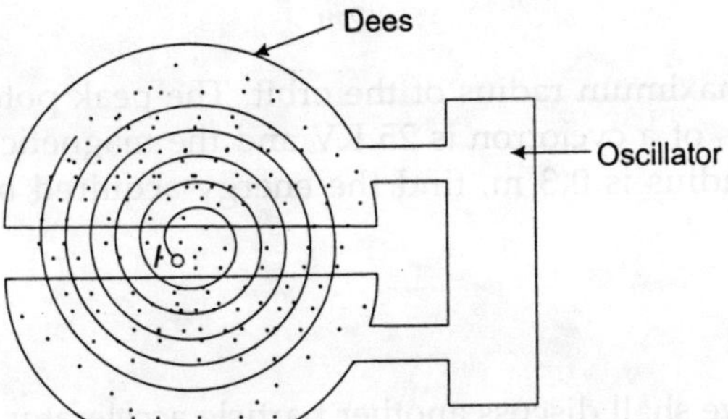

Fig. 10.10 (Top view)

$$r = \frac{mv}{qB} \tag{1}$$

As it enters the gap between the dees, the oscillating voltage has changed sign so that the lower dee is negative with respect to the upper one. The particle is accelerated across the gap. It has a greater energy when it enters the lower dee. So it describes a semicircle of larger radius in the magnetic field after it enters the lower dee. This process is repeated and the radius of the orbit increases. The energy of the particle also increases. The time required for the particle to describe a semicircle is

$$\frac{\pi r}{v} = \frac{\pi m}{qB}$$

This must be equal to half the time period T_0 of the oscillating voltage so that the voltage reverses in polarity just at the right time as the particle emerges from the dee. Thus

$$\frac{T_0}{2} = \frac{\pi m}{qB}$$

or,
$$T_0 = \frac{2\pi m}{qB}$$

In other words, the frequency ν_0 of the oscillating voltage must be

$$\nu_0 = \frac{1}{T_0} = \frac{qB}{2\pi m} \tag{2}$$

The maximum kinetic energy of the charged particle is

$$K_{\max} = \frac{1}{2}mv^2 = \frac{1}{2}m.\frac{q^2R^2B^2}{m^2}$$

$$\therefore \qquad K_{\max} = \frac{q^2R^2B^2}{2m} \tag{3}$$

where R is the maximum radius of the orbit. The peak potential difference between the dees of a cyclotron is 25 KV and the magnetic field is 1.6 T. If the maximum radius is 0.3 m, find the energy acquired by the proton in electron volts.

Tutorial 2*

In this tutorial, we shall discuss another particle accelerator — the betatron. The betatron accelerates electrons. Figs. 10.11(a) and (b) are a top view and a side view of the betatron. Here the electrons are made to move in an orbit

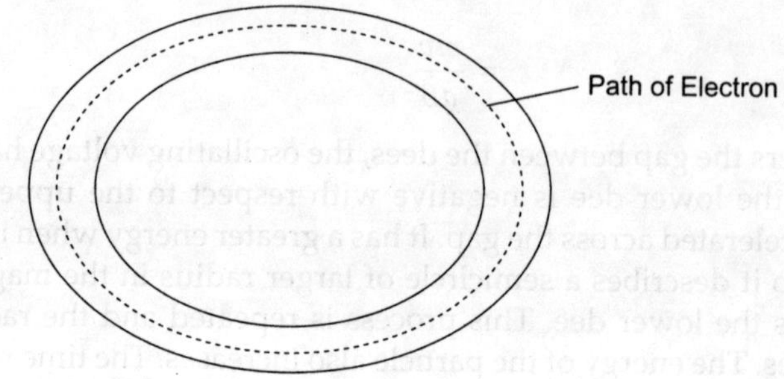

Fig. 10.11(a)

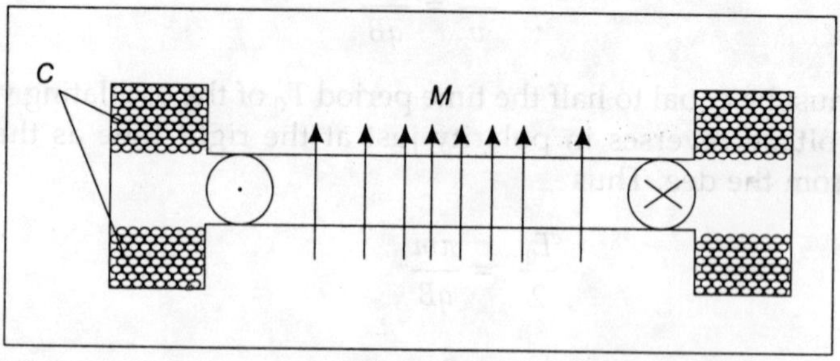

Fig. 10.11(b)

*This tutorial should be read after going through Sec. 11.1 in the next chapter.

of fixed radius in a doughnut-shaped vacuum chamber. This is done by using a time-varying magnetic field which is axially symmetric. In other words, the magnetic field is a function of r, the distance from the axis. There is a steel magnet M. Also, alternating current flows through the coils C to produce a changing flux. The principle of operation is Faraday's law of electromagnetic induction. A changing magnetic field produces a changing flux through the orbit. Thus an electric field is induced. If r is the radius of the orbit and B_{av} is the average value of the magnetic field in the circular region enclosed by the orbit, we have, from Faraday's law,

$$\oint \mathbf{E}.d\mathbf{l} = \frac{-d\phi}{dt}$$

or,

$$E.2\pi r = \frac{-d}{dt}(B_{av}\pi r^2)$$

This electric field is tangential to the orbit, in the clockwise direction. Since r is fixed,

$$E = \frac{-r}{2}\frac{dB_{av}}{dt}$$

The force on an electron is

$$F = -eE = \frac{er}{2}\frac{dB_{av}}{dt}$$

This force equals the rate of change of linear momentum of the electron.[12]

$$\therefore \quad \frac{dp}{dt} = \frac{er}{2}\frac{dB_{av}}{dt}$$

The change in the linear momentum of the electron is

$$\Delta p = \frac{er}{2}\Delta B_{av} \tag{1}$$

in a small time interval Δt. The magnetic field at the location of the orbit B_{orbit} must also increase to keep the electron moving in an orbit of fixed radius. Thus we must have

$$\frac{mv^2}{r} = qvB_{\text{orbit}}$$

or,

$$mv = qrB_{\text{orbit}}$$

or,

$$p = qrB_{\text{orbit}}$$

∴ A change Δp in the linear momentum in time Δt due to a change ΔB_{orbit} in the magnetic field at the orbit in this same time is

$$\Delta p = qr\Delta B_{\text{orbit}} \quad (2)$$

∴ For the electron to keep moving in an orbit of fixed radius, (1) and (2) must be equal.

$$\therefore \quad 2\Delta B_{\text{orbit}} = \Delta B_{av} \quad (3)$$

In other words, the average magnetic field must increase at twice the rate at which the field at the orbit increases.

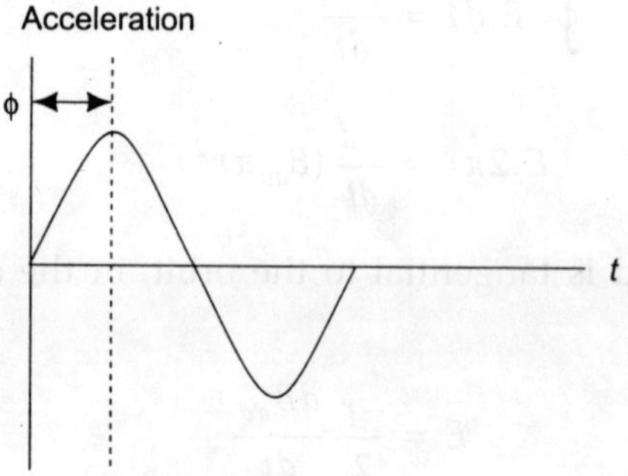

Fig. 10.12

Acceleration of the electron occurs during the half of the positive cycle when the flux is increasing with time (Fig. 10.12). The time of acceleration of the electrons is $\pi/2\omega$ where ω is the angular frequency of the oscillating flux. The electron is moving at close to the speed of light c. ∴ The distance travelled by the electron during this time is $\pi c/2\omega$. If R is the radius of the orbit, the number of revolutions which the electron makes is $\dfrac{\pi c}{2\omega}/2\pi R = \dfrac{c}{4\omega R}$. Take $\omega = 2\pi \times 50$ rad/s and orbit radius = 25 cm. Calculate the number of revolutions.

Problems

1. A mass spectrometer uses parallel electric and magnetic fields, both perpendicular to the path of the narrow incident beam of positive ions. The magnetic field is 0.282 T and the electric field is 5×10^4 V/m and both extend in the direction of the incident beam over a distance of 5 cm. If the photographic plate is 20 cm from the centre of the deflecting fields, at what point with respect to the axis of the system does a singly charged oxygen atom, which has fallen through a potential difference of 10 kV in the discharge tube, strike the photographic plate? (mass of $^{16}O = 2.56 \times 10^{-26}$ kg)

   [Ans. x = 4.98 cm, y = 2.5 cm]

2. A cathode-ray tube has electrostatic deflection plates 2 cm square, separated by 0.5 cm and 25 cm from the fluorescent screen. If the electrons have 1500 eV of kinetic energy when they pass between the plates, calculate the approximate beam deflection on the screen if 20 V is applied between the deflection plates. **[Ans.** 0.7 cm]
3. An electron is accelerated through a potential difference V. It then passes between two long parallel plates with electric field E between them. Perpendicular to E is a magnetic field B such that the electron is not deflected as it passes between the plates. Show that $e/m = E^2/2VB^2$.
4. What is the mass of singly charged ions which follow a circular path of radius 0.41 m when placed in a transverse magnetic field of 0.223 T, the initial energy being 100 keV? What electric field must be superimposed if the ions are to pass undeflected through the magnetic field? **[Ans.** $m = 6.55 \times 10^{-27}$ kg, $E = 4.9 \times 10^5$ V/m]
5. Calculate the maximum energy of protons obtainable from a cyclotron of 1.2 m dee diameter and 1.5 T magnetic field. At what frequency must the cyclotron be operated? If the average energy gain per dee passage is 50 keV, how many revolutions do the protons make? **[Ans.** K_{max} = 39 MeV, ν_0 = 23 MHz, n = 390]

11

Maxwell's Equations

11.1 FARADAY'S LAW OF ELECTROMAGNETIC INDUCTION

After a series of experiments in 1831, Faraday concluded that an *emf* is induced in an electric circuit whenever the flux through the circuit changes. Thereafter Lenz observed that the direction of induced emf is such as to oppose the change producing it. In the SI system of units, the induced *emf* ε is given by

$$\varepsilon = \frac{-d\phi}{dt} \tag{11.1.1}$$

where ϕ is the flux. The negative sign incorporates Lenz's observation. The emf in a circuit is defined as the work done in talking a unit positive charge once around the circuit. Since work done is the line integral of the force with respect to the displacement and the force per unit charge is the electric field, we can write (11.1.1) as

$$\oint \mathbf{E}.d\mathbf{l} = \frac{-d\phi}{dt}$$

This electric field is not an electrostatic field, for which the right hand side would be zero. The flux is the surface integral of the magnetic field over the area enclosed by the circuit.

$$\therefore \qquad \oint \mathbf{E}.d\mathbf{l} = \frac{-d}{dt}\int_s \mathbf{B}.d\,\mathbf{a} \tag{11.1.2}$$

where S is the area enclosed by the circuit. When we take the time derivative within the integral, it becomes a partial derivative with respect to time because **B** is a function of both space and time. Thus we have

$$\oint \mathbf{E}.d\mathbf{l} = -\int_s \frac{\partial \mathbf{B}}{\partial t}.d\,\mathbf{a}$$

Using Stokes' theorem, we can transform the line integral into a surface integral over *S*.

$$\int_S (\nabla \times \mathbf{E}).d\mathbf{a} = -\int_S \frac{\partial \mathbf{B}}{\partial t}.d\mathbf{a}$$

or, $$\int_S \left(\nabla \times \mathbf{E} + \frac{\partial \mathbf{B}}{\partial t}\right).d\mathbf{a} = 0$$

Since this is true for any surface *S* bound by the circuit, the integrand is zero.

$$\therefore \quad \nabla \times \mathbf{E} + \frac{\partial \mathbf{B}}{\partial t} = 0 \tag{11.1.3}$$

This equation is the differential form of Faraday's law. Equation (11.1.2) was the integral form of the law.

Example 1: Consider the *U*-shaped conductor in the figure below. AC is conducting rod which moves to the right with a constant speed *v*. There is a magnetic field *B* perpendicular to the plane of the diagram, pointing away from the reader. The distance between the two points at which the rod AC touches the *U* shaped conductor is *l*. Find the induced *emf*.

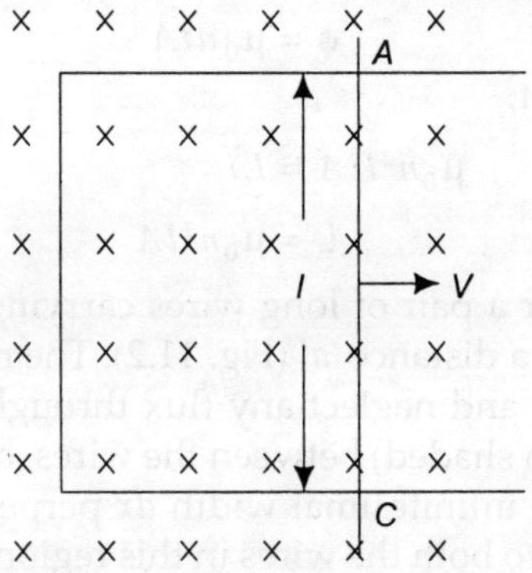

Fig. 11.1

Solution: As the rod moves to the right, the flux through the closed circuit formed by the *U* shaped conductor and the rod AC increases. An *emf* is induced in the circuit so as to reduce the flux. In an infinitesimal time dt, the rod AC moves to the right by an infinitesimal distance $dx = vdt$. The area enclosed is ldx and so the increase in flux is $d\phi = Bldx = Blvdt$. $\therefore$ The magnitude of the emf induced is $\varepsilon = Blv$. A current flows from *C* to *A*. This *emf* is called motional emf. A magnetic field pointing towards the reader appears due to this current.

11.2 SELF AND MUTUAL INDUCTANCE

The flux through a circuit is proportional to the current in the circuit.⊕ This is because, from Biot-Savart's law, the magnetic field is proportional to the current. The constant of proportionality is called the self-inductance L.

$$\phi = Li \tag{11.2.1}$$

The *SI* unit of flux is Weber (*Wb*). If we now use Faraday's law (11.1.1), then we get,

$$\varepsilon = \frac{-Ldi}{dt} \tag{11.2.2}$$

The *SI* unit of inductance is Henry (H). We can use (11.2.2) to define it. A coil is said to have a self-inductance of 1 H if the emf induced in it is 1 volt when the current through it is changing at the rate of 1 A/s.

Let us now calculate the self-inductance in some simple cases. First we shall consider a solenoid of length l having n turns per unit length. The magnetic field inside the solenoid is along its axis and has a magnitude $B = \mu_0 ni$. If ϕ is the flux through each turn, then we can writer

$$nl\phi = Li \tag{11.2.3}$$

where nl is the total number of turns. Let A be the area of a coil. ∴ The flux through each turn of the coil is

$$\phi = \mu_0 niA$$

∴ Using (11.2.3), we get,

$$\mu_0 n^2 liA = Li$$

$$\therefore \quad L = \mu_0 n^2 lA \tag{11.2.4}$$

Next let us consider a pair of long wires carrying currents I in opposite directions separated by a distance 'a' (Fig. 11.2). The radius of the wires is 'r'. We assume that $r << a$, and neglect any flux through the wires. Consider a rectangular area (shown shaded) between the wires, of length l parallel to the length of the wires and infinitesimal width dx perpendicular to this length. The magnetic field due to both the wires in this region is perpendicular to the plane of the diagram pointing away from the reader. The field due to the upper wire at the location of the shaded region is

$$B_1 = \frac{\mu_0 I}{2\pi x}$$

and that due to the lower wire is

$$B_2 = \frac{\mu_0 I}{2\pi(a-x)}$$

The total field is thus

⊕This is on the assumption that there are no nonlinear materials in the vicinity.

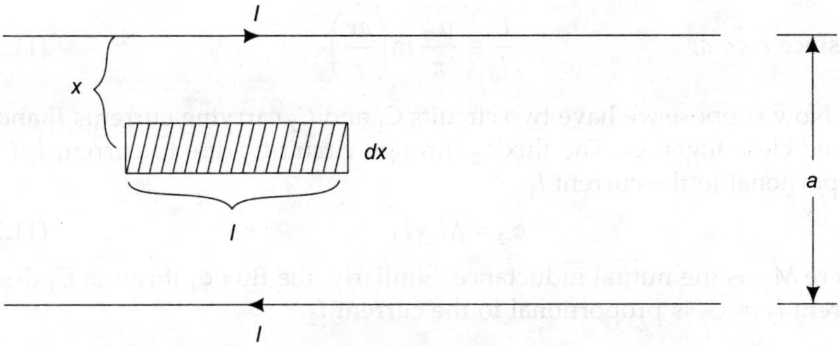

Fig. 11.2

$$B = B + B_2 = \frac{\mu_0 I}{2\pi}\left(\frac{1}{x} + \frac{1}{a-x}\right)$$

The flux through the shaded region is

$$d\phi = Bldx = \frac{\mu_0 Il}{2\pi}\left(\frac{1}{x} + \frac{1}{a-x}\right)dx$$

The flux through the entire region between the wires is obtained by integrating the above from $x = r$ to $x = a - r$ Thus, we have

$$\phi = \frac{\mu_0 Il}{2\pi}\left[\int_r^{a-r} \frac{dx}{x} + \int_r^{a-r} \frac{dx}{a-x}\right]$$

$$= \frac{\mu_0 Il}{2\pi}\left[\ln x\,\right]_r^{a-r} - \ln(a-r)\Big]_r^{a-r}$$

$$= \frac{\mu_0 Il}{2\pi}\left[\ln\left(\frac{a-r}{r}\right) - \ln\left(\frac{r}{a-r}\right)\right]$$

$$\phi = \frac{\mu_0 Il}{\pi}\ln\left(\frac{a-r}{r}\right)$$

$\therefore$ The inductance of a length l of the pair of wires is $L = \frac{\mu_0 l}{\pi}\ln\left(\frac{a-r}{r}\right)$. The inductance per unit length is

$$\frac{L}{l} = \frac{\mu_0}{\pi}\ln\left(\frac{a-r}{r}\right)$$

or, since $r << a$, $$\frac{L}{l} \cong \frac{\mu_0}{\pi} \ln\left(\frac{ar}{r}\right) \tag{11.2.5}$$

Now suppose we have two circuits C_1 and C_2 carrying currents I_1 and I_2, placed close together. The flux ϕ_2 through circuit C_2 due to current I_1 C_1 is proportional to the current I_1

$$\phi_2 = M_{21} I_1 \tag{11.2.6}$$

where M_{21} is the nutual inductance. Similarly, the flux ϕ_1 through C_1 due to current I_2 in C_2 is proportional to the current I_2

$$\phi_1 = M_{12} I_2 \tag{11.2.7}$$

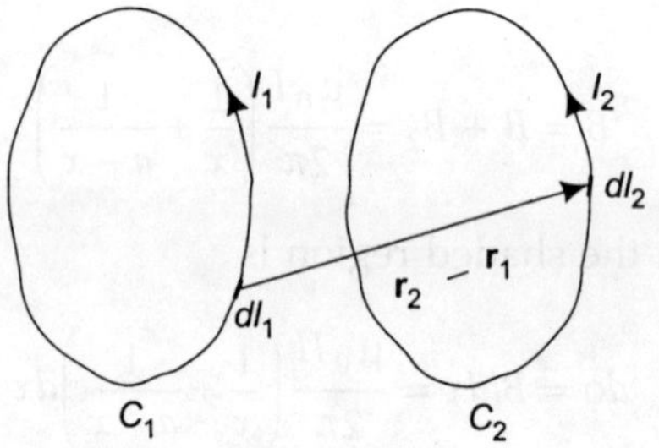

Fig. 11.3

Biot-Savart's law tells us that the magnetic field $\mathbf{B}_1$ at the location of $d\mathbf{I}_2$ due to current I_1 in C_1 is

$$\mathbf{B}_1 = \frac{\mu_0 \mathbf{I}_1}{4\pi} \oint_{C_1} \frac{d\mathbf{I} \times (\mathbf{r}_2 - \mathbf{r}_1)}{|\mathbf{r}_2 - \mathbf{r}_1|^3}$$

where $\mathbf{r}_1$ and $\mathbf{r}_2$ are the position vectors of the infinitesimal elements of length dl_1 and dl_2 respectively (with respect to some origin).

$$\mathbf{B}_1 = \frac{-\mu_0 \mathbf{I}_1}{4\pi} \oint_{C_1} d\mathbf{I}_1 \times \nabla_2 \left(\frac{1}{|\mathbf{r}_2 - \mathbf{r}_1|}\right)$$

where ∇_2 is the gradient with respect to $\mathbf{r}_2$.

$$\therefore \quad \mathbf{B}_1 = \frac{\mu_0 \mathbf{I}_1}{4\pi} \oint_{C_1} \nabla_2 \left(\frac{1}{|\mathbf{r}_2 - \mathbf{r}_1|}\right) \times d\mathbf{l}_1$$

$$= \frac{\mu_0 \mathbf{I}_1}{4\pi} \nabla_2 \times \left\{ \oint_{C_1} \frac{d\mathbf{l}_1}{|\mathbf{r}_2 - \mathbf{r}_1|} \right\} \tag{11.2.8}$$

using a vector identity. We have taken the gradient outside the integral because the integration is with respect to $\mathbf{r}_1$. The flux ϕ_2 through C_2 is

$$\phi_2 = \int_{S_2} \mathbf{B}_1 . d\mathbf{a}_2$$

where S_2 is the area of C_2.

$$\therefore \qquad \phi_2 = \frac{\mu_0 I_1}{4\pi} \int_{S_2} \left[\nabla_2 \times \left\{ \oint_{C_1} \frac{d\mathbf{l}_1}{|\mathbf{r}_2 - \mathbf{r}_1|} \right\} \right] . d\mathbf{a}_2$$

$$= \frac{\mu_0 I_1}{4\pi} \oint_{C_2} \oint_{C_1} \frac{d\mathbf{l}_1 . d\mathbf{l}_2}{|\mathbf{r}_2 - \mathbf{r}_1|}$$

using Stokes' theorem.

$$\therefore \qquad M_{21} = \frac{\mu_0}{4\pi} \oint_{C_1} \oint_{C_2} \frac{d\mathbf{l}_1 . d\mathbf{l}_2}{|\mathbf{r}_2 - \mathbf{r}_1|} \qquad (11.2.9)$$

This is *Neumann's formula* for mutual inductance. We see that $M_{21} = M_{12}$, on interchanging 1 and 2 in the right side. We drop the subscripts and cail M the mutual inductance between the coils.

Let us calculate the mutual inductance in a simple case. Consider a short solenoid of length l with n_1 turns per unit length and radius 'a' placed along the axis of a larger, longer solenoid with n_2 turns per unit length and radius 'b' ($b > a$). See Fig. 11.4.

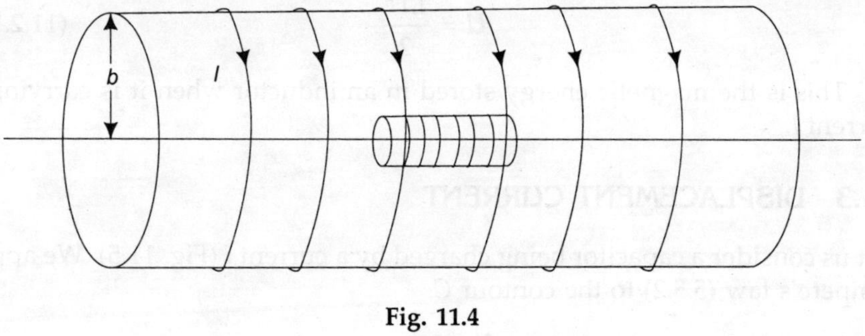

Fig. 11.4

Suppose a current I flows through the larger soleniod. The magnetic field along the axis is

$$B = \mu_0 n_2 I$$

The flux ϕ through each turn of the short solenoid is $\mathbf{B}.\pi a^2$ and it has $n_1 l$ turns. $\therefore$ The flux through the solenoid is

$$n_1 l\phi = n_1 lB\pi a^2$$
$$= n_1 l\mu_0 n_2 I\pi a^2$$

$\therefore$ The mutual inductance M between the two solenoids is

$$M = \mu_0 n_1 n_2 \pi a^2 l \tag{11.2.10}$$

Let us now calculate the energy stored in an inductor when a current I flows through it. Let a source of voltage V be applied to a circuit of resistance R and let ε be the induced *emf*. Then

$$V = iR - \varepsilon$$

The work done by the source in pushing a charge dq around the circuit in time dt is

$$Vdq = Vidt = -\varepsilon idt + i^2Rdt$$

From Faraday's law, we have $\varepsilon = -d\phi/dt$. So the above equation becomes

$$Vdq = id\phi + i^2Rdt \tag{11.2.11}$$

The second term represents the conversion of electrical energy into heat, which is irreversible. The first term is the work done against the induced emf in the circuit. It is the term responsible for setting up of the magnetic field. If the circuits are stationary, no mechanical work is assiociated with the change in flux and the first term is the change dU in the magnetic energy of the circuit.

$\therefore$ $$dU = id\phi$$

For linear media, $d\phi = Ldi$

$\therefore$ $$dU = Lidi$$

Integrating from $i = 0$ to $i = I$, we get,

$$U = \frac{LI^2}{2} \tag{11.2.12}$$

This is the magnetic energy stored in an inductor when it is carrying a current I.

11.3 DISPLACEMENT CURRENT

Let us consider a capacitor being charged by a current i (Fig. 11.5). We apply Ampere's law (5.5.2) to the contour C.

$$\oint \mathbf{B}.d\mathbf{l} = \mu_0 i \tag{5.5.2}$$

Here i is the current enclosed by the contour. Since $I_{enc.} = \int_S \mathbf{J}.d\mathbf{a}$ where $\mathbf{J}$ is the current density and S is a surface with C as its boundary, we could use either surface S_1 which is pierced by the conductor, or surface S_2 which lies partly in the space between the capacitor plates. Clearly S_1 encloses current

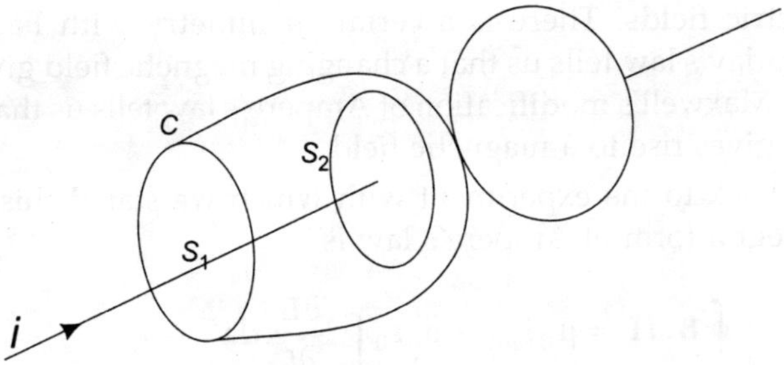

Fig. 11.5

i but S_2 does not. So $I_{enc.} = 0$ if we use S_2. This is an inconsistency. We have to remember that the current density **J** is nonsteady in this case. Charge is piling up on the capacitor plates. ∴ **J** and ρ are functions of both space and time and satisfy the continuity equation.

$$\nabla.\mathbf{J} + \frac{\partial \rho}{\partial t} = 0 \tag{7.1.2}$$

The differential form of Ampere's law is $\nabla \times \mathbf{B} = \mu_0 \mathbf{J}$. Taking the divergence of both sides we have $\nabla.(\nabla \times \mathbf{B}) = \mu_0 \nabla.\mathbf{J}$. The left need side is zero identically, but the right hand side is not zero in this case. Let us rewrite (7.1.2) using the differential form of Gauss' law $\nabla.\mathbf{E} = \frac{\rho}{\varepsilon_0}$. $\therefore \frac{\varepsilon_0 \partial}{\partial t}(\nabla.\mathbf{E}) = \frac{\partial \rho}{\partial t}$. We can interchange space and time derivatives. So, $\varepsilon_0 \nabla.\left(\frac{\partial \mathbf{E}}{\partial t}\right) = \frac{\partial \rho}{\partial t}$. ∴ The continuity equation becomes

$$\nabla.\left(\mathbf{J} + \frac{\varepsilon_0 \partial \mathbf{E}}{\partial t}\right) = 0$$

So if we rewrite the differential form of Ampere's law as

$$\nabla \times \mathbf{B} = \mu_0 \left(\mathbf{J} + \frac{\varepsilon_0 \partial \mathbf{E}}{\partial t}\right) \tag{11.3.1}$$

we get zero on both sides on applying the divergence operator. Eqn. (11.3.1) is Maxwell's modification to Ampere's law. The term $\mu_0 \varepsilon_0 \frac{\partial \mathbf{E}}{\partial t}$ is called displacement current.[⊕] It also produces a magnetic field like a conduction current. But it is not due to physical transport of change carriers. This term is negligibly small compared to the first term except for extremely rapidly

---

[⊕]More precisely, it is a displacement current density.

varying electric fields. There is a certain symmetry with Faraday's law. Whereas Faraday's law tells us that a changing magnetic field gives rise to an electric field, Maxwell's modification of Ampere's law tells us that a changing electric field gives rise to a magnetic field.

Coming back to the experiment with which we stated this section, the modified integral form of Ampere's law is

$$\oint \mathbf{B}.d\mathbf{l} = \mu_0 I_{enc} + \mu_0 \varepsilon_0 \int_S \frac{\partial \mathbf{E}}{\partial t}.\mathbf{da} \tag{11.3.2}$$

For the surface S_1, the second term is zero because $E = 0$. For the surface S_2, $I_{enc} = 0$, but $\varepsilon_0 \int_S \frac{\partial \mathbf{E}}{\partial t}.d\mathbf{a} = i$.

Example 2: Consider a discharging capacitor with circular plates of radius b (Fig. 11.6) and separation s. Calculate the magnetic field at a point P distant r from the axis of symmetry.

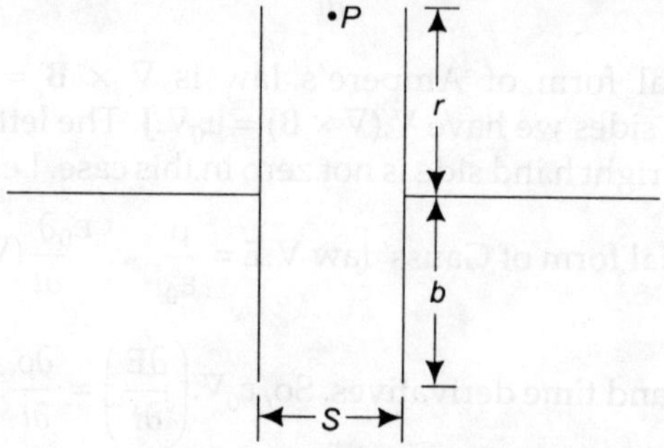

Fig. 11.6

Solution: Consider a circular loop C of radius r passing through P and centered on the axis of symmetry (Fig. 11.7). The magnetic field is tangential

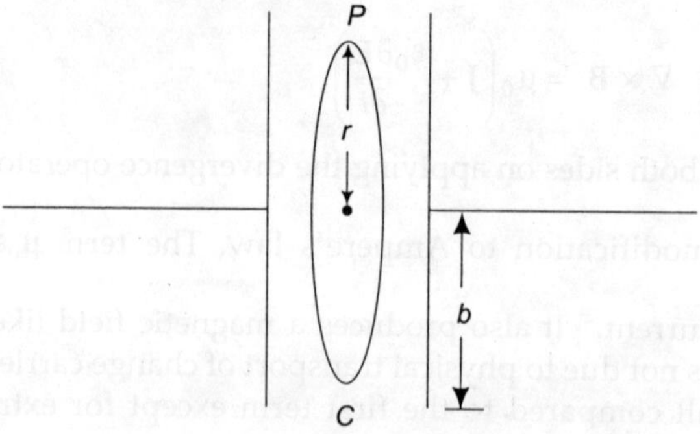

Fig. 11.7

to the circle. $I_{enc} = 0$. The electric field at any instant t is $E(t) = \dfrac{q(t)}{\varepsilon_0 \pi b^2}$ since πb^2 is the area of the plates and $q(t)$ is the charge on the plates at that instant.

$\therefore \quad \dfrac{\partial E}{\partial t} = \dfrac{1}{\varepsilon_0 \pi b^2}\dfrac{dq}{dt} = \dfrac{I}{\varepsilon_0 \pi b^2}$ where I is the magnitude of the conduction current flowing in the wires attached to the plates. This is directed normal to the plates.

$$\therefore \qquad \int_S \frac{\partial \mathbf{E}}{\partial t}.d\mathbf{a} = \frac{I}{\varepsilon_0 \pi b^2}.\pi r^2$$

S is the area of circle C. From (11.3.2), we have

$$B.2\pi r = \frac{I}{\varepsilon_0 \pi b^2}.\pi r^2$$

$$\therefore \qquad B = \frac{Ir}{2\pi\varepsilon_0 b^2}$$

11.4 MAXWELL'S EQUATIONS

We will now recall equations from our study of electric and magnetic fields from previous chapters and the two equations derived in this chapter.

$$\nabla.\mathbf{E} = \frac{\rho}{\varepsilon_0} \qquad (1.8.11)$$

$$\nabla.\mathbf{B} = 0 \qquad (5.5.1)$$

$$\nabla \times \mathbf{E} = \frac{-\partial \mathbf{B}}{\partial t} \qquad (11.1.3)$$

$$\nabla \times \mathbf{B} = \mu_0 \mathbf{J} + \mu_0\varepsilon_0 \frac{\partial \mathbf{E}}{\partial t} \qquad (11.3.1)$$

Equation (1.8.11) is the differential form of Gauss' law. Eqn. (5.5.1) indicates absence of magnetic monopoles. Eqn. (11.1.3) is Faraday's law. Eqn. (11.3.1) is Maxwell's modified form of Ampere's law. The last equation has far-reaching consequences. As we shall learn later, it predicts existence of electromagnetic waves. Moreover, Maxwell also showed that these waves propagate at the speed of light. In other words, light is an electromagnetic wave. Maxwell deduced the modification of Ampere's law in 1865 and because of these outstanding achievements, the set of four equations above are known as Maxwell's equations.

In free space there are no charges ($\rho = 0$) and no currents ($J = 0$). Maxwell's equations then become

$$\nabla.\mathbf{E} = 0 \tag{11.4.1}$$

$$\nabla.\mathbf{B} = 0 \tag{5.5.1}$$

$$\nabla \times \mathbf{E} = \frac{-\partial \mathbf{B}}{\partial t} \tag{11.1.3}$$

$$\nabla \times \mathbf{B} = \mu_0 \varepsilon_0 \frac{\partial \mathrm{E}}{\partial t} \tag{11.4.2}$$

When we consider material media, we introduce two vectors: the auxiliary field **H** (Chapter 6) and the electric displacement **D** (Chapter 4). Gausss' law then takes the form

$$\nabla.\mathbf{D} = \rho \tag{4.3.6}$$

and Ampere's law took the form

$$\nabla \times \mathbf{H} = \mathbf{J}_f \tag{6.2.2}$$

In order to satisfy the continuity equation, eqn. (6.2.2) must be modified to

$$\nabla \times \mathbf{H} = \mathbf{J}_f + \frac{\partial \mathbf{D}}{\partial t} \tag{11.4.3}$$

Maxwell's equations inside matter are thus

$$\nabla.\mathbf{D} = \rho \tag{4.3.6}$$

$$\nabla.\mathbf{B} = 0 \tag{5.5.1}$$

$$\nabla \times \mathbf{E} = \frac{-\partial \mathbf{B}}{\partial t} \tag{11.1.3}$$

$$\nabla \times \mathbf{H} = \mathbf{J}_f + \frac{\partial \mathbf{D}}{\partial t} \tag{11.4.3}$$

11.5 BOUNDARY CONDITIONS

We shall now look at the interface between two media 1 and 2 and find the relation between the field vectors in them. We will use the integral form of Maxwell's equations inside matter.

$$\int_S \mathbf{D}.d\mathbf{a} = q \tag{4.3.5}$$

Consider a Gaussian pillbox of base area A and infinitesimal thickness at the interface between two media (Fig. 11.8). The surface S refers to the top and bottom faces and the curved surface. The contribution to the surface integral from the curved surface goes to zero as the thickness tends to zero. If σ is the surface charge density at the interface, then we get,

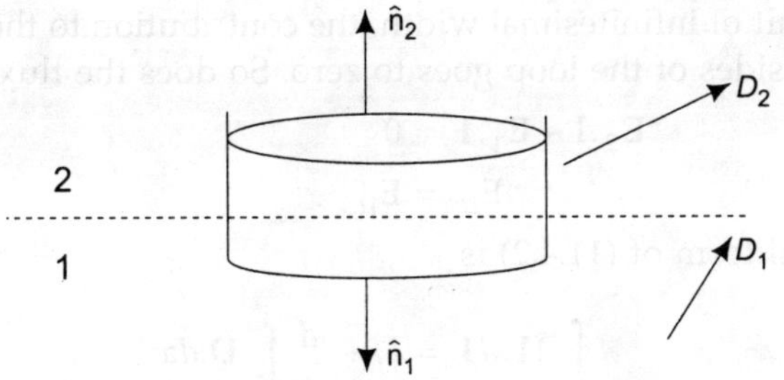

Fig. 11.8

$$\mathbf{D}_2.\hat{\mathbf{n}}_2 A + \mathbf{D}_1.\hat{\mathbf{n}}_1 A = \sigma A$$

or, since $\hat{\mathbf{n}}_1 = -\hat{\mathbf{n}}_2$

$$(\mathbf{D}_2 - \mathbf{D}_1).\hat{\mathbf{n}}_2 = \sigma$$

or, $$D_{2n} - D_{1n} = \sigma \tag{11.5.1}$$

where the subscript n denotes the normal components.

The integral form of (5.5.1) is

$$\int_S \mathbf{B}.d\mathbf{a} = 0$$

Using the same pillbox and the same reasoning, we find

$$\mathbf{B}_2.\hat{\mathbf{n}}_2 A + \mathbf{B}_1.\hat{\mathbf{n}}_1 \mathrm{A} = 0$$

or, $$\mathbf{B}_2.\hat{\mathbf{n}}_2 + \mathbf{B}_1.\hat{\mathbf{n}}_1 = 0$$

or, $$(\mathbf{B}_2 - \mathbf{B}_1).\hat{\mathbf{n}}_2 = 0$$

or, $$B_{2n} = B_{1n} \tag{11.5.2}$$

Next let us consider an Amperian loop of length l and infinitesimal width straddling the interface between 1 and 2 (Fig. 11.9). The integral form of (11.1.3) is

$$\oint \mathbf{E}.d\mathbf{l} = \frac{-d}{dt}\int_S \mathbf{B}.d\mathbf{a} \tag{11.1.2}$$

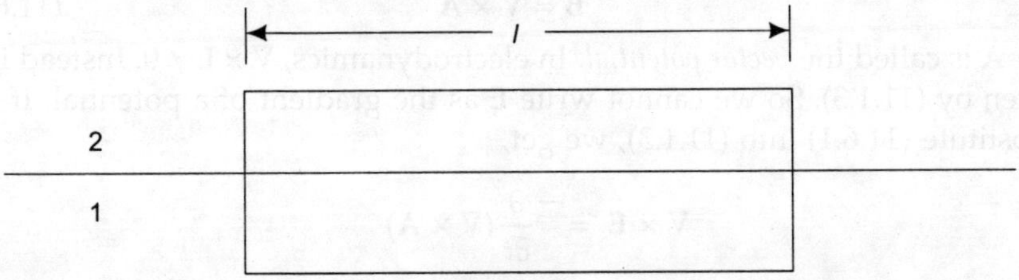

Fig. 11.9

In the limit of infinitesimal width, the contribution to the integral on the left from the sides of the loop goes to zero. So does the flux. ∴ We get

$$\mathbf{E}_2.\mathbf{l} - \mathbf{E}_1.\mathbf{l} = 0$$

or,
$$\mathbf{E}_{2t} = \mathbf{E}_{1t} \tag{11.5.3}$$

The integral form of (11.4.2) is

$$\oint \mathbf{H}.d\mathbf{l} = I_f + \frac{d}{dt}\int_S \mathbf{D}.d\mathbf{a} \tag{11.5.4}$$

As the width of the loop goes to zero, the flux of **D** goes to zero and the contribution to the line integral from the two sides goes to zero. So we have

$$\mathbf{H}_2.\mathbf{l} - \mathbf{H}_1.\mathbf{l} = I_f$$

If we take $\hat{\mathbf{n}}$ as the unit vector pointing from 1 to 2, we can write the free current I_f as

$$I_f = \mathbf{K}_f.(\hat{\mathbf{n}} \times \mathbf{l})$$

where $\mathbf{K}_f$ is the surface current density.

∴
$$I_f = (\mathbf{K}_f \times \hat{\mathbf{n}}).\mathbf{l}$$

∴
$$\mathbf{H}_{2t} - \mathbf{H}_{1t} = \mathbf{K}_f \times \hat{\mathbf{n}} \tag{11.5.5}$$

For linear media, $\mathbf{D} = \varepsilon\mathbf{E}$ and $\mathbf{B} = \mu\mathbf{H}$. So we can express the boundary conditions in terms of **E** and **B**:

$$\varepsilon_2 E_{2n} - \varepsilon_1 E_{1n} = \sigma \tag{11.5.1a}$$

$$B_{2n} = B_{1n} \tag{11.5.2}$$

$$E_{2t} = E_{1t} \tag{11.5.3}$$

$$\frac{\mathbf{B}_{2t}}{\mu_2} - \frac{\mathbf{B}_{1t}}{\mu_1} = \mathbf{K}_f \times \hat{\mathbf{n}} \tag{11.5.5a}$$

11.6 VECTOR AND SCALAR POTENTIALS

Let us come back to eqn. (5.5.1). Since the divergence of the curl of a vector is identically zero, we can write

$$\mathbf{B} = \nabla \times \mathbf{A} \tag{11.6.1}$$

A is called the *vector potential.* In electrodynamics, $\nabla \times \mathbf{E} \neq 0$. Instead it is given by (11.1.3). So we cannot write **E** as the gradient of a potential. If we substitute (11.6.1) into (11.1.3), we get,

$$\nabla \times \mathbf{E} = \frac{-\partial}{\partial t}(\nabla \times \mathbf{A})$$

or,
$$\nabla \times \mathbf{E} + \frac{\partial}{\partial t}(\nabla \times \mathbf{A}) = 0$$

We can interchange space and time derivatives

$$\therefore \qquad \nabla \times \mathbf{E} + \nabla \times \frac{\partial \mathbf{A}}{\partial t} = 0$$

So we may write

$$\mathbf{E} + \frac{\partial \mathbf{A}}{\partial t} = -\nabla\phi$$

or,

$$\mathbf{E} = -\nabla\phi - \frac{\partial \mathbf{A}}{\partial t} \tag{11.6.2}$$

Here ϕ is the scalar potential. Let us rewrite eqns. (1.8.11) and (11.3.1) in terms of vector and scalar potentials. First let us take eqn. (1.8.11). Substituting (11.6.2) into (1.8.11), we get,

$$\nabla^2\phi + \frac{\partial}{\partial t}(\nabla.\mathbf{A}) = \frac{-\rho}{\varepsilon_0} \tag{11.6.3}$$

Now consider (11.3.1). We get

$$\nabla \times (\nabla \times \mathbf{A}) = \mu_0 \mathbf{J} - \mu_0 \varepsilon_0 \frac{\partial}{\partial t}\left(\nabla\phi + \frac{\partial A}{\partial t}\right)$$

or, using a vector identity

$$\nabla(\nabla.\mathbf{A}) - \nabla^2\mathbf{A} = \mu_0 \mathbf{J} - \mu_0 \varepsilon_0 \frac{\partial}{\partial t}(\nabla\phi) - \mu_0 \varepsilon_0 \frac{\partial^2 \mathbf{A}}{\partial t^2}$$

or,

$$\left(\nabla^2\mathbf{A} - \mu_0 \varepsilon_0 \frac{\partial^2 \mathbf{A}}{\partial t^2}\right) - \nabla\left(\nabla.\mathbf{A} + \mu_0 \varepsilon_0 \frac{\partial \phi}{\partial t}\right) = -\mu_0 \mathbf{J} \tag{11.6.4}$$

Equation (11.6.1) and (11.6.2) do not uniquely define the vector and scalar potentials. Let us choose another set ($\mathbf{A}'$, ϕ') which produce the same **B** and **E**.

$$\therefore \qquad \mathbf{B} = \nabla \times \mathbf{A}' \tag{11.6.5}$$

$$\mathbf{E} = -\nabla\phi' - \frac{\partial \mathbf{A}'}{\partial t} \tag{11.6.6}$$

Let us take

and

$$\left.\begin{aligned} \mathbf{A}' &= \mathbf{A} + \mathbf{C} \\ \phi' &= \phi + V \end{aligned}\right\} \tag{11.6.7}$$

and find equations that **C** and V must satisfy.

$$\nabla \times \mathbf{A}' = \nabla \times \mathbf{A} + \nabla \times \mathbf{C}$$

$\therefore$ For (11.6.1) and (11.6.5) to be simultaneously true, we must have

$$\nabla \times \mathbf{C} = 0$$

$$\therefore \qquad \mathbf{C} = \nabla f \qquad (11.6.8)$$

$$\nabla \phi' = \nabla \phi + \nabla V$$

$$\frac{\partial \mathbf{A}'}{\partial t} = \frac{\partial \mathbf{A}}{\partial t} + \frac{\partial \mathbf{C}}{\partial t}$$

$\therefore$ For (11.6.2) and (11.6.6) to be simultaneously true, we must have

$$\nabla V + \frac{\partial \mathbf{C}}{\partial t} = 0$$

or,
$$\nabla V + \frac{\partial}{\partial t}(\nabla f) = 0 \qquad \text{using (11.6.8)}$$

or,
$$\nabla\left(V + \frac{\partial f}{\partial t}\right) = 0$$

or,
$$V + \frac{\partial f}{\partial t} = a_1(t)$$

We can absorb $a_1(t)$ into $f(t)$ by suitably redefining it. With this redefinition, we rewrite

$$V = -\frac{\partial f}{\partial t} \qquad (11.6.9)$$

This redefinition will leave ∇f unchanged. So substituting these into (11.6.7), we get

$$\mathbf{A}' = \mathbf{A} + \nabla f \qquad (11.6.10)$$

$$\phi' = \phi - \frac{\partial f}{\partial t} \qquad (11.6.11)$$

The vector potential $\mathbf{A}'$ and ϕ' will give the same $\mathbf{E}$ and $\mathbf{B}$ as $\mathbf{A}$ and ϕ. The transformations (11.6.10) and (11.6.11) are called *gauge transformations.*

We shall look at two possible choices of the quantity $\nabla.\mathbf{A}$. One of these choices is called the *Coulomb gauge*. Here, we take

$$\nabla.\mathbf{A} = 0 \qquad (11.6.12)$$

With this substitution, eqn. (11.6.3) becomes

$$\nabla^2 \phi = -\frac{\rho}{\varepsilon_0} \qquad (11.6.13)$$

This is Poisson's equation. Eqn. (11.6.4) becomes

$$\nabla^2 \mathbf{A} - \mu_0 \varepsilon_0 \frac{\partial^2 \mathbf{A}}{\partial t^2} - \mu_0 \varepsilon_0 \nabla\left(\frac{\partial \phi}{\partial t}\right) = -\mu_0 \mathbf{J} \qquad (11.6.14)$$

The other choice is the Lorentz gauge. In this gauge, we take

$$\nabla . \mathbf{A} = -\mu_0 \varepsilon_0 \frac{\partial \phi}{\partial t} \tag{11.6.15}$$

Then eqn. (11.6.3) becomes

$$\nabla^2 \phi - \mu_0 \varepsilon_0 \frac{\partial^2 \phi}{\partial t^2} = \frac{-\rho}{\varepsilon_0} \tag{11.6.16}$$

and eqn. (11.6.4) becomes

$$\nabla^2 \mathbf{A} - \mu_0 \varepsilon_0 \frac{\partial^2 \mathbf{A}}{\partial t^2} = -\mu_0 \mathbf{J} \tag{11.6.17}$$

So in this gauge **A** and ϕ satisfy similar equations. We will look at the form of the solutions in Chapter 13.

Example 3: Let us come back to magnetostatics. Here we can also define a vector potential **A** in the same way. $\mathbf{B} = \nabla \times \mathbf{A}$. Here **A** is independent of time. Also ϕ is independent of time. With a choice $\nabla . \mathbf{A} = 0$, eqn. (11.6.14) becomes

$$\nabla^2 \mathbf{A} = -\mu_0 \mathbf{J}$$

and the solution is

$$\mathbf{A} = \frac{\mu_0}{4\pi} \int \frac{\mathbf{J}(\mathbf{r}')}{r_1} dv' \tag{11.6.18}$$

Consider a long, straight conductor carrying a current *I*. We know the magnetic field *B* at a distance *r* due to this conductor. Find the vector potential **A**.

Solution: The magnetic field due to a straight conductor carrying a current *I* at a distance *r* from it is

$$\mathbf{B} = \frac{\mu_0 I}{2\pi r} \hat{\phi}$$

Let us use the equation $\mathbf{B} = \nabla \times \mathbf{A}$. We expect from symmetry, that **A** should be a function of *r* alone. Since the problem has cylindrical symmetry, we shall use cylindrical co-ordinates. From the formula for curl in these co-ordinates, we have,

$$\frac{\mu_0 I}{2\pi r} = \frac{-d A_z}{dr}$$

$$\therefore \quad A_z = \frac{-\mu_0 I}{2\pi r} \ln r + c_1$$

c_1 is a constant of integration.

Note: The vector potential for a linear current distribution, in analogy with (11.6.18) is given by

$$\mathbf{A} = \frac{\mu_0}{4\pi}\int \frac{\mathbf{I}}{r}\,dl \qquad (11.6.19)$$

but cannot be used for a current distribution that extends to infinity, like this one.

Example 4: For surface currents, the vector potential is given by $A = \frac{\mu_0}{4\pi}\int \frac{\mathbf{K}\,da}{r_1}$ where **K** is the surface current density. Consider a spherical shell of radius R carrying a uniform surface charge σ. It is spinning at an angular speed ω. Find the vector potential at a point P (Fig. 11.10a).

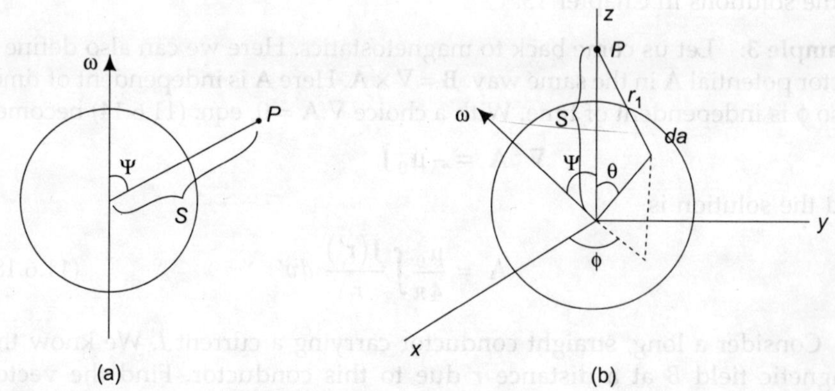

Fig. 11.10

Solution: Let us choose the point P to lie on the z axis. ω is inclined at an angle ψ with respect to the position vector **s** of P (Fig. 11.10b). ω is taken to lie in the XZ plane. We will use the formula

$$\mathbf{A} = \frac{\mu_0}{4\pi}\int \frac{\mathbf{K}}{r_1}\,da$$

Here $\mathbf{K} = \sigma\mathbf{v}$, $r_1 = \sqrt{(R^2 + s^2 - 2sR\cos\theta)}$, $da = R^2 \sin\theta\, d\theta\, d\phi$,

$$\mathbf{v} = \boldsymbol{\omega}\times\mathbf{r} = \begin{vmatrix} \hat{\mathbf{i}} & \hat{\mathbf{j}} & \hat{\mathbf{k}} \\ \omega\sin\psi & 0 & \omega\cos\psi \\ R\sin\theta\cos\phi & R\sin\theta\sin\phi & R\cos\theta \end{vmatrix}$$

$= R\omega[-\cos\psi\sin\theta\sin\phi\,\hat{\mathbf{i}} + (\cos\psi\sin\theta\cos\phi - \sin\psi\cos\theta)\,\hat{\mathbf{j}} + \sin\psi\sin\theta\sin\phi\,\hat{\mathbf{k}}]$. When we integrate over φ, the terms involving cos φ or sin φ will give zero because

$$\int_0^{2\pi} \sin\phi \, d\phi = 0, \int_0^{2\pi} \cos\phi \, d\phi = 0$$

$$\therefore \qquad \mathbf{A} = \frac{-\mu_0 R^3 \omega\sigma}{4\pi} \hat{\mathbf{j}} \sin\psi \int_0^{\pi} \frac{\cos\theta \sin\theta \, d\theta}{\sqrt{R^2 + s^2 - 2Rs\cos\theta}} \int_0^{2\pi} d\phi$$

$$= \frac{-\mu_0 R^3 \omega\sigma \sin\psi}{2} \hat{\mathbf{j}} \int_0^{\pi} \frac{\cos\theta \sin\theta \, d\theta}{\sqrt{R^2 + s^2 - 2Rs\cos\theta}}$$

To do the integration, let $u = \cos\theta$

$$\int_0^{\pi} \frac{\cos\theta \sin\theta \, d\theta}{\sqrt{R^2 + s^2 - 2Rs\cos\theta}} = \int_{-1}^{+1} \frac{u\,du}{\sqrt{R^2 + s^2 - 2sRu}}$$

$$= \frac{-\left(R^2 + s^2 - Rsu\right)}{3R^2 s^2} \sqrt{R^2 + s^2 - 2Rsu}\,\Big|_{-1}^{+1}$$

$$= \frac{-1}{3R^2 s^2} [(R^2 + s^2 + Rs)\ |R - s| - (R^2 + s^2 - Rs)(R + s)]$$

If the point P lies inside the sphere, then $R > s$ and this expression reduces to $2s/3R^2$. If P lies outside the sphere so that $R < s$, it reduces to $2R/3s^2$. Nothing that $\omega \times s = -\omega s \sin\psi\, \hat{\mathbf{j}}$, we have, finally, for points inside the sphere

$$\mathbf{A} = \frac{-\mu_0 R^3 \omega\sigma \sin\psi}{2} \hat{\mathbf{j}} \frac{2s}{3R^2} = \frac{\mu_0 R\sigma}{3} (\omega \times \mathbf{s})$$

and for points outside the sphere,

$$\mathbf{A} = \frac{-\mu_0 R^3 \omega\sigma \sin\psi}{2} \hat{\mathbf{j}} \frac{2R}{3s^2} = \frac{\mu_0 R^4 \sigma}{3S^2} (\omega \times \mathbf{s})$$

Let us now revert to the co-ordinates of Fig. 11.10(a) where ω is along the z axis and P has co-ordinates r, θ, ϕ. So replace s by r, ψ by θ. Then, for points inside the sphere ($r \le R$)

$$\mathbf{A} = \frac{\mu_0 R\omega\sigma}{3} r \sin\theta\, \hat{\phi}$$

and for points outside the sphere ($r \ge R$)

$$\mathbf{A} = \frac{\mu_0 R^4 \omega\sigma}{3r^2} \sin\theta\, \hat{\phi}$$

Let us now rewrite the Lorentz force law using the vector and scalar potentials. From (11.6.1) and (11.6.2), we have,

$$\mathbf{F} = q\left\{-\nabla\phi - \frac{\partial \mathbf{A}}{\partial t} + \mathbf{v}\times(\nabla\times\mathbf{A})\right\} \tag{11.6.18}$$

We will now use a vector identity:

$$\nabla(\mathbf{A}_1.\mathbf{A}_2) = \mathbf{A}_1\times(\nabla\times\mathbf{A}_2) + \mathbf{A}_2\times(\nabla\times\mathbf{A}_1) + (\mathbf{A}_1.\nabla)\mathbf{A}_2 + (\mathbf{A}_2.\nabla)\mathbf{A}_1$$

Take $\mathbf{A}_1 = \mathbf{v}$ and $\mathbf{A}_2 = \mathbf{A}$. $\mathbf{v}$ is only a function of time. $\therefore\ \nabla\times\mathbf{v} = 0$. Also $(\mathbf{A}.\nabla)\mathbf{v} = 0$.

$$\therefore \qquad \nabla(\mathbf{v}.\mathbf{A}) = \mathbf{v}\times(\nabla\times\mathbf{A}) + (\mathbf{v}.\nabla)\mathbf{A}$$

$$\therefore \qquad \mathbf{v}\times(\nabla\times\mathbf{A}) = \nabla(\mathbf{v}.\mathbf{A}) - (\mathbf{v}.\nabla)\mathbf{A}$$

Substituting this in (11.6.18), we get,

$$\mathbf{F} = q\left\{-\nabla\phi - \frac{\partial \mathbf{A}}{\partial t} + \nabla(\mathbf{v}.\mathbf{A}) - (\mathbf{v}.\nabla)\mathbf{A}\right\} \tag{11.6.19}$$

We define the *convective derivative* of **A** as

$$\frac{d\mathbf{A}}{dt} = \frac{\partial \mathbf{A}}{\partial t} + (\mathbf{v}.\nabla)\mathbf{A} \tag{11.6.20}$$

It represents the time rate of change of **A** at the location of the moving particle. Let the position vector of the particle at time t be $\mathbf{r}$, where the potential is $\mathbf{A}(\mathbf{r}, t)$. At time $t + dt$, the position vector of the particle is $\mathbf{r} + \mathbf{v}dt$ and the potential there is $\mathbf{A}(\mathbf{r} + \mathbf{v}dt, t + dt)$. The change in **A** is,

$$d\mathbf{A} = \mathbf{A}(\mathbf{r} + \mathbf{v}dt, t + dt) - \mathbf{A}(\mathbf{r}, t)$$

$$= \left(\frac{\partial \mathbf{A}}{\partial x}\right)v_x dt + \left(\frac{\partial \mathbf{A}}{\partial y}\right)v_y dt + \left(\frac{\partial \mathbf{A}}{\partial z}\right)v_z dt + \frac{\partial \mathbf{A}}{\partial t}dt$$

$$\therefore \qquad \frac{d\mathbf{A}}{dt} = (\mathbf{v}.\nabla)\mathbf{A} + \frac{\partial \mathbf{A}}{\partial t}$$

Using this definition, we can rewrite (11.6.19) as

$$\mathbf{F} = q\left\{-\nabla\phi + \nabla(\mathbf{v}.\mathbf{A}) - \frac{d\mathbf{A}}{dt}\right\}$$

Since $\mathbf{F} = \dfrac{d\mathbf{p}}{dt}$ where **p** is the linear momentum, we have,

$$\frac{d\mathbf{p}}{dt} = \frac{-q\,d\mathbf{A}}{dt} - q\nabla(\phi - \mathbf{v}.\mathbf{A})$$

or, $$\frac{d}{dt}(\mathbf{p} + q\mathbf{A}) = -\nabla\{q(\phi - \mathbf{v}.\mathbf{A})\}$$

This equation is of the form

$$\frac{d\,\mathbf{p}_{can}}{dt} = -\nabla V$$

where $\mathbf{p}_{can}$ is the canonical momentum, given in this case, by

$$\mathbf{p}_{can} = \mathbf{p} + q\mathbf{A} \tag{11.6.21}$$

and V is the velocity-dependent quantity

$$V = q(\phi - \mathbf{v}.\mathbf{A})$$

11.7 ENERGY IN MAGNETIC FIELDS

In section 11.2, we had calculated the energy stored in an inductor when it carries a current i. Now we shall use that result to derive the energy stored in a magnetic field. We will assume that the magnetic field is set up fairly slowly so that the equations of magnetostatics can be used. The flux through a surface S (in a magnetic field) bound by a curve C is

$$\phi = \int_S \mathbf{B}.d\mathbf{a} = \int_S (\nabla \times \mathbf{A}).d\mathbf{a} = \oint_C \mathbf{A}.d\mathbf{l}$$

using the vector potential and Stokes' theorem. Now, using the definition (11.2.1) of self inductance, we have,

$$Li = \oint_C \mathbf{A}.d\mathbf{l}$$

From (11.2.12), energy U stored in the magnetic field is

$$U = \frac{1}{2}Li^2 = \frac{1}{2}\oint_C \mathbf{A}.i\,d\mathbf{l}$$

$$= \frac{1}{2}\oint_C \mathbf{A}.i\,d\mathbf{l}$$

transferring the vector sign to the current i since $d\mathbf{l}$ is taken along the direction of i. For a volume current distribution, the above expression is generalized to

$$U = \frac{1}{2}\int_V \mathbf{A}.\mathbf{J}\,dv \tag{11.7.1}$$

We will now use Ampere's law in magnetostatics $\nabla \times \mathbf{H} = \mathbf{J}$.

$$U = \frac{1}{2}\int_V \mathbf{A}.(\nabla \times \mathbf{H})\,dv$$

There is a vector identity

$$\nabla.(\mathbf{A}\times\mathbf{H}) = \mathbf{H}.(\nabla\times\mathbf{A}) - \mathbf{A}.(\nabla\times\mathbf{H})$$
$$= \mathbf{H}.\mathbf{B} - \mathbf{A}.(\nabla\times\mathbf{H})$$

$$\therefore \quad U = \frac{1}{2}\int_V \mathbf{H}.\mathbf{B}\,dv - \frac{1}{2}\int_V \nabla.(\mathbf{A}\times\mathbf{H})\,dv$$

$$= \frac{1}{2}\int_V \mathbf{H}.\mathbf{B}\,dv - \frac{1}{2}\int_S (\mathbf{A}\times\mathbf{H})\,d\mathbf{a} \qquad (11.7.2)$$

Now we can extend the surface S far enough from the current distribution. The current distribution occupies a finite region of space. If r is the distance from an origin near the middle of the current distribution to the surface S, $H \sim 1/r^2$, $A \sim 1/r$, $da \sim r^2$. $\therefore$ The surface integral falls off as $1/r$. As S is moved to infinity, this contribution vanishes. $\therefore$ In this limit,

$$U = \frac{1}{2}\int_{\text{all space}} \mathbf{H}.\mathbf{B}\,dv \qquad (11.7.3)$$

For, linear magnetic materials, $\mathbf{B} = \mu\mathbf{H}$ and we have

$$U = \frac{1}{2\mu}\int_{\text{all space}} B^2\,dv \qquad (11.7.4)$$

The table below summarizes some of the important results of this chapter in *SI* and Gaussian units.

Physical quantity/equation	*SI*	*Gaussian*
Faraday's law (11.1.3)	$\nabla\times\mathbf{E} = \frac{-\partial\mathbf{B}}{\partial t}$	$\nabla\times\mathbf{E} = -\frac{1}{c}\frac{\partial\mathbf{B}}{\partial t}$
(11.3.1)	$\nabla\times\mathbf{B} = \mu_0\mathbf{J} + \mu_0\varepsilon_0\frac{\partial\mathbf{E}}{\partial t}$	$\nabla\times\mathbf{B} = \frac{4\pi}{c}\mathbf{J} + \frac{1}{c}\frac{\partial\mathbf{E}}{\partial t}$
(11.4.2)	$\nabla\times\mathbf{H} = \mathbf{J} + \frac{\partial\mathbf{D}}{\partial t}$	$\nabla\times\mathbf{H} = \frac{4\pi}{c}\mathbf{J} + \frac{1}{c}\frac{\partial\mathbf{E}}{\partial t}$
Lorentz gauge	$\nabla.\mathbf{A} + \mu_0\varepsilon_0\frac{\partial\phi}{\partial t} = 0$	$\nabla.\mathbf{A} + \frac{1}{c}\frac{\partial\phi}{\partial t} = 0$

Tutorial 1

A small loop of wire (radius a) lies a distance z above the centre of a large loop (radius b) as shown in the figure below. The planes of the loops are parallel, and perpendicular to a common axis. Suppose current I flows in the big loop. Find the flux through the little loop. (The little loop is so small that you may

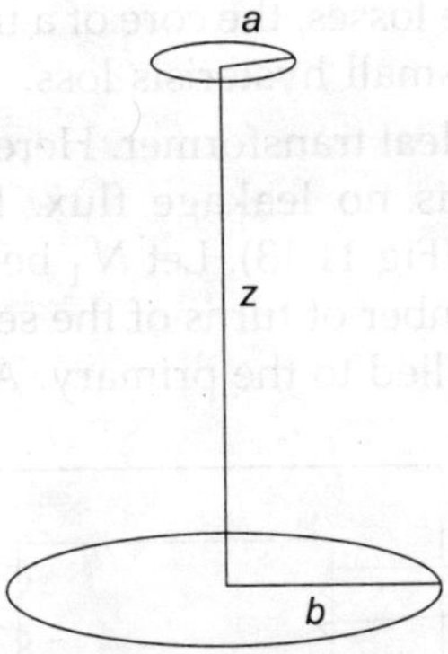

Fig. 11.11

consider the field of the big loop to be essentially constant). Calculate the mutual inductance between the loops.

Tutorial 2

In this tutorial, we shall study the transformer. A transformer consists of two pairs of coils wound on the same iron core (Fig. 11.12). An alternating voltage is applied to one winding, called the primary and a load is connected to the other winding, called the secondary. A transformer is used for providing higher voltages at smaller currents or lower voltages at larger currents to a load. An alternating flux is set up in the core due to the alternating voltage applied to the primary. This flux links with the secondary winding. In practice, a small amount of flux passes through the surrounding air instead of the core. The flux which links the primary only is called the primary leakage flux. The flux which links the secondary only is called the secondary leakage flux. There will also be some heating in the windings (I^2R losses). These are called copper losses. In addition, there may be heating of the core due to hysterisis and eddy current circulation in the core. These are called

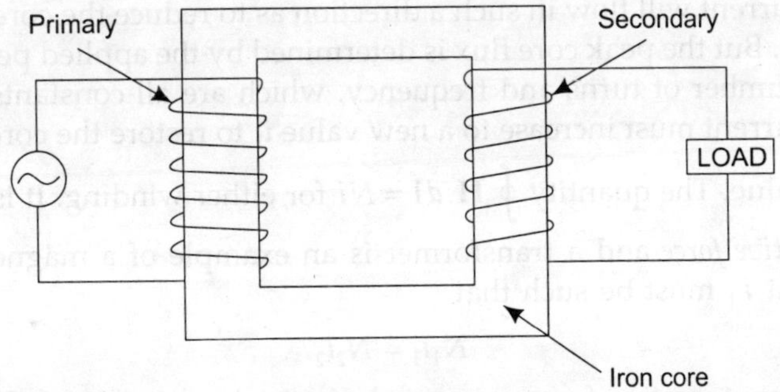

Fig. 11.12

core losses. To reduce these losses, the core of a transformer is laminated and made of materials with a small hysterisis loss.

We shall consider an ideal transformer. Here it is assumed that all losses are negligible and there is no leakage flux. First we shall consider an open-circuited secondary (Fig 11.13). Let N_1 be the number of turns of the primary and N_2 be the number of turns of the secondary winding. Let v_1 be the alternating voltage applied to the primary. A small magnetizing current

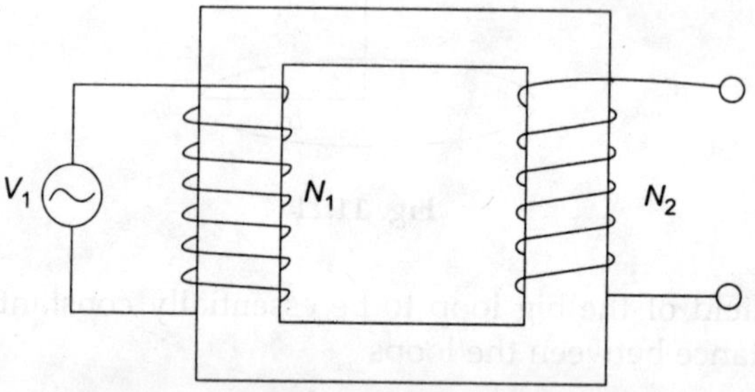

Fig. 11.13

i_φ flows through the primary. So if r_1 is the resistance of the primary, and e_1 be the back emf in the primary, then

$$V_1 = r_1 i_\varphi + e_1 \tag{1}$$

where
$$e_1 = -\frac{N_1 d\phi}{dt} \tag{2}$$

In the ideal case, $|v_1| \cong |e_1|$, because both r_1 and i_φ are extremely small. Now suppose a load is connected to the secondary (Fig. 11.12). Then a current i_2 flows in the secondary and the emf induced in the secondary is

$$e_2 = -\frac{N_2 d\phi}{dt} \tag{3}$$

This current will flow in such a direction as to reduce the core flux, from Lenz's law. But the peak core flux is determined by the applied peak voltage, primary number of turns and frequency, which are all constants. Thus the primary current must increase to a new value i_1 to restore the core flux to its no-load value. The quantity $\oint \mathbf{H}.d\mathbf{l} = Ni$ for either winding. It is called the *magnetomotive force* and a transformer is an example of a magnetic circuit. The current i_1 must be such that

$$N_1 i_1 = N_2 i_2 \tag{4}$$

i.e. the magnetomotive forces are equal. If we take the ratio of (3) to (2), we get,

$$\frac{e_2}{e_1} = \frac{N_2}{N_1} = \frac{v_2}{v_1} \tag{5}$$

where $|v_2| \equiv |e_2|$ is the terminal voltage of the secondary. Thus a transformer transforms voltages in direct ratio as the number of turns and currents in inverse ratio. If $N_2 > N_1$, $e_2 > e_1$ and we have a step-up transformer. If $N_2 < N_1$, $e_2 < e_1$ and we have a step-down transformer. Also, for an ideal transformer, combining (4) and (5), we have,

$$v_1 i_1 = v_2 i_2$$

i.e. input power = output power, as we have neglected all losses. In practice output power is less than input power.

Taking $\phi = \phi_m \sin \omega t$, show that

$$\phi_m = \frac{V_1}{4.44\, f N_1}$$

where f is the frequency of the alternating supply voltage and V_1 is its rms value.

Problems

1. A copper rod of length L rotates at angular speed ω in a plane perpendicular to a uniform magnetic field B. Find the emf developed between the two ends of the rod. $\left[\textbf{Ans. } \varepsilon = \frac{B\omega L^2}{2}\right]$

2. A long coaxial cable carries a current I. The current flows down the surface of the inner cylinder of radius 'a' and back along the outer cylinder of radius 'b'. Find the energy stored in a section l. Hence find

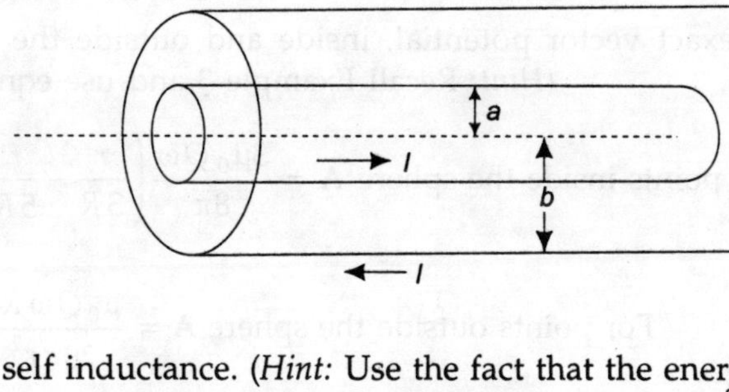

the self inductance. (*Hint:* Use the fact that the energy density of the magnetic field in given, according to (11.7.4), by $B^2/2\mu_0$.)

$$\left[\textbf{Ans. } U = \frac{\mu_0 I^2 l}{4\pi}\ln\left(\frac{b}{a}\right), L = \frac{\mu_0 l}{2\pi}\ln\left(\frac{b}{a}\right)\right]$$

3. Given two circuits: a very long straight wire and a rectangle of dimensions h and d. The rectangle lies in a plane through the wire, the sides of length h being parallel to the wire and at distances r and $r + d$ from it. Calculate the mutual inductance between the two circuits.

$$\left[\textbf{Ans. } M = \frac{\mu_0 h}{2\pi} \ln\left(1 + \frac{d}{r}\right)\right]$$

4. Show that the following are all possible vector potentials of the uniform field $\mathbf{B} = B\hat{\mathbf{k}}$: $\mathbf{A}_1 = -By\hat{\mathbf{i}}$, $\mathbf{A}_2 = Bx\hat{\mathbf{j}}$, $\mathbf{A}_3 = -\frac{1}{2}\,\mathbf{r} \times \mathbf{B}$.

5. Find the magnetic vector potential of a finite segment of straight wire carrying a current I. By calculating $\nabla \times \mathbf{B}$, show that is leads to the magnetic field due to a finite element of length of a wire.

(*Hint:* Use cylindrical coordinates)

$$\left[\textbf{Ans. } \mathbf{A} = \frac{\mu_0 I}{4\pi}\hat{\mathbf{k}}\ \{\ln|z + \sqrt{r^2 + z^2}| - \ln|z - z_1 + \sqrt{r^2 + (z - z_1)^2}|\right.$$

It is assumed that the wire extends from $z = 0$ to $z = z_1$ along the z axis. The current flows in the direction of positive z. The point P has coordinates (r, ϕ, z)]

6. A uniformly charged solid sphere of radius R carries a total charge Q and is set spinning with angular speed ω about the z axis. What is the magnetic dipole moment of the sphere? The vector potential due to a dipole of moment $\mathbf{m}$ at a distance r is given by $\mathbf{A} = \frac{\mu_0}{4\pi}\frac{\mathbf{m} \times \hat{\mathbf{r}}}{r^2}$. Find the approximate vector potential due to the sphere at large distance $r >> R$ from the sphere. $\left[\textbf{Ans. } \mathbf{m} = \frac{Q\omega R^2}{5}\hat{\mathbf{k}},\ \mathbf{A} = \frac{\mu_0 Q\omega \mathbf{R}^2 \sin\theta}{20\pi r^2}\hat{\phi}\right]$

7. Find the exact vector potential, inside and outside the sphere, in problem 6. (*Hint:* Recall Example 3 and use eqn. (11.6.18))

$$\left[\textbf{Ans. } \text{For points inside the sphere } \mathbf{A} = \frac{3\mu_0 Q\omega}{8\pi}\left(\frac{r}{3R} - \frac{r^3}{5R^3}\right)\sin\theta\,\hat{\phi}\right.$$

$$\left.\text{For points outside the sphere } \mathbf{A} = \frac{\mu_0 Q\omega R^2}{20\pi r^2}\sin\theta\,\hat{\phi}\right]$$

8. Suppose potentials $(\mathbf{A}, \phi)$ satisfy the Lorentz gauge condition (11.6.15). Consider a gauge transformation (11.6.10), (11.6.11) to new potentials $(\mathbf{A}', \phi')$. What equation must f satisfy if $(\mathbf{A}', \phi')$ is to satisfy the Lorentz gauge condition also? $\left[\textbf{Ans.} = \nabla^2 f - \mu_0\varepsilon_0\frac{\partial^2 f}{\partial t^2} = 0\right]$

9. Calculate the magnetic field inside and outside the spherical shell of Example 3.

$$\left[\textbf{Ans.} \text{ Inside, } \mathbf{B} = \frac{2}{3}\mu_0 \sigma R \omega \hat{\mathbf{k}} \text{ Outside, } \mathbf{B} = \frac{\mu_0 \sigma R^4 \omega}{3r^3} \right.$$

$$\left. (2 \cos \theta\, \hat{\mathbf{r}} + \sin \theta\, \hat{\theta}) \right]$$

12

Electromagnetic Waves

12.1 PLANE WAVE SOLUTIONS OF MAXWELL'S EQUATIONS

In this chapter, we shall study plane wave solutions of Maxwell's equations. Let us recall Maxwell's equations in vacuum.

$$\nabla . \mathbf{E} = 0 \tag{11.4.1}$$

$$\nabla . \mathbf{B} = 0 \tag{5.5.1}$$

$$\nabla \times \mathbf{E} = \frac{-\partial \mathbf{B}}{\partial t} \tag{11.1.3}$$

$$\nabla \times \mathbf{B} = \mu_0 \varepsilon_0 \frac{\partial \mathbf{E}}{\partial t} \tag{11.4.2}$$

Take the curl of eqn. (11.1.3). We get

$$\nabla \times (\nabla \times \mathbf{E}) = -\nabla \times \left(\frac{\partial \mathbf{B}}{\partial t} \right)$$

or, using a vector identity,

$$\nabla (\nabla . \mathbf{E}) - \nabla^2 \mathbf{E} = \frac{-\partial}{\partial t} (\nabla \times \mathbf{B})$$

Now use (11.4.1) and (11.4.2):

$$\nabla^2 \mathbf{E} = \mu_0 \varepsilon_0 \frac{\partial^2 \mathbf{E}}{\partial t^2} \tag{12.1.1}$$

The components of the electric field thus satisfy the wave equation. The speed of the wave c is given by

$$c = \frac{1}{\sqrt{\mu_0 \varepsilon_0}} \tag{12.1.2}$$

This value of c turns out to be the measured speed of light.[⊕] Take the curl of eqn. (11.4.2). We get

$$\nabla(\nabla.\mathbf{B}) - \nabla^2\mathbf{B} = \mu_0\varepsilon_0\nabla \times \left(\frac{\partial \mathbf{E}}{\partial t}\right)$$

Using (5.5.1), we get

$$-\nabla^2\mathbf{B} = \mu_0\varepsilon_0\frac{\partial}{\partial t}(\nabla \times \mathbf{E}) = -\mu_0\varepsilon_0\frac{\partial^2\mathbf{B}}{\partial t^2}$$

$$\therefore \quad \nabla^2\mathbf{B} = \mu_0\varepsilon_0\frac{\partial^2\mathbf{B}}{\partial t^2} \tag{12.1.3}$$

∴ The components of the magnetic field also satisfy the same equation. Let us take

$$\mathbf{E} = \tilde{\mathbf{E}}_0 e^{i(\mathbf{k}.\mathbf{r}-\omega t)} \tag{12.1.4}$$

$$\mathbf{B} = \tilde{\mathbf{B}}_0 e^{i(\mathbf{k}.\mathbf{r}-\omega t)} \tag{12.1.5}$$

If we substitute this in the wave equation, we get $k^2 = \omega^2/c^2$. Here $\tilde{\mathbf{E}}_0$ and $\tilde{\mathbf{B}}_0$ are complex vector amplitudes of the electric and magnetic fields. Eqns. (12.1.4) and (12.1.5) represent plane waves since the phase is the same, at a given instant, in each plane perpendicular to some specified direction. This direction is the direction of **k**. We can write, in general,

$$\tilde{\mathbf{E}}_0 = \tilde{E}_{0x}\hat{\mathbf{i}} + \tilde{E}_{0y}\hat{\mathbf{j}} + \tilde{E}_{0z}\hat{\mathbf{k}}$$

Where $\tilde{E}_{0x} = E_{0x}e^{i\phi_1}$, etc. and E_{0x} is real, ϕ_1 being the phase. The physical electric field along the x axis would then be the real part of (12.1.4)

$$Re[E_{0x}e^{i(\mathbf{k}.\mathbf{r}-\omega t+\phi_1)}]$$
$$= E_{0x}\cos(\mathbf{k}.\mathbf{r} - \omega t + \phi_1)$$

We can treat the magnetic field similarly. The vector **k** is called the wavevector or the propagation vector.

Maxwell's equations impose some conditions on the directions of **E**, **B** and **k**. If we substitute (12.1.4) into (11.4.1), we get,

$$\mathbf{k}.\tilde{\mathbf{E}}_0 = 0 \tag{12.1.6}$$

This implies that the electric field is perpendicular to **k**. Similarly, substituting (12.1.5) into (5.5.1), we get,

$$\mathbf{k}.\tilde{\mathbf{B}}_0 = 0 \tag{12.1.7}$$

⊕The current value is: C = 2.99792457×10^8 m/s.[13]

The magnetic field is also perpendicular to **k**. ∴ Electromagnetic waves are transverse in nature. Substituting (12.1.4) and (12.1.5) into (11.1.3), we get,

$$\mathbf{k} \times \tilde{\mathbf{E}}_0 = \omega \tilde{\mathbf{B}}_0$$

or,

$$\tilde{\mathbf{B}}_0 = \frac{\mathbf{k}}{\omega} \times \tilde{\mathbf{E}}_0 \tag{12.1.8}$$

If we take the dot product of both sides with $\mathbf{E}_0$, we find $\tilde{\mathbf{E}}_0 . \tilde{\mathbf{B}}_0 = 0$. Thus the magnetic field is also perpendicular to the electric field. The electric field, magnetic field, and the propagation vector form an orthogonal, right handed system.

Let us choose the vector **k** to be along the z axis. ∴ $\mathbf{k} = k\hat{\mathbf{k}}$. Then $\tilde{\mathbf{E}}_0 = \tilde{E}_0 \hat{\mathbf{i}}$ and $\tilde{\mathbf{B}}_0 = \tilde{\mathbf{B}}_0 \hat{\mathbf{j}}$ is a possible choice.

$$\mathbf{E} = \tilde{E}_0 \hat{\mathbf{i}}\, e^{i(kz - \omega t)} \tag{12.1.9}$$

$$\mathbf{B} = \tilde{B}_0 \hat{\mathbf{j}}\, e^{i(kz - \omega t)} \tag{12.1.10}$$

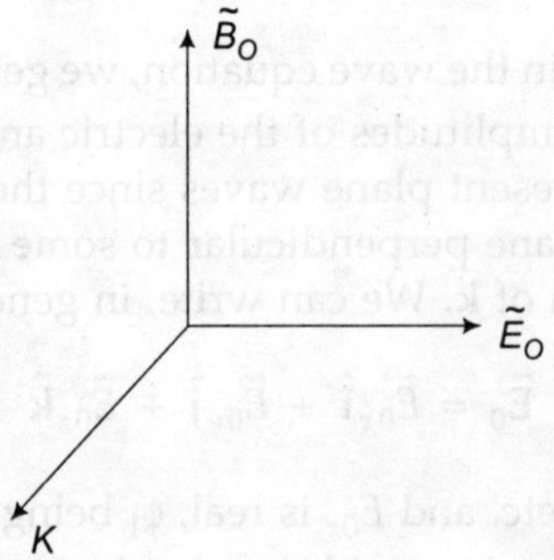

Fig. 12.1

Let us take $\tilde{E}_0 = E_0 e^{i\phi}$ where E_0 is real.

∴ $$Re(\mathbf{E}) = E_0 \cos(kz - \omega t + \phi)\,\hat{\mathbf{i}}$$

The electric field vector vibrates along the x axis with amplitude E_0. The magnetic field vector vibrates along the y axis with amplitude B_0. From eqn. (12.1.8)

$$\tilde{B}_0 = \frac{k}{\omega} \tilde{E}_0 = \frac{\tilde{E}_0}{c} \tag{12.1.11}$$

By convention, the direction of vibration of the electric field vector is taken as the direction of polarization and (12.1.9) is said to represent a linearly polarized wave. In general, with $\mathbf{k} = k\hat{\mathbf{k}}$, we could choose $\tilde{\mathbf{E}}_0$ arbitrarily

$$\tilde{\mathbf{E}}_0 = \tilde{E}_{0x}\hat{\mathbf{i}} + \tilde{E}_{0y}\hat{\mathbf{j}} \tag{12.1.12}$$

and
$$\tilde{\mathbf{B}}_0 = \tilde{B}_{0x}\hat{\mathbf{i}} + \tilde{B}_{0y}\hat{\mathbf{j}} \tag{12.1.13}$$

Then (12.1.8) gives

$$\tilde{B}_{0x} = \frac{-\tilde{E}_{0y}}{c} \quad \tilde{B}_{0y} = \frac{\tilde{E}_{0x}}{c} \tag{12.1.14}$$

Taking $\tilde{E}_{0x} = \tilde{E}_{0x}e^{i\phi}$ and $\tilde{E}_{0y} = E_{0y}$ where E_{0x} and E_{0y} are real, we have

$$Re(\mathbf{E}) = E_{0x}\cos(kz - \omega t + \phi)\,\hat{\mathbf{i}} + E_{0y}\cos(kz - \omega t)\,\hat{\mathbf{j}}$$

For $\phi = 0$, we have

$$Re(\mathbf{E}) = E_{0x}\cos(kz - \omega t)\,\hat{\mathbf{i}} + E_{0y}\cos(kz - \omega t)\,\hat{\mathbf{j}}$$

This represents a linearly polarized wave of amplitude $\sqrt{(E_{0x}{}^2 + E_{0y}{}^2)}$ making an angle of $\tan^{-1}\left(\frac{E_{0y}}{E_{0x}}\right)$ with the x axis.

For $\phi = \pi/2$.

$$Re(\mathbf{E}) = -E_{0x}\sin(kz - \omega t)\,\hat{\mathbf{i}} + E_{0y}\cos(kz - \omega t)\,\hat{\mathbf{j}}$$

The tip of the electric field vector traces out an ellipse. The wave is said to be elliptically polarized. The direction of rotation is clockwise as the wave advances towards the reader along the z axis (Fig. 12.2). This is a case of right handed elliptical polarization. If the direction of rotation is anticlockwise, the wave is left handed elliptically polarized.

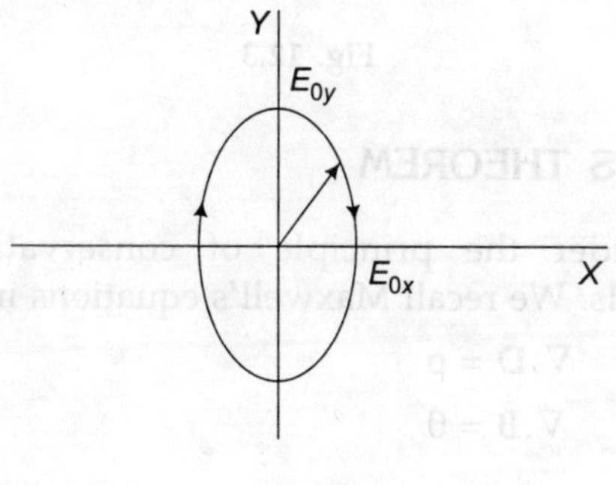

Fig. 12.2

Example 1: For $\phi = \pi/2$, what happens if $E_{0x} = E_{0y}$? Explain with a diagram

Solution: For $\phi = \pi/2$ and $E_{0x} = E_{0y} = E_0$, say, this is a case of circular polarization. We have

$$Re(\mathbf{E}) = -E_0 \sin(kz - \omega t)\,\hat{\mathbf{i}} + E_0 \cos(kz - \omega t)\,\hat{\mathbf{j}}$$

The tip of the electric vector traces out a circle of radius E_0. To see this, consider snapshots of the electric field vector at $z = 0$, at different times (Fig. 12.3 a, b, c, d.

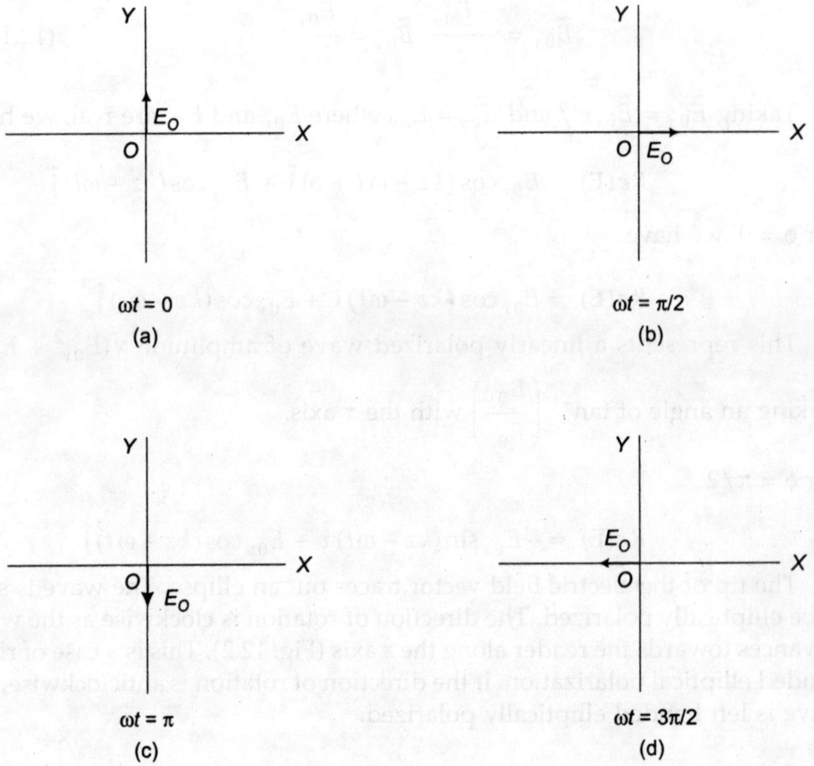

Fig. 12.3

12.2 POYNTING'S THEOREM

Let us now consider the principle of conservation of energy for electromagnetic fields. We recall Maxwell's equations inside matter.

$$\nabla . \mathbf{D} = \rho \tag{4.3.6}$$

$$\nabla . \mathbf{B} = 0 \tag{5.5.1}$$

$$\nabla \times \mathbf{E} = \frac{-\partial \mathbf{B}}{\partial t} \tag{11.1.3}$$

$$\nabla \times \mathbf{H} = \mathbf{J}_f + \frac{\partial \mathbf{D}}{\partial t} \tag{11.4.3}$$

Taking the dot product of (11.1.3) with **H**, we get

$$\mathbf{H}.(\nabla \times \mathbf{E}) = -\frac{\mathbf{H}.\partial \mathbf{B}}{\partial t} \tag{12.2.1}$$

and the dot product of (11.4.3) with **E** gives

$$\mathbf{E}.(\nabla \times \mathbf{H}) = \mathbf{E}.\mathbf{J}_f + \frac{\mathbf{E}.\partial \mathbf{D}}{\partial t} \tag{12.2.2}$$

Subtracting (12.2.2) from (12.2.1), we have,

$$\mathbf{H}.(\nabla \times \mathbf{E}) - \mathbf{E}.(\nabla \times \mathbf{H}) = -\mathbf{E}.\mathbf{J}_f - \frac{\mathbf{H}.\partial \mathbf{B}}{\partial t} - \frac{\mathbf{E}.\partial \mathbf{D}}{\partial t}$$

The left hand side can be rewritten using a vector identity. Let us now assume the medium to be linear. Thus we have the constitutive equations $\mathbf{D} = \varepsilon \mathbf{E}$, $\mathbf{B} = \mu \mathbf{H}$.

$$\therefore \quad \frac{\mathbf{H}.\partial \mathbf{B}}{\partial t} = \frac{\mathbf{H}.\partial(\mu \mathbf{H})}{\partial t} = \frac{\mu \partial\left(\mathbf{H}^2\right)}{2\,\partial t} = \frac{\partial(\mathbf{H}.\mathbf{B})}{2\,\partial t}$$

$$\frac{\mathbf{E}.\partial \mathbf{D}}{\partial t} = \frac{\mathbf{E}.\partial(\varepsilon \mathbf{E})}{\partial t} = \frac{\varepsilon \partial\left(\mathbf{E}^2\right)}{2\,\partial t} = \frac{\partial(\mathbf{E}.\mathbf{D})}{2\,\partial t}$$

∴ The above equations can be rewritten as

$$\nabla.(\mathbf{E} \times \mathbf{H}) = -\mathbf{E}.\mathbf{J}_f - \frac{1}{2}\frac{\partial}{\partial t}(\mathbf{H}.\mathbf{B} + \mathbf{E}.\mathbf{D}) \tag{12.2.3}$$

Integrate this over a volume V, containing the fields and matter.

$$\int_V \nabla.(\mathbf{E} \times \mathbf{H})\, dv = -\int_V \mathbf{E}.\mathbf{J}_f\, dv - \frac{d}{dt}\int_V \frac{1}{2}(\mathbf{H}.\mathbf{B} + \mathbf{E}.\mathbf{D})\, dv$$

or, using Stokes' theorem,

$$\int_S (\mathbf{E} \times \mathbf{H}).d\mathbf{a} = -\int_V \mathbf{E}.\mathbf{J}_f\, dv - \frac{d}{dt}\int_V \frac{1}{2}(\mathbf{H}.\mathbf{B} + \mathbf{E}.\mathbf{D})\, dv$$

or,

$$-\int_V \mathbf{E}.\mathbf{J}_f\, d\mathbf{v} = \frac{d}{dt}\int_V \frac{1}{2}(\mathbf{H}.\mathbf{B} + \mathbf{E}.\mathbf{D})\, dv + \int_V (\mathbf{E} \times \mathbf{H}).d\mathbf{a} \tag{12.2.4}$$

Consider a charge q moving with a velocity **v**. The work done on the charge by the electric field **E** in the dt is $q\mathbf{E}.\mathbf{v}dt$. The magnetic field does not do any work. ∴ The rate at which work is done by the electric field is $q\mathbf{E}.\mathbf{v}$. A continuous volume charge distribution is characterized by a charge density ρ. An elementary volume dv then contains charge $\rho\, dv$. If this charge is moving with a velocity **v**, the current density $\mathbf{J} = \rho\mathbf{v}$ and the work done per unit time is $\rho\mathbf{v}.\mathbf{E}dv = \mathbf{J}.\mathbf{E}dv$. ∴ The rate at which work at done by the fields on the charged particles within volume V is $\int_V \mathbf{E}.\mathbf{J}dv$. This represents a conversion of electromagnetic energy into mechanical energy or thermal energy. The left hand side of eqn. (12.2.4) is the negative of this. ∴ It is the energy transferred to the electromagnetic field. The first term on the right hand side is the rate of change of energy stored in the electric and magnetic fields. The second term is the rate at which energy is flowing out through the surface S bounding the volume V. Eqn. (12.2.4) is Poynting's theorem. It says that the negative of the rate at which work is done by the electromagnetic fields on the charged particles in a volume V is equal to the sum of the rate of change of energy of the system of electromagnetic fields in the volume and the rate at which energy flows out of the volume. The Poynting vector **S** is defined as

$$\mathbf{S} = \mathbf{E} \times \mathbf{H} \tag{12.2.5}$$

It is the energy flowing per unit area per second. Let $u = \frac{1}{2}(\mathbf{H}.\mathbf{B} + \mathbf{E}.\mathbf{D})$.

Then eqn. (12.2.3) can be rewritten as

$$\nabla.\mathbf{S} + \frac{\partial u}{\partial t} = -\mathbf{E}.\mathbf{J}_f \tag{12.2.6}$$

The term on the right hand side can be written as $\frac{-\partial u_{\text{particle}}}{\partial t}$ and so we have

$$\nabla.\mathbf{S} + \frac{\partial}{\partial t}(u + u_{\text{particle}}) = 0 \tag{12.2.6a}$$

This equation is similar to the continuity equation (7.1.2). Here the energy density $u + u_{\text{particle}}$ plays the role of charge density and the Poynting vector plays the role of the current density. This is an energy conservation principle for the system of particles and fields.

Let us calculate the energy density and the Poynting vector for plane electromagnetic waves in vacuum. We shall use the real parts of the complex fields. For the linearly polarized wave (12.2.12), (12.2.13), with $\phi = 0$, the energy density is

$$u = \frac{1}{2}\left(\frac{B^2}{\mu_0} + \varepsilon_0 E^2\right)$$

$$= \frac{1}{2}\left(\frac{B_0^2}{\mu_0} + \varepsilon_0 E^2\right)\cos^2(kz - \omega t)$$

where $E_0^2 = E_{0x}^2 + E_{0y}^2$, $B_0^2 = B_{0x}^2 + B_{0y}^2 = \frac{E_{0y}^2 + E_{0x}^2}{c^2} = \frac{E_0^2}{c^2}$ using (12.1.14)

$$\therefore \quad u = \frac{1}{2}\left(\frac{E_0^2}{\mu_0 c^2} + \varepsilon_0 E_0^2\right)\cos^2(kz - \omega t)$$

$$= \frac{1}{2}\varepsilon_0 E_0^2\left(\frac{1}{\mu_0 \varepsilon_0 c^2} + 1\right)\cos^2(kz - \omega t)$$

$$\therefore \quad u = \varepsilon_0 E_0^2 \cos^2(kz - \omega t) \tag{12.2.8}$$

and the Poynting vector is

$$\mathbf{S} = \frac{E_0^2}{\mu_0 c}\cos^2(kz - \omega t)\hat{\mathbf{k}} = \varepsilon_0 c E_0^2 \cos^2(kz - \omega t)\hat{\mathbf{k}} \tag{12.2.9}$$

$$\therefore \quad \mathbf{S} = uc\hat{\mathbf{k}} \tag{12.2.9a}$$

The direction of energy flow is the same as the direction of propagation.

At high frequencies typical of light, it is the time average that is measurable. Since

$$\frac{1}{T}\int_0^T \cos^2(kz - \omega t)\,dt = \frac{1}{2}$$

the time averages of u and S are

$$\overline{u} = \frac{\varepsilon_0 c E_0^2}{T}\int_0^T \cos^2(kz - \omega t)\,dt = \frac{1}{2}\varepsilon_0 E_0^2 \tag{12.2.10}$$

$$\overline{S} = \frac{\varepsilon_0 c E_0^2}{T}\int_0^T \cos^2(kz - \omega t)\,dt = \frac{1}{2}\varepsilon_0 c E_0^2 \tag{12.2.11}$$

We can also work with complex fields. We have to use the formulae

$$\overline{u} = \frac{1}{4}\left(\varepsilon_0 \mathbf{E}.\mathbf{E}^* + \frac{1}{\mu_0}\mathbf{B}.\mathbf{B}^*\right) \tag{12.2.12}$$

where * denotes complex conjugate.

and $$\overline{\mathbf{S}} = \frac{1}{2}Re(\mathbf{E} \times \mathbf{H}^*) \tag{12.2.13}$$

Let us take the case of linear polarization considered above.

$$\mathbf{E}.\mathbf{E}^* = |\tilde{E}_0|^2 = E_{0x}^2 + E_{0y}^2 = E_0^2$$

$$\mathbf{B}.\mathbf{B}^* = |\tilde{B}_0|^2 = B_{0x}^2 + B_{0y}^2 = \left(\frac{E_{0y}^2 + E_{0x}^2}{c^2}\right) = \frac{E_0^2}{c^2}$$

$$\therefore \quad \overline{u} = \frac{1}{4}\left(\varepsilon_0 E_0^2 + \frac{E_0^2}{\mu_0 c^2}\right) = \frac{\varepsilon_0 E_0^2}{4}\left(1 + \frac{1}{\mu_0 \varepsilon_0 c^2}\right) = \frac{\varepsilon_0 E_0^2}{2}$$

$$\mathbf{E} \times \mathbf{H}^* = (E_{0x}\hat{\mathbf{i}} + E_{0y}\hat{\mathbf{j}})e^{i(\mathbf{k}z-\omega t)} \times \left(\frac{\mathbf{B}_{0x}\hat{\mathbf{i}}}{\mu_0} + \frac{\mathbf{B}_{0y}\hat{\mathbf{j}}}{\mu_0}\right)e^{-i(\mathbf{k}z-\omega t)}$$

$$= (E_{0x}\hat{\mathbf{i}} + E_{0y}\hat{\mathbf{j}}) \times \left(\frac{-\mathbf{E}_{0y}\hat{\mathbf{i}}}{\mu_0 c} + \frac{\mathbf{E}_{0x}\hat{\mathbf{j}}}{\mu_0 c}\right) \qquad \text{using (12.1.4)}$$

$$= \left(\frac{\mathbf{E}_{0x}^2 + \mathbf{E}_{0y}^2}{\mu_0 c}\right)\hat{\mathbf{k}} = \frac{\mathbf{E}_0^2}{\mu_0 c}\hat{\mathbf{k}} = \varepsilon_0 c E_0^2 \hat{\mathbf{k}}$$

$$\therefore \qquad \mathbf{S} = \frac{1}{2}\varepsilon_0 c E_0^2 \hat{\mathbf{k}}$$

12.3 ELECTROMAGNETIC MOMENTUM

The rate of change of linear momentum for a particle of charge q in a region of space where there are electric and magnetic fields, is

$$\frac{d\mathbf{p}_{\text{mechanical}}}{dt} = q\mathbf{E} + q\mathbf{v} \times \mathbf{B}$$

For a continuous charge distribution, the charge in linear momentum is

$$\frac{d\mathbf{p}_{\text{mechanical}}}{dt} = \int_V (\rho\mathbf{E} + \mathbf{J} \times \mathbf{B})dv \qquad (12.3.1)$$

We shall use the equations (1.8.11) and (11.3.1) to eliminate ρ and $\mathbf{J}$.

$$\therefore \qquad \rho = \varepsilon_0(\nabla.\mathbf{E})$$

and

$$\mathbf{J} = \frac{1}{\mu_0}(\nabla \times \mathbf{B}) - \frac{\varepsilon_0 \partial \mathbf{E}}{\partial t}$$

$\therefore$ We get

$$\frac{d\mathbf{p}_{\text{mechanical}}}{dt} = \int_V \left\{\varepsilon_0 \mathbf{E}\,(\nabla.\mathbf{E}) - \frac{1}{\mu_0}\mathbf{B} \times (\nabla \times \mathbf{B}) + \varepsilon_0 \mathbf{B} \times \frac{\partial \mathbf{E}}{\partial t}\right\}dv$$

$$= \varepsilon_0 \int_V \left\{\mathbf{E}\,(\nabla.\mathbf{E}) - c^2\mathbf{B} \times (\nabla \times \mathbf{B}) + \mathbf{B} \times \frac{\partial \mathbf{E}}{\partial t}\right\}dv$$

Now

$$\frac{\partial}{\partial t}(\mathbf{E} \times \mathbf{B}) = \frac{\partial \mathbf{E}}{\partial t} \times \mathbf{B} + \mathbf{E} \times \frac{\partial \mathbf{B}}{\partial t}$$

$$\therefore \quad \mathbf{B} \times \frac{\partial \mathbf{E}}{\partial t} = \frac{-\partial}{\partial t}(\mathbf{E} \times \mathbf{B}) + \mathbf{E} \times \frac{\partial \mathbf{B}}{\partial t}$$

$$= \frac{-\partial}{\partial t}(\mathbf{E} \times \mathbf{B}) - \mathbf{E} \times (\nabla \times \mathbf{E}) \qquad \text{using (11.1.3)}$$

Adding $c^2\mathbf{B}(\nabla.\mathbf{B}) = 0$ to the right hand side of the above, we get

$$\frac{d\mathbf{p}_{\text{mechanical}}}{dt} = \varepsilon_0 \int_V \{\mathbf{E}(\nabla.\mathbf{E}) - c^2\mathbf{B} \times (\nabla \times \mathbf{B}) - \frac{\partial}{\partial t}(\mathbf{E} \times \mathbf{B}) - \mathbf{E} \times (\nabla \times \mathbf{E})$$

$$+ c^2\mathbf{B}(\nabla.\mathbf{B})\}\, dv$$

or $$\frac{d\mathbf{p}_{\text{mechanical}}}{dt} + \frac{d}{dt}\int_V \varepsilon_0(\mathbf{E} \times \mathbf{B})\, dv$$

$$= \varepsilon_0 \int_V \{\mathbf{E}(\nabla.\mathbf{E}) - \mathbf{E} \times (\nabla \times \mathbf{E}) + c^2\mathbf{B}(\nabla.\mathbf{B}) - c^2\mathbf{B} \times (\nabla \times \mathbf{B})\}\, dv \qquad (12.3.2)$$

Let us call the volume integral on the left the total electromagnetic momentum in the volume V.[3]

$$\mathbf{P}_{\text{field}} = \varepsilon_0 \int_V (\mathbf{E} \times \mathbf{B})\, dv = \mu_0 \varepsilon_0 \int_V (\mathbf{E} \times \mathbf{H})\, dv$$

$$\therefore \quad \mathbf{P}_{\text{field}} = \frac{1}{c^2}\int_V (\mathbf{E} \times \mathbf{H})\, dv \qquad (12.3.3)$$

Let $$\mathbf{g} = \frac{1}{c^2}(\mathbf{E} \times \mathbf{H}) \qquad (12.3.4)$$

In order to identify the volume integral of **g** as the electromagnetic momentum and to establish eqn. (12.3.2) as a momentum conservation law, we have to convert the volume integral into a surface integral of something that can regarded as a momentum flow. Let us call the Cartesian co-ordinates $x = x_1$, $y = x_2$ and $z = x_3$. Then the x-component of the right hand side of (12.3.2), involving the electric field only, is

$$E_1\left(\frac{\partial E_1}{\partial x_1} + \frac{\partial E_2}{\partial x_2} + \frac{\partial E_3}{\partial x_3}\right) - E_2\left(\frac{\partial E_2}{\partial x_1} - \frac{\partial E_1}{\partial x_2}\right) + E_3\left(\frac{\partial E_1}{\partial x_3} - \frac{\partial E_3}{\partial x_1}\right)$$

$$= E_1\frac{\partial E_1}{\partial x_1} + E_1\frac{\partial E_2}{\partial x_2} + E_1\frac{\partial E_3}{\partial x_3} - E_2\frac{\partial E_2}{\partial x_1} + E_2\frac{\partial E_1}{\partial x_2} + E_3\frac{\partial E_1}{\partial x_3} - E_3\frac{\partial E_3}{\partial x_1}$$

$$= \frac{\partial}{\partial x_1}(E_1^2) - E_1\frac{\partial E_1}{\partial x_1} - E_2\frac{\partial E_2}{\partial x_1} - E_3\frac{\partial E_3}{\partial x_1} + \frac{\partial}{\partial x_2}(E_1E_2) + \frac{\partial}{\partial x_3}(E_1E_3)$$

$$= \frac{\partial}{\partial x_1}(E_1^2) - \frac{1}{2}\frac{\partial}{\partial x_1}(E_1^2) - \frac{1}{2}\frac{\partial}{\partial x_1}(E_2^2) - \frac{1}{2}\frac{\partial}{\partial x_1}(E_3^2) + \frac{\partial}{\partial x_2}(E_1E_2)$$

$$+ \frac{\partial}{\partial x_3}(E_1E_3) \qquad (12.3.5)$$

Introducing an index α ($\alpha = 1, 2$ or 3) and the Kronecker delta symbol $\delta_{\alpha\beta}$ where

$$\delta_{\alpha\beta} = 1 \text{ for } \alpha = \beta$$
$$= 0 \text{ for } \alpha \neq \beta$$

We can write the α component of the electric field part of the integrand as

$$[\mathbf{E}(\nabla.\mathbf{E}) - \mathbf{E} \times (\nabla \times \mathbf{E})]_\alpha = \sum_\beta \frac{\partial}{\partial x_\beta}\left(E_\alpha E_\beta - \frac{1}{2}\mathbf{E}.\mathbf{E}\,\delta_{\alpha\beta}\right) \quad (12.3.6)$$

This is the divergence of a second rank tensor. This tensor is called *Maxwell's stress tensor* $T_{\alpha\beta}$ and is defined as:

$$T_{\alpha\beta} = \varepsilon_0[E_\alpha E_\beta + c^2 B_\alpha B_\beta - \frac{1}{2}(\mathbf{E}.\mathbf{E} + c^2\mathbf{B}.\mathbf{B})\,\delta_{\alpha\beta}] \quad (12.3.7)$$

We can therefore write eqn. (12.3.2) in component form as

$$\frac{d}{dt}(\mathbf{p}_{\text{mechanical}} + \mathbf{P}_{\text{field}}) = \sum_\beta \int_V \frac{\partial T_{\alpha\beta}}{\partial x_\beta}\,dv \quad (12.3.8)$$

Applying the divergence theorem to the right of this equation, we get,

$$\frac{d}{dt}(\mathbf{p}_{\text{mechanical}} + \mathbf{P}_{\text{field}}) = \oint_S T_{\alpha\beta}\, n_\beta\, da \quad (12.3.9)$$

where $\hat{\mathbf{n}}$ is the outward normal to the closed surface S. If eqn. (12.3.9) represents the principle of conservation of linear momentum, $\sum_\beta T_{\alpha\beta} n_\beta$ is the αth component of the flow per unit area of momentum across the surface S into volume V.

12.4 REFLECTION AND REFRACTION OF PLANE WAVES IN NONCONDUCTING MEDIA: NORMAL INCIDENCE

Let us consider propagation of electromagnetic waves in linear nonconducting media. In absence of free charge density and free current density, Maxwell's equations inside matter are:

$$\nabla.\mathbf{D} = 0 \quad (12.4.1)$$

$$\nabla.\mathbf{B} = 0 \quad (5.5.1)$$

$$\nabla \times \mathbf{E} = -\frac{\partial \mathbf{B}}{\partial t} \quad (11.1.3)$$

$$\nabla \times \mathbf{H} = \frac{\partial \mathbf{D}}{\partial t} \quad (11.4.3)$$

For linear media, we have the constitutive equations $\mathbf{D} = \varepsilon\mathbf{E}$, and $\mathbf{B} = \mu\mathbf{H}$. Then, in terms of **E** and **B**, the above equations become

$$\nabla.\mathbf{E} = 0 \tag{12.4.1a}$$

$$\nabla.\mathbf{B} = 0 \tag{5.5.1}$$

$$\nabla \times \mathbf{E} = -\frac{\partial \mathbf{B}}{\partial t} \tag{11.1.3}$$

$$\nabla \times \mathbf{B} = \mu\varepsilon\frac{\partial \mathbf{E}}{\partial t} \tag{11.4.3a}$$

These equations are similar to those in Sec. 12.1 with μ, ε replacing μ_0, ε_0. Thus the electric and magnetic fields **E** and **B** satisfy the wave equation

$$\nabla^2\mathbf{E} = \mu\varepsilon\frac{\partial^2 \mathbf{E}}{\partial t^2} \tag{12.4.2a}$$

$$\nabla^2\mathbf{B} = \mu\varepsilon\frac{\partial^2 \mathbf{B}}{\partial t^2} \tag{12.4.2b}$$

These waves travel with a speed $v = 1/\sqrt{\mu\varepsilon}$. This is related to the speed of light in vacuum by the equation

$$v = \frac{c}{n} \tag{12.4.3}$$

where n is the refractive index of the medium. The waves thus travel slower in the medium than in free space.

We will now consider the reflection and transmission of linearly polarized waves at the interface of two media 1 and 2. We will be using the boundary conditions derived in Sec. 11.5 for the case $\sigma = 0$, $\mathbf{K}_f = 0$, i.e. there are no free surface charges or surface currents.

$$\varepsilon_2\mathbf{E}_{2n} = \varepsilon_1\mathbf{E}_{1n} \tag{12.4.4}$$

$$\mathbf{B}_{2n} = \mathbf{B}_{1n} \tag{11.5.2}$$

$$\mathbf{E}_{2t} = \mathbf{E}_{1t} \tag{11.5.3}$$

$$\frac{\mathbf{B}_{2t}}{\mu_2} = \frac{\mathbf{B}_{1t}}{\mu_1} \tag{11.5.5a}$$

We will take the electric field to be polarized along the x axis and the wave to be propagating along the z axis. interface between the two media is the XY plane (Fig. 12.4). Thus

$$\mathbf{E}_1 = E_{01}\hat{\mathbf{i}}e^{i(\mathbf{k}_1\mathbf{z} - \omega t)} \tag{12.4.5}$$

$$\mathbf{B}_1 = B_{01}\hat{\mathbf{j}}e^{i(\mathbf{k}_1\mathbf{z} - \omega t)} \tag{12.4.6}$$

where we have taken the initial phase to be zero so that the amplitudes are real. For the reflected wave,

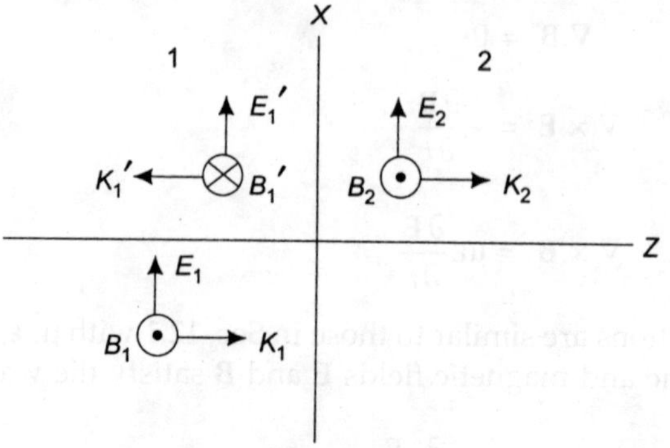

Fig. 12.4

$$\mathbf{E}_1' = E_{01}'\hat{\mathbf{i}}e^{-i(\mathbf{k}_1 z + \omega t)} \tag{12.4.7}$$

$$\mathbf{B}_1' = -\mathrm{B}_{01}'\hat{\mathbf{j}}e^{-i(\mathbf{k}_1 z + \omega t)} \tag{12.4.8}$$

B_{01}' is determined by eqn. (12.1.8).

$$\therefore \quad B_{01}' = \frac{E_{01}'}{v_1} \tag{12.4.9}$$

For the transmitted wave,

$$\mathbf{E}_2 = E_{02}\hat{\mathbf{i}}\,e^{i(\mathbf{k}_2 z - \omega t)} \tag{12.4.10}$$

$$\mathbf{B}_2 = B_{02}\hat{\mathbf{j}}\,e^{i(\mathbf{k}_2 z - \omega t)} \tag{12.4.11}$$

$$B_{02} = \frac{E_{02}}{v_2} \tag{12.4.12}$$

The reflected and transmitted waves must have the same frequency as the incident wave, if the boundary conditions are to be satisfied at the interface ($z = 0$) for all time t.

From the boundary condition (11.5.3), we get

$$E_{01} + E_{01}' = E_{02} \tag{12.4.13}$$

From the boundary condition (11.5.5a) for nonmagnetic media ($\mu_1 = \mu_2 \cong \mu_0$), we get

$$B_{01} - B_{01}' = B_{02}$$

or

$$\frac{E_{01}}{v_1} - \frac{E_{01}'}{v_1} = \frac{E_{02}}{v_2}$$

or $$E_{01} - E_{01}' = \frac{v_1}{v_2} E_{02} = \frac{n_2}{n_1} E_{02} \tag{12.4.14}$$

Solving eqns. (12.4.13) and (12.4.14) for E_{01}' and E_{02} we get,

$$E_{01}' = \left(\frac{n_1 - n_2}{n_1 + n_2}\right) E_{01} \tag{12.4.15}$$

$$E_{02} = \left(\frac{2n_1}{n_1 + n_2}\right) E_{01} \tag{12.4.16}$$

If $n_2 > n_1$, E_{01}'/E_{01} is negative. In other words, there is a phase change of π in the reflected beam with respect to the incident beam, when the wave is going from a rarer to a denser medium.

The intensities of the waves, given by the time averages of the magnitudes of the Poynting vectors, are $I_1 = \frac{1}{2}\varepsilon_1 v_1 E_{01}^2$, $I_1' = \frac{1}{2}\varepsilon_1 v_1 E_{01}'^2$, $I_2 = \frac{1}{2}\varepsilon_2 v_2 E_{02}^2$.

The ratio of the intensity of the reflected wave to the intensity of the incident wave is called the *reflectance* R. In this case, it is

$$R = \frac{I_1'}{I_1} = \left(\frac{E_{01}'}{E_{01}}\right)^2 = \left(\frac{n_1 - n_2}{n_1 + n_2}\right)^2 \tag{12.4.17}$$

The ratio of the intensity of the transmitted wave to the intensity of the incident wave is called the *transmittance* T. In this case, it is

$$T = \frac{I_2}{I_1} = \frac{\varepsilon_2 v_2}{\varepsilon_1 v_1}\left(\frac{E_{02}}{E_{01}}\right)^2 = \frac{n_2}{n_1}\left(\frac{2n_1}{n_1 + n_2}\right)^2 \tag{12.4.18}$$

For the air-glass interface, $n_1 = 1$, $n_2 = 1.5$, $R = 0.04$, $T = 0.96$.
From (12.4.17) and (12.4.18),

$$R + T = 1 \tag{12.4.19}$$

This equation expresses conservation of energy

12.5 REFLECTION AND REFRACTION OF PLANE WAVES IN NONCONDUCTING MEDIA: OBLIQUE INCIDENCE

We shall now consider plane electromagnetic waves incident at an arbitrary angle on the interface between two media. The interface will be taken to be the XY plane. First we will take the polarization to be in the plane of incidence. θ_1 is the angle of incidence (Fig. 12.5). θ_1' is the angle of reflection. θ_2 is the angle of refraction. All the **B** vectors are directed along the y axis. Let

$$\mathbf{E}_1 = \widetilde{\mathbf{E}}_{01}\, e^{i(\mathbf{k}_1.\mathbf{r}-\omega t)} \tag{12.5.1}$$

$$\mathbf{B}_1 = B_{01}\hat{\mathbf{j}}\, e^{i(\mathbf{k}_1.\mathbf{r}-\omega t)} \tag{12.5.2}$$

be the fields in the incident wave. The fields in the reflected wave are

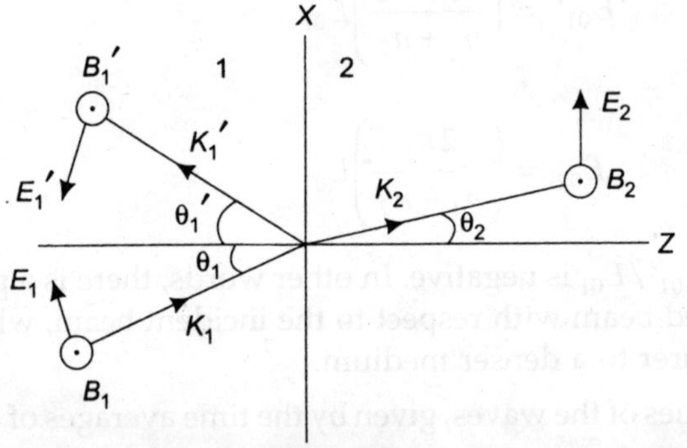

Fig. 12.5

$$\mathbf{E}_1{}' = \widetilde{\mathbf{E}}_{01}{}' e^{i(\mathbf{k}_1{}'.\mathbf{r}-\omega t)} \tag{12.5.3}$$

$$\mathbf{B}_1{}' = B_{01}{}'\hat{\mathbf{j}}\, e^{i(\mathbf{k}_1{}'.\mathbf{r}-\omega t)} \tag{12.5.4}$$

and the fields in the transmitted wave are

$$\mathbf{E}_2 = \widetilde{\mathbf{E}}_{02}\, e^{i(\mathbf{k}_2.\mathbf{r}-\omega t)} \tag{12.5.5}$$

$$\mathbf{B}_2 = B_{02}\hat{\mathbf{j}}\, e^{i(\mathbf{k}_2.\mathbf{r}-\omega t)} \tag{12.5.6}$$

We have the relations

$$E_{01} = v_1 B_{01} \tag{12.5.7}$$

$$E_{01}{}' = v_1 B_{01}{}' \tag{12.5.8}$$

$$E_{02} = v_2 B_{02} \tag{12.5.9}$$

[The E_0's turn out to be real as we have taken the B_0's to be real.]

We shall now use the boundary conditions. Eqn. (12.4.4) gives, in this case,

$$\varepsilon_1(\mathbf{E}_1)_z\, e^{i(\mathbf{k}_1.\mathbf{r}-\omega t)} + \varepsilon_1(\mathbf{E}_1{}')_z e^{i(\mathbf{k}_1{}'.\mathbf{r}-\omega t)} = \varepsilon_2(\mathbf{E}_2)_z e^{i(\mathbf{k}_2.\mathbf{r}-\omega t)}$$

For this to be true (and in fact for the other boundary conditions also), the exponents must have the same value at all times and at all points on the interface. ∴ We must have, at the interface,

$$\mathbf{k}_1.\mathbf{r} = \mathbf{k}_1{}'\mathbf{r} = \mathbf{k}_2.\mathbf{r}$$

The unit normal to the interface is $\hat{\mathbf{n}} = \hat{\mathbf{k}}$. The interface is given by the equation $z = 0$.

With $\quad \mathbf{r} = x\hat{\mathbf{i}} + y\hat{\mathbf{j}} + z\hat{\mathbf{k}}, \; \hat{\mathbf{n}}.\mathbf{r} = z = 0.$

Now $\quad \hat{\mathbf{n}} \times (\hat{\mathbf{n}} \times \mathbf{r}) = (\hat{\mathbf{n}}.\mathbf{r})\hat{\mathbf{n}} - \mathbf{r} = -\mathbf{r}$

∴ Using this in the above equation, we get,

$$\mathbf{k}_1.\{\hat{\mathbf{n}} \times (\hat{\mathbf{n}} \times \mathbf{r})\} = \mathbf{k}_1'.\{\hat{\mathbf{n}} \times (\hat{\mathbf{n}} \times \mathbf{r})\} = \mathbf{k}_2.\{\hat{\mathbf{n}} \times (\hat{\mathbf{n}} \times \mathbf{r})\}$$

or, interchanging the dot and the cross products, we get,

$$(\mathbf{k}_1 \times \hat{\mathbf{n}}).(\hat{\mathbf{n}} \times \mathbf{r}) = (\mathbf{k}_1' \times \hat{\mathbf{n}}).(\hat{\mathbf{n}} \times \mathbf{r}) = (\mathbf{k}_2 \times \hat{\mathbf{n}}).(\hat{\mathbf{n}} \times \mathbf{r})$$

Since $\mathbf{r}$ is an arbitrary vector,

$$\mathbf{k}_1 \times \hat{\mathbf{n}} = \mathbf{k}_1' \times \hat{\mathbf{n}} = \mathbf{k}_2 \times \hat{\mathbf{n}} \qquad (12.5.10)$$

The plane defined by $\mathbf{k}_1$ and $\hat{\mathbf{n}}$ is the plane of incidence. $\mathbf{k}_1 \times \hat{\mathbf{n}}$ is perpendicular to this plane. Also, $\mathbf{k}_1'$ is perpendicular to $\mathbf{k}_1 \times \hat{\mathbf{n}}$. ∴ $\mathbf{k}_1'$ also lies in the plane of incidence. Similarly $\mathbf{k}_2$ is also in the plane of incidence. this proves the first law of refraction, i.e., the incident ray, the refracted ray and the normal lie in the same plane. Further, taking the magnitudes in eqn. (12.5.10), we get,

$$k_1 \sin\theta_1 = k_1' \sin\theta_1' = k_2 \sin\theta_2 \qquad (12.5.11)$$

$$k_1 = k_1' = \frac{\omega}{v_1} \qquad (12.5.12)$$

$$k_2 = \frac{\omega}{v_2} \qquad (12.5.13)$$

$$\therefore \quad \theta_1 = \theta_1' \qquad (12.5.14)$$

The angle of reflection is equal to the angle of incidence. Also

$$\frac{\sin\theta_1}{v_1} = \frac{\sin\theta_2}{v_2}$$

or, $$n_1 \sin\theta_1 = n_2 \sin\theta_2 \qquad (12.5.15)$$

This is Snell's second law of refraction. Note that the deduction of the laws of reflection and refraction did not require the boundary conditions on the electric and magnetic field vectors. We obtained them simply by requiring the phases of the waves to be the same at all points on the interface.

Having derived the laws of reflection and refraction, let us now find the relations between the amplitudes of the fields. We shall be using the boundary conditions. Eqn. (12.4.4) gives

$$\varepsilon_1(-E_{01}\sin\theta_1 - E_{01}'\sin\theta_1') = \varepsilon_2(-E_{02}\sin\theta_2)$$

or, $$\varepsilon_1(E_{01} + E_{01}')\sin\theta_1 = \varepsilon_2 E_{02}\sin\theta_2 \qquad \text{...using (12.5.14)}$$

or, $$E_{01} + E_{01}' = \frac{\varepsilon_2 n_1}{\varepsilon_1 n_2} E_{02} \quad \text{...using (12.5.15)}$$

We shall be considering non-magnetic media. $\mu_1 \cong \mu_2 = \mu_0$. Then the refractive index is

$$n = \sqrt{\frac{\varepsilon\mu}{\varepsilon_0 \mu_0}} \cong \sqrt{\frac{\varepsilon}{\varepsilon_0}}$$

∴ The above equation becomes

$$n_1(E_{01} + E_{01}') = n_2 E_{02} \tag{12.5.16}$$

The boundary condition (11.5.3) gives

$$E_{01} \cos\theta_1 - E_{01}' \cos\theta_1 = E_{02} \cos\theta_2$$

or, using (12.5.14),

$$(E_{01} - E_{01}') \cos\theta_1 = E_{02} \cos\theta_2 \tag{12.5.17}$$

The boundary condition (11.5.5a) gives, with $\mu_1 = \mu_2$

$$B_{01} + B_{01}' = B_{02}$$

or, $$\frac{E_{01} + E_{01}'}{v_1} = \frac{E_{02}}{v_2} \quad \text{using (12.5.7) to (12.5.9)}$$

or, $$n_1 (E_{01} + E_{01}') = n_2 E_{02}$$

which is the same as eqn. (12.5.16). Solving eqns. (12.5.16) and (12.5.17) for E_{01}' and E_{02} we get

$$E_{01}' = \left(\frac{n_2 \cos\theta_1 - n_1 \cos\theta_2}{n_2 \cos\theta_1 + n_1 \cos\theta_2}\right) E_{01} \tag{12.5.18}$$

$$E_{02} = \left(\frac{2n_1 \cos\theta_1}{n_2 \cos\theta_1 + n_1 \cos\theta_2}\right) E_{01} \tag{12.5.19}$$

The quantities in brackets are called the *Fresnel coefficients* (i.e. the ratios E_{01}'/E_{01} and E_{02}/E_{01} as given by the above relations). For $\theta_1 = 0$ (normal incidence), these formulae reduce to (12.4.15) and (12.4.16). For $\theta_1 = \pi/2$ (grazing incidence), $E_{02} = 0$ and $E_{01}'/E_{01} = 1$. ∴ The reflectance $\left|\frac{E_{01}'}{E_{01}}\right|^2 = 1$.

If the numerator of eqn. (12.5.18) vanishes, i.e.

$$n_2 \cos\theta_1 = n_1 \cos\theta_2$$

then $E_{01}' = 0$. There is no reflected wave. Let this occur for an angle of incidence $\theta_1 = \theta_B$. Then

$$\cos\theta_B = \frac{n_1}{n_2}\cos\theta_2$$

and it can be shown, using Snell's law that

$$\tan\theta_B = \frac{n_2}{n_1} \qquad (12.5.20)$$

The angle θ_B is called *Brewster's angle* and eqn. (12.5.20) is called *Brewter's law.*

Now we shall take the polarization to be perpendicular to the plane of incidence (Fig. 12.6). Let

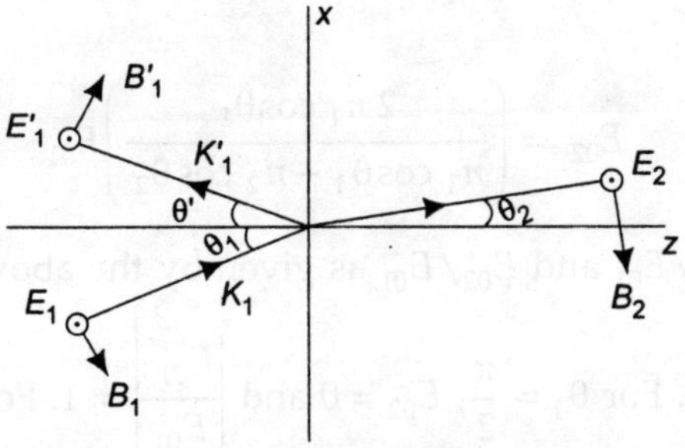

Fig. 12.6

$$\mathbf{E}_1 = E_{01}\hat{\mathbf{j}}\, e^{i(\mathbf{k}_1.\mathbf{r} - \omega t)} \qquad (12.5.21)$$

$$\mathbf{B}_1 = \mathbf{B}_{01} e^{i(\mathbf{k}_1.\mathbf{r} - \omega t)} \qquad (12.5.22)$$

represent the incident wave.

Let

$$\mathbf{E}_1' = E_{01}'\hat{\mathbf{j}}\, e^{i\left(\mathbf{k}_1'.\mathbf{r} - \omega t\right)} \qquad (12.5.23)$$

$$\mathbf{B}_1' = \mathbf{B}_{01} e^{i\left(\mathbf{k}_1'.\mathbf{r} - \omega t\right)} \qquad (12.5.24)$$

represent the reflected wave.

Let

$$\mathbf{E}_2 = E_{02}\hat{\mathbf{j}}\, e^{i(\mathbf{k}_2.\mathbf{r} - \omega t)} \qquad (12.5.25)$$

$$\mathbf{B}_2 = \mathbf{B}_{02} e^{i(\mathbf{k}_2.\mathbf{r} - \omega t)} \qquad (12.5.26)$$

represent the refracted wave. Then the relations (12.5.7)-(12.5.9) hold true. Let us use the boundary conditions. From eqn. (11.5.2), we get,

$$B_{01} \sin \theta_1 + B_{01}' \sin \theta_1 = B_{02} \sin \theta_2$$

or, $$(B_{01} + B_{01}') \sin \theta_1 = B_{02} \sin \theta_2$$

or, using Snell' law and the relations (12.5.7)-(12.5.9) we get,

$$E_{01} + E_{01}' = E_{02} \tag{12.5.27}$$

From the boundary condition (11.5.3), we get the same equation. From eqn. (11.5.5a), for non-magnetic media, we get,

$$-B_{01} \cos \theta_1 + B_{01}' \cos \theta_1' = -B_{02} \cos \theta_2$$

or, $$(-B_{01} + B_{01}') \cos \theta_1 = -B_{02} \cos \theta_2$$

or, using the relations (12.5.7)-(12.5.9), we have

$$n_1(-E_{01} + E_{01}') \cos \theta_1 = -n_2 E_{02} \cos \theta_2 \tag{12.5.28}$$

Solving eqn. (12.5.27) and (12.5.28) for E_{01}' and E_{02} we get,

$$E_{01}' = \left(\frac{n_1 \cos \theta_1 - n_2 \cos \theta_2}{n_1 \cos \theta_1 + n_2 \cos \theta_2} \right) E_{01} \tag{12.5.29}$$

$$E_{02} = \left(\frac{2 n_1 \cos \theta_1}{n_1 \cos \theta_1 + n_2 \cos \theta_2} \right) E_{01} \tag{12.5.30}$$

The ratios E_{01}'/E_{01} and E_{02}/E_{01} as given by the above relations are the Fresnel coefficients. For $\theta_1 = \frac{\pi}{2}$, $E_{02} = 0$ and $\left| \frac{E_{01}'}{E_{01}} \right| = 1$. For $\theta_2 = \frac{\pi}{2}$, $\frac{E_{01}'}{E_{01}} = 1$ in both cases. The wave is totally reflected. This is the case of total internal reflection and occurs for $n_1 > n_2$. Let $\theta_1 = \theta_c$, where θ_c is called the critical angle. Then from Snell's law, $\sin \theta_c = \frac{n_2}{n_1}$.

Example 2: Calculate the reflectance and transmittance for electromagnetic waves polarized parallel to and perpendicular to the plane of incidence.

Solution: The reflectance R is given by

$$R = \frac{\overline{\mathbf{S}}_1{}' . \hat{\mathbf{n}}}{\overline{\mathbf{S}}_1 . \hat{\mathbf{n}}}$$

and the transmittance T is given by

$$T = \frac{\overline{\mathbf{S}}_2 . \hat{\mathbf{n}}}{\overline{\mathbf{S}}_1 . \hat{\mathbf{n}}}$$

where $\mathbf{S}_1$, $\mathbf{S}_1'$ and $\mathbf{S}_2$ are the Poynting vectors of the incident wave, reflected wave and transmitted wave. $\hat{\mathbf{n}}$ is the unit vector normal to the interface. I shall work out the case of polarization parallel to the plane of incidence. The case of polarization perpendicular to the plane of incidence is similar.

$$\overline{\mathbf{S}}_1 = \frac{1}{2} Re(\mathbf{E}_1 \times \mathbf{H}_1^*)$$

$$\mathbf{E}_1 \times \mathbf{H}_1^* = \widetilde{\mathbf{E}}_{01} e^{i(\mathbf{k}_1.\mathbf{r}-\omega t)} \times \frac{B_{01}}{\mu_1} \hat{\mathbf{j}}\, e^{-i(k_1.r-\omega t)}$$

$$= \frac{\widetilde{\mathbf{E}}_{01} B_{01}}{\mu_0} (\cos\theta_1 \hat{\mathbf{i}} - \sin\theta_1 \hat{\mathbf{k}}) \times \hat{\mathbf{j}}$$

$$= \frac{E_{01}^2}{\mu_0 v_1} (\cos\theta_1 \hat{\mathbf{k}} + \sin\theta_1 \hat{\mathbf{i}})$$

$$= \varepsilon_1 v_1 E_{01}^2 (\cos\theta_1 \hat{\mathbf{k}} + \sin\theta_1 \hat{\mathbf{i}})$$

$$\therefore \quad \overline{\mathbf{S}}_1 = \frac{\varepsilon_1 v_1}{2} E_{01}^2 (\cos\theta_1 \hat{\mathbf{k}} + \sin\theta_1 \hat{\mathbf{i}})$$

$$\therefore \quad \overline{\mathbf{S}}_1.\hat{\mathbf{n}} = \overline{\mathbf{S}}_1.\hat{\mathbf{k}} = \frac{\varepsilon_1 v_1}{2} E_{01}^2 \cos\theta_1$$

$$\mathbf{E}_1' \times \mathbf{H}_1'^* = \frac{E_{01}' B_{01}'}{\mu_0} (-\cos\theta_1 \hat{\mathbf{i}} - \sin\theta_1 \hat{\mathbf{k}}) \times \hat{\mathbf{j}}$$

$$= \frac{E_{01}'^2}{\mu_0 v_1} (-\cos\theta_1 \hat{\mathbf{k}} + \sin\theta_1 \hat{\mathbf{i}})$$

$$= \varepsilon_1 v_1 E_{01}'^2 (-\cos\theta_1 \hat{\mathbf{k}} + \sin\theta_1 \hat{\mathbf{i}})$$

$$\therefore \quad \overline{\mathbf{S}}_1' = \frac{\varepsilon_1 v_1}{2} E_{01}'^2 (-\cos\theta_1 \hat{\mathbf{k}} + \sin\theta_1 \hat{\mathbf{i}})$$

For the reflected wave, $\hat{\mathbf{n}} = -\hat{\mathbf{k}}$.

$$\therefore \quad \overline{\mathbf{S}}_1'.\hat{\mathbf{n}} = \frac{\varepsilon_1 v_1}{2} E_{01}'^2 \cos\theta_1$$

$$\mathbf{E}_2 \times \mathbf{H}_2^* = \frac{E_{02} B_{02}}{\mu_0} (\cos\theta_2 \hat{\mathbf{i}} - \sin\theta_2 \hat{\mathbf{k}}) \times \hat{\mathbf{j}}$$

$$= \frac{E_{02}^2}{\mu_0 v_2} (\cos\theta_2 \hat{\mathbf{k}} + \sin\theta_2 \hat{\mathbf{i}})$$

$$= \varepsilon_2 v_2 E_{02}^2 (\cos\theta_2 \hat{\mathbf{k}} + \sin\theta_2 \hat{\mathbf{i}})$$

$$\therefore \quad \overline{\mathbf{S}}_2 = \frac{\varepsilon_2 v_2}{2} E_{02}^2 (\cos\theta_2 \hat{\mathbf{k}} + \sin\theta_2 \hat{\mathbf{i}})$$

$$\therefore \quad \overline{\mathbf{S}}_2 . \hat{\mathbf{n}} = \frac{\varepsilon_2 v_2}{2} E_{02}^2 \cos\theta_2$$

$$\therefore \quad R = \left(\frac{E_{01}'}{E_{01}}\right)^2 = \left(\frac{n_2 \cos\theta_1 - n_1 \cos\theta_2}{n_2 \cos\theta_1 + n_1 \cos\theta_2}\right)^2$$

$$T = \frac{\varepsilon_2 v_2 \cos\theta_2}{\varepsilon_1 v_1 \cos\theta_1}\left(\frac{E_{02}}{E_{01}}\right)^2 = \frac{n_2 \cos\theta_2}{n_1 \cos\theta_1}\left(\frac{E_{02}}{E_{01}}\right)^2$$

$$= \frac{n_2 \cos\theta_2}{n_1 \cos\theta_1}\left(\frac{2n_1}{n_2 \cos\theta_1 + n_1 \cos\theta_2}\right)^2$$

$$R + T = 1$$

12.6 ELECTROMAGNETIC WAVES IN CONDUCTORS

We shall now study the propagation of electromagnetic waves in conductors. Here, there is no free charge. But there is a current $\mathbf{J} = \sigma\mathbf{E}$ (Ohm's law) where σ is the conductivity. Inside the conductor, Maxwell's equations become (for linear media)

$$\nabla.\mathbf{E} = 0 \qquad (11.4.1)$$

$$\nabla.\mathbf{B} = 0 \qquad (5.5.1)$$

$$\nabla \times \mathbf{E} = -\frac{\partial \mathbf{B}}{\partial t} \qquad (11.1.3)$$

$$\nabla \times \mathbf{B} = \mu\sigma\mathbf{E} + \mu\varepsilon\frac{\partial \mathbf{E}}{\partial t} \qquad (12.6.1)$$

Taking the curl of both sides of eqn. (11.1.3), we get,

$$\nabla \times (\nabla \times \mathbf{E}) = -\frac{\partial}{\partial t}(\nabla \times \mathbf{B})$$

or, using a vector identity and eqns. (11.4.1) and (12.6.1), we get,

$$\nabla^2 \mathbf{E} = \mu\sigma\frac{\partial \mathbf{E}}{\partial t} + \mu\varepsilon\frac{\partial^2 \mathbf{E}}{\partial t^2} \qquad (12.6.2)$$

This equation is the modified wave equation in a conductor. **B** satisfies the same equation. We will consider plane wave solutions of this equation. So let us choose

$$\mathbf{E} = \tilde{E}_0 e^{i(\mathbf{k}_1 . \mathbf{r} - \omega t)} \tag{12.6.3}$$

Substituting in (12.6.2), we get,

$$-k^2 \mathbf{E} = -i\omega\mu\sigma \mathbf{E} - \mu\varepsilon\omega^2 \mathbf{E}$$

$$\therefore \quad k^2 = i\omega\mu\sigma + \mu\varepsilon\omega^2 \tag{12.6.4}$$

We shall be concerned with nonmagnetic media $\mu \cong \mu_0$. $\therefore \mu_0\varepsilon_0 \cong \mu\varepsilon_0 = 1/c^2$.

$$\therefore \quad k^2 = \frac{i\omega\sigma}{\varepsilon_0 c^2} + \frac{\varepsilon\omega^2}{\varepsilon_0 c^2}$$

$$= \frac{i\omega\sigma}{\varepsilon_0 c^2} + \frac{K\omega^2}{c^2}$$

$$\therefore \quad k^2 = \frac{\omega^2}{c^2}\left(K + \frac{i\sigma}{\omega\varepsilon_0}\right) \tag{12.6.5}$$

where K is the dielectric constant of the medium. We define a complex dielectric constant $\tilde{K}$ as

$$\tilde{K} = K + \frac{i\sigma}{\omega\varepsilon_0} \tag{12.6.6}$$

We also define a complex refractive index $\tilde{n}$ such that

$$\tilde{n}^2 = \tilde{K} \tag{12.6.7}$$

The propagation vector has to be taken complex. Let $\mathbf{k} = \mathbf{k}_1 + i\mathbf{k}_2$ where $\mathbf{k}_1$ and $\mathbf{k}_2$ are real vectors. We will assume them to be in the same direction. Let $\hat{\mathbf{m}}$ be a unit vector in this direction. Then

$$\mathbf{k} = (k_1 + ik_2)\hat{\mathbf{m}} = \tilde{K}\hat{\mathbf{m}} \tag{12.6.8}$$

$\therefore$ From (12.6.5), we get,

$$\tilde{k} = \frac{\omega}{c}\tilde{n} \tag{12.6.9}$$

With a complex propagation vector, the electric field now becomes

$$\mathbf{E} = \tilde{\mathbf{E}}_0 e^{-\mathbf{k}_2 \mathbf{r}} e^{i(\mathbf{k}_1 . \mathbf{r} - \omega t)} \tag{12.6.10}$$

Thus the amplitude of the electric field vector decays with distance in the conductor. Substituting (12.6.3) and its magnetic counterpart $\mathbf{B} = \tilde{\mathbf{B}}_0 e^{i(k.r-\omega t)}$ into (11.1.3), we get,

$$\frac{\mathbf{k}}{\omega} \times \tilde{\mathbf{E}}_0 = \tilde{\mathbf{B}}_0$$

or, using eqns. (12.6.8) and (12.6.9),

$$\frac{\tilde{n}}{c}\left(\tilde{\mathbf{m}} \times \tilde{\mathbf{E}}_0\right) = \tilde{\mathbf{B}}_0 \tag{12.6.11}$$

Since $\tilde{n}$ is complex, $\tilde{\mathbf{E}}$ and $\tilde{\mathbf{B}}$ are not in phase with each other. From eqns. (11.4.1) and (5.5.1), we get, on using eqn. (12.6.3) and the magnetic field,

$$\mathbf{k}.\tilde{\mathbf{E}}_0 = 0, \mathbf{k}.\tilde{\mathbf{B}}_0 = 0 \tag{12.6.12}$$

∴ The electromagnetic wave is transverse.

Let us now solve for the complex refractive index $\tilde{n}$ introduced in (12.6.7). Let

$$\tilde{n} = n + ik$$

where n and k are called the optical constants.

Squaring this equation, we get,

$$\tilde{n}^2 = n^2 - k^2 + 2\,ink$$

From eqn. (12.6.6), the complex dielectric constant is

$$\tilde{K} = K_r + iK_i$$

Where $K_r = K$ and $K_i = \dfrac{\sigma}{\omega \varepsilon_0}$

∴ From eqn. (12.6.7), on equating real and imaginary parts, we get,

$$n^2 - k^2 = K_r$$

$$2nk = K_i$$

Solving for n and k, we have

$$n = \sqrt{\frac{1}{2}\left\{K_r + \sqrt{K_r^2 + K_i^2}\right\}} \tag{12.6.14}$$

$$k = \sqrt{\frac{1}{2}\left\{-K_r + \sqrt{K_r^2 + K_i^2}\right\}} \tag{12.6.15}$$

where we have taken the positive sign before the square root.

If $|K_r| >> K_i$, $K_r > 0$ then

$$n \cong \sqrt{K_r} \tag{12.6.16}$$

$$k \cong \sqrt{\frac{1}{2}\left\{-K_r + K_r\left(1+\frac{K_i^2}{2K_r^2}\right)\right\}} = \frac{K_i}{2\sqrt{K_r}}$$

$$\therefore \quad k \cong \frac{K_i}{2n} \tag{12.6.17}$$

In other words, if $K >> \frac{\sigma}{\omega\varepsilon_0}$, i.e. if $\omega >> \frac{\sigma}{\varepsilon}$, $n \cong \sqrt{K}$ and $k \cong \frac{\sigma}{\omega\sqrt{\varepsilon\varepsilon_0}}$

This holds for a good insulator.

If $|K_f| << K_i$, i.e. $\omega << \frac{\sigma}{\varepsilon}$, then

$$n \cong \sqrt{\frac{1}{2}\{K_r + K_i\}} \cong \sqrt{\frac{K_i}{2}}$$

$$k \cong \sqrt{\frac{1}{2}\{-K_r + K_i\}} \cong \sqrt{\frac{K_i}{2}}$$

$$\therefore \quad n \cong k \cong \sqrt{\frac{\sigma}{2we_0}} \tag{12.6.18}$$

This holds in metals at microwave frequencies (~ 10^{10} to 10^{11} Hz) and below. Coming back to eqn. (12.6.10), suppose the propagation direction is the z axis. Then

$$\mathbf{E} = \tilde{\mathbf{E}}_0 e^{-k_2 z} e^{i(k_1 z - \omega t)}$$

where $$k_2 = \frac{\omega k}{c} \text{ and } k_1 = \frac{\omega n}{c}$$

$$\therefore \quad \mathbf{E} = \tilde{\mathbf{E}}_0 e^{-\omega k z/c}\, e^{i\omega(nz/c - t)}$$

The distanced after which the amplitude of the electric field decays to $1/e$ of its initial value (i.e. the value at the point where the wave enters the medium) is given by

$$d = \frac{c}{\omega k} \tag{12.6.19}$$

and is called the skin depth. Since the wavelength is $\lambda = \frac{2\pi c}{n\omega}$, we have

$$d = \frac{n\lambda}{2\pi k} \tag{12.6.20}$$

or, for a good conductor, $d \cong \dfrac{\lambda}{2\pi}$

Eqn. (12.6.18) is usually valid when the skin depth is important in electrical problems. Then we have, substituting (12.6.18) into (12.6.19),

$$d \cong \sqrt{\frac{2}{\mu_0 \omega_0 \sigma}} \tag{12.6.21}$$

For sea water, $\sigma \cong 5\ (\Omega\text{m})^{-1}$, $\varepsilon = 6 \times 10^{-10}\ C^2/Nm^2$, $\sigma/\varepsilon \cong 10^{10}$ Hz. So sea water is a poor conductor for frequencies far above 10^{10} Hz. If we consider a frequency of say $f = 60$ kHz, where $\sigma = 4.3\ (\Omega\text{m})^{-1}$, sea water is a good conductor and the skin depth is[2]

$$d \cong 1\text{m}$$

For silver, $\sigma \cong 10^7 (\Omega\text{m})^{-1}$, $\varepsilon \sim 10^{-11}\ \text{C}^2/\text{Nm}^2$. $\therefore\ \dfrac{\sigma}{\varepsilon} = 10^{18}$ Hz.

For a frequency ~ 10^{10} Hz, silver is a good conductor and the skin depth is $d \cong 10^{-8}$ m.[14]

Now we shall briefly consider reflection and transmission at the conducting surface. We will take medium 1 to be nonconducting and medium 2 to be conducting and consider normal incidence only. Let us take the incident wave to be polarized along the x axis and the interface to be the XY plane (Fig. 12.4).

$$\mathbf{E}_1 = \tilde{E}_{01}\hat{\mathbf{i}}\, e^{i(k_1 z - \omega t)} \tag{12.6.22}$$

$$\mathbf{B}_1 = \tilde{B}_{01}\hat{\mathbf{j}}\, e^{i(k_1 z - \omega t)} \tag{12.6.23}$$

with $\tilde{E}_{01} = v_1 \tilde{B}_{01}$

where the $\tilde{E}_0$ s are now complex.

The reflected wave is

$$\mathbf{E}_1' = \tilde{E}_{01}'\hat{\mathbf{i}}\, e^{-i(k_1 z - \omega t)} \tag{12.6.24}$$

$$\mathbf{B}_1' = \tilde{B}_{01}\hat{\mathbf{j}}\, e^{-i(k_1 z - \omega t)} \tag{12.6.25}$$

with $\tilde{E}_{01}' = v_1 \hat{B}_{01}'$

The transmitted wave is

$$\mathbf{E}_2 = \tilde{E}_{02}\hat{\mathbf{i}}\, e^{i(\tilde{k}_2 z - \omega t)} \quad (12.6.26)$$

$$\mathbf{B}_2 = \tilde{B}_{02}\hat{\mathbf{j}}\, e^{i(\tilde{k}_2 z - \omega t)} \quad (12.6.27)$$

with
$$\tilde{E}_{02} = \frac{\tilde{B}_{02}\,\omega}{\tilde{k}_2}$$

We have to remember that $\tilde{k}_2$ is complex because the second medium is conducting. We have to use the general boundary conditions

$$\varepsilon_2 E_{2n} - \varepsilon_1 E_{1n} = \sigma \quad (11.5.1a)$$

$$B_{2n} = B_{1n} \quad (11.5.2)$$

$$E_{2t} = E_{1t} \quad (11.5.3)$$

$$\frac{\mathbf{B}_{2t}}{\mu_2} - \frac{\mathbf{B}_{1t}}{\mu_1} = \mathbf{K}_f \times \hat{\mathbf{n}} \quad (11.5.5a)$$

Here $\hat{\mathbf{n}} = \hat{\mathbf{k}}$ For conductors obeying Ohm's law, there can be no free surface currents since this would require the field to be infinite at the boundary. $\therefore K_f = 0$. From eqn. (11.5.1a), there is no normal component of E (i.e. $E_z = 0$). $\therefore \sigma = 0$. Eqn. (11.5.2) is automatically satisfied since there is no normal component of B (i.e. $B_z = 0$). From eqn. (11.5.3), we get,

$$\tilde{E}_{01} + \tilde{E}_{01}' = \tilde{E}_{02} \quad (12.6.28)$$

From eqn. (11.5.5a), we get

$$\tilde{B}_{01} - \tilde{B}_{01}' = \tilde{B}_{02}$$

or,
$$\frac{\tilde{E}_{01} - \tilde{E}_{01}'}{\mu_1 v_1} = \frac{\tilde{k}_2 \tilde{E}_{02}}{\omega}$$

or,
$$\tilde{E}_{01} - \tilde{E}_{01}' = \frac{\mu_1 v_1}{\mu_2 \omega}\tilde{k}_1 \tilde{E}_{02} \quad (12.6.29)$$

Solving (12.6.28) and (12.6.29) for $\tilde{E}_{01}'$ and $\tilde{E}_{02}$ we get,

$$\tilde{E}_{01}' = \frac{\left(1 - \dfrac{\mu_1 v_1 \tilde{k}_2}{\mu_2 \omega}\right)}{\left(1 + \dfrac{\mu_1 v_1 \tilde{k}_2}{\mu_2 \omega}\right)}\tilde{E}_{01} \quad (12.6.30)$$

$$\tilde{E}_{02} = \left(\frac{2}{1 + \dfrac{\mu_1 v_1 \tilde{k}_2}{\mu_2 \omega}} \right) \tilde{E}_{01} \tag{12.6.31}$$

Consider a good conductor where $\sigma >> \omega\varepsilon_2$. Take $\mu_1 = \mu_2 \cong \mu_0$. Then

$$\tilde{E}_{01}' = \frac{\left(1 - \dfrac{v_1 \tilde{k}_2}{\omega}\right)}{\left(1 + \dfrac{v_1 \tilde{k}_2}{\omega}\right)} \tilde{E}_{01} \tag{12.6.32}$$

$$\tilde{E}_{02} = \left(\frac{2}{1 + \dfrac{v_1 \tilde{k}_2}{\omega}} \right) \tilde{E}_{01} \tag{12.6.33}$$

Since $\tilde{k}_2$ is complex, the electric fields of the reflected and transmitted waves are phase shifted relative to the incident wave.

$$\frac{v_1 \tilde{k}_2}{\omega} = \frac{v_1}{c}(n_2 + ik_2) = \frac{1}{n_1}(n_2 + ik_2)$$

In this case, $n_2 \cong k_2 = \sqrt{\dfrac{\sigma}{2\omega\varepsilon_0}}$. $\quad \therefore \dfrac{v_1 \tilde{k}_2}{\omega} = \dfrac{1}{n_1}\sqrt{\dfrac{\sigma}{2\omega\varepsilon_0}}(1 + i)$

Take the first medium to be air. $\therefore n_1 = 1$

$$\frac{\tilde{E}_{01}'}{\tilde{E}_{01}} = \frac{1 - \sqrt{\dfrac{\sigma}{2\omega\varepsilon_0}}(1+i)}{1 + \sqrt{\dfrac{\sigma}{2\omega\varepsilon_0}}(1+i)}$$

$$= \frac{\sqrt{\dfrac{2\omega\varepsilon_0}{\sigma}} - 1 - i}{\sqrt{\dfrac{2\omega\varepsilon_0}{\sigma}} + 1 + i}$$

The reflectance is

$$R = \left|\frac{E_{01}{}'}{E_{01}}\right|^2 = \frac{\left(\sqrt{\frac{2\,\omega\varepsilon_0}{\sigma}}-1\right)^2+1}{\left(\sqrt{\frac{2\,\omega\varepsilon_0}{\sigma}}+1\right)^2+1} = \frac{1-\sqrt{\frac{2\,\omega\varepsilon_0}{\sigma}}+\frac{\omega\varepsilon_0}{\sigma}}{1+\sqrt{\frac{2\,\omega\varepsilon_0}{\sigma}}+\frac{\omega\varepsilon_0}{\sigma}}$$

$$\cong \left(1-\sqrt{\frac{2\,\omega\varepsilon_0}{\sigma}}+\frac{\omega\varepsilon_0}{\sigma}\right)\left(1-\sqrt{\frac{2\,\omega\varepsilon_0}{\sigma}}-\frac{\omega\varepsilon_0}{\sigma}\right)$$

$R \cong 1-2\sqrt{\frac{2\,\omega\varepsilon_0}{\sigma}}$, keeping terms of first order in $\sqrt{\frac{\omega\varepsilon_0}{\sigma}}$ only. For silver[(2)] at $f = 10^{10}$ Hz, $R = 0.9996$. To find the transmittance, we have to first calculate the average Poynting vector of the transmitted wave.

$$\overline{\mathbf{S}}_2 = \frac{1}{2}\,Re(\mathbf{E}_2\times\mathbf{H}_2{}^*)$$

$$\mathbf{H}_2 = \frac{\mathbf{B}_2}{\mu_2} = \frac{\widetilde{B}_{02}}{\mu_2}\hat{\mathbf{j}}e^{i(k_2 z-\omega t)}$$

$$= \frac{\widetilde{B}_{02}}{\mu_2}\hat{\mathbf{j}}e^{i\left\{\frac{\omega}{c}(n+ik_2)z-\omega t\right\}}$$

$$\mathbf{H}_2 = \frac{\widetilde{B}_{02}}{\mu_2}\hat{\mathbf{j}}\,e^{\frac{-\omega k_2 z}{c}}\,e^{i\left(\frac{\omega n_2 z}{c}-\omega t\right)}$$

$$\mathbf{E}_2\times\mathbf{H}_2{}^* = \widetilde{E}_{02}\hat{\mathbf{i}}\,e^{\frac{-\omega k_2 z}{c}}\,e^{i\left(\frac{\omega n_2 z}{c}-\omega t\right)}\times\frac{\widetilde{B}_{02}{}^*}{\mu_2}\hat{\mathbf{j}}\,e^{\frac{-\omega k_2 z}{c}}\,e^{-i\left(\frac{\omega n_2 z}{c}-\omega t\right)}$$

$$= \frac{\widetilde{E}_{02}\widetilde{B}_{02}{}^*}{\mu_2}\hat{\mathbf{k}}\,e^{\frac{-2\omega k_2 z}{c}} = \frac{\left|\widetilde{E}_{02}\right|^2}{\mu_2\omega}\widetilde{k}_2{}^*\hat{\mathbf{k}}\,e^{\frac{-2\omega k_2 z}{c}}$$

$$= \frac{\left|\widetilde{E}_{02}\right|^2}{\mu_2 c}\hat{\mathbf{k}}\,(n_2-ik_2)\,e^{\frac{-2\omega k_2 z}{c}}$$

$$\therefore \quad \overline{\mathbf{S}}_2 = \frac{\left|\widetilde{E}_{02}\right|^2}{2\mu_2 c}n_2\hat{\mathbf{k}}\,e^{\frac{-2\omega k_2 z}{c}}$$

At the interface, $(z = 0)$ $\bar{\mathbf{S}}_2 = \dfrac{\left|\tilde{E}_{02}\right|^2}{2\mu_2 c} n_2 \hat{\mathbf{k}} \cong \dfrac{\left|\tilde{E}_{02}\right|^2}{2\mu_0 c} n_2 \hat{\mathbf{k}}$

$$\therefore \qquad T = \frac{\bar{\mathbf{S}}_2 . \hat{\mathbf{k}}}{\bar{\mathbf{S}}_1 . \hat{\mathbf{k}}} = \frac{\dfrac{\left|\tilde{E}_{02}\right|^2 n_2}{2\mu_0 c}}{\dfrac{\left|\tilde{E}_{01}\right|^2}{2\mu_0 c}} = n_2 \left|\frac{\tilde{E}_{02}}{\tilde{E}_{01}}\right|^2$$

$$\therefore \qquad T = \frac{4n_2}{(1+n_2)^2 + k_2^2}$$

The table below summarizes some of the important results of this chapter in *SI* and Guassian units:

Physical quantity/equation	*SI*	*Gaussian*
Poynting vector **S** (12.2.5)	$\mathbf{S} = \mathbf{E} \times \mathbf{H}$	$\mathbf{S} = \dfrac{c}{4\pi} (\mathbf{E} \times \mathbf{H})$
Maxwell's Strees Tensor	$T_{\alpha\beta} = \varepsilon_0 \{E_\alpha E_\beta + c^2 B_\alpha$	$T_{\alpha\beta} = \dfrac{1}{4\pi} \{E_\alpha E_\beta + B_\alpha B_\beta$
$T_{\alpha\beta}$ (12.3.7)	$B_\beta - \dfrac{1}{2}(\mathbf{E.E} + c^2 \mathbf{B.B}) \delta_{\alpha\beta}\}$	$- \dfrac{1}{2}(\mathbf{E.E} + \mathbf{B.B}) \delta_{\alpha\beta}\}$

Tutorial 1

In this tutorial, we will briefly recall the early theories of light.⊕ It was Francesco Maria Grimaldi, a professor at the Jesuit College in Bologna, Italy, who first observed the phenomenon of diffraction of light. [15] He observed light in the geometrical shadow of a rod illuminated by a small source (1665). Robert Hooke and Robert Boyle independently observed the phenomenon of interference in thin films. [16] Hooke also observed diffraction. Hooke was the first to propose that light consists of rapid vibrations propagated instantaneously, or with a very great speed, over any distance, and believed that in a homogeneous medium, every vibration will generate a sphere which will grow steadily. [16] He was thus the first to propose the wave theory. In 1666, Newton observed the phenomenon of dispersion of light by a prism. The wave theory proposed by Hooke could not explain the rectilinear propagation of light. Newton thus proposed a corpuscular theory of light, according to which "light is propagated from a luminous body in the form of

⊕René Descartes, who put the law of refraction in its familiar form in terms of sines, proposed (1637) that light was a pressure transmitted through an elastic medium.[16]

minute particles."[16] In the meantime, Huygens extended the wave theory. At that time it was supposed that light waves were logitudinal and propagated in a medium called the "aether". Huygens' principle was that each point of the aether upon which the luminous disturbance is incident may be regarded as the centre of a new disturbance propagated in the form of spherical waves; these secondary waves combine in such a manner that their envelope determines the wavefront at a later time. He was able to explain the laws of reflection and refraction using this principle. He also discovered the phenomenon of polarization. The explanation of polarization was given by Newton. However, he assumed that the rays had "sides"[16] and this transversality seemed to him to be an impedement to the acceptance of the wave theory.

The wave theory of light was revived by Thomas Young in 1801-03 when he proposed his Principle of Interference. In 1814, Augustin Fresnel placed the wave theory on a secure foundation when he combined Huygens' principle and Young's principle of interference. He explained both rectilinear propagation of light and diffraction. He also calculated the diffraction caused by straight edges, small apertures and screens. In 1816, Dominique Arago and Fresnel found that two rays polarized at right angles never interface. This fact could not be reconciled with the longitudinal wave picture. In 1817, Young showed that the phenomenon could be explained by assuming light waves to be transverse.

Tutorial 2

We have learnt in this chapter that electromagnetic waves travel with a speed c in vacuum. They can have a wide range of frequencies varying by several orders of magnitude. They have been classified accordingly. For example, radio frequency waves extend form about a few kHz to several hundred MHz (1 MHz = 10^6 Hz). That corresponds to a wavelength range ($\lambda = c/f$) of 3 m to 300 km. Electromagnetic waves having frequencies from a few GHz (1 GHz = 10^9 Hz) to about 3×10^{11} Hz are called microwaves.[15] What range of wavelengths do they correspond to? The frequency range from 3×10^{11} Hz to about 4×10^{14} Hz is the infrared region. What wavelength range does that correspond to? Visible light has a wavelength range of 400 to 700 nm (1 nm = 10^{-9} m). What frequency range does that correspond to? Beyond this, from 8×10^{14} Hz to 3.4×10^{16} Hz we have the ultraviolet region.[15] What is the corresponding range of wavelengths? From about 2.4 $\times 10^{16}$ Hz to 5×10^{19} Hz, we have X-rays. What is the range of wavelengths?

Tutorial 3

In this tutorial, we shall define anisotropic media and look at some properties of wave propagation in them. Anisotropic media are those for which the electric displacement **D** and the electric field **E** are not in the same direction. (We assume however that $\mathbf{B} = \mu_0 \mathbf{H}$ so that they are nonmagnetic and magnetically isotropic.[16]) Thus we may write

$$D_x = \varepsilon_{xx} E_x + \varepsilon_{xy} E_y + \varepsilon_{xz} E_z$$
$$D_y = \varepsilon_{yx} E_x + \varepsilon_{yy} E_y + \varepsilon_{yz} E_z$$
$$D_z = \varepsilon_{zx} E_x + \varepsilon_{zy} E_y + \varepsilon_{zz} E_z \quad ^{(16)}$$

It can be shown that[16] $\varepsilon_{xy} = \varepsilon_{yx}$, $\varepsilon_{xz} = \varepsilon_{zx}$, $\varepsilon_{yz} = \varepsilon_{zy}$. The coeffiecients ε_{ij} constitute the dielectric tensor. It is possible to transform to a new set of axes $x'\ y'\ z'$ for which we have

$$D_x' = \varepsilon_x' E_x', D_y' = \varepsilon_y' E_y' D_z' = \varepsilon_z' E_z'$$

These axes $x'\ y'\ z'$ are called principal axes and the constants $\varepsilon_x', \varepsilon_y', \varepsilon_z'$ are called principal dielectric constants. The Poynting vector is still defined as

$$\mathbf{S} = \mathbf{E} \times \mathbf{H}$$

We assume that there are no free charges ($\rho = 0$) and no free currents ($\mathbf{J}_f = 0$). Then Maxwell's equations inside matter become

$$\nabla.\mathbf{D} = 0 \tag{12.4.1}$$

$$\nabla.\mathbf{B} = 0 \tag{5.5.1}$$

$$\nabla \times \mathbf{E} = -\frac{\partial \mathbf{B}}{\partial t} \tag{11.1.3}$$

$$\nabla \times \mathbf{H} = \frac{\partial \mathbf{D}}{\partial t} \tag{11.4.3}$$

For wave propagation in anisotropic media, we consider plane wave solutions

$$\mathbf{E} = \tilde{\mathbf{E}}_0 e^{i(\mathbf{k}.\mathbf{r} - \omega t)} \tag{1}$$

$$\mathbf{D} = \tilde{\mathbf{D}}_0 e^{i(\mathbf{k}.\mathbf{r} - \omega t)} \tag{2}$$

$$\mathbf{H} = \tilde{\mathbf{H}}_0 e^{i(\mathbf{k}.\mathbf{r} - \omega t)} \tag{3}$$

$$\mathbf{B} = \tilde{\mathbf{B}}_0 e^{i(\mathbf{k}.\mathbf{r} - \omega t)} \tag{4}$$

where $\mathbf{H} = \mathbf{B}/\mu_0$. Substituting these in eqn.(11.1.3), we get,

$$\mathbf{H} = \frac{\mathbf{k} \times \mathbf{E}}{\mu_0 \omega} \tag{5}$$

and from eqn. (11.4.3), we get,

$$\mathbf{D} = \frac{\mathbf{H} \times \mathbf{k}}{\omega} \tag{6}$$

From eqn. (5) and (6), we find that **E**, **D** and **k** are coplanar and **H** is perpendicular to this plane. Further eqn. (12.4.1), we get,

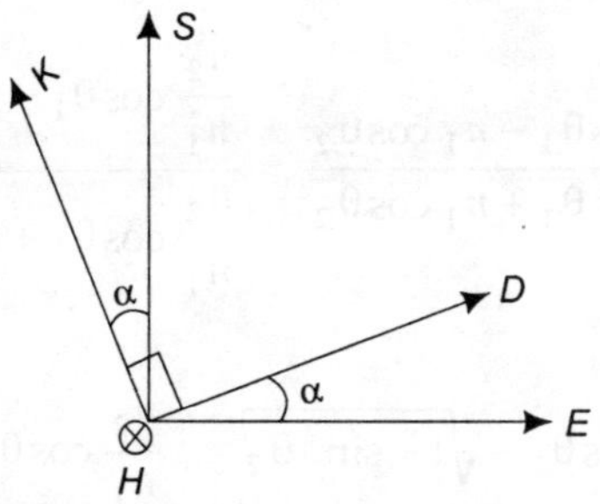

Fig. 12.7

$$\mathbf{k}.\mathbf{D} = 0 \tag{7}$$

Thus **D** and **H** are transverse to **k** but **E** is not. In fig. 12.7, **S** is the Poynting vector. The angle between **k** and **S** is the same as the angle between **E** and **D**. The direction of propagation of energy (**S**) is different from the direction of propagation of the wave (**k**). The phase velocity v_P is defined to be in the direction of **k** and has a magnitude $v_P = c/n$. The ray velocity is defined to be in the direction of **S**. Its magnitude is the ratio of the energy flow per unit time per unit area perpendicular to the flow direction, to the energy density. Substituting for **H** from (5) into (6), we get,

$$\mathbf{D} = \frac{1}{\mu_0 \omega^2}\{(\mathbf{k} \times \mathbf{E}) \times \mathbf{k}\}$$

or, using a vector identity,

$$\mathbf{D} = \frac{1}{\mu_0 \omega^2}\{k^2 \mathbf{E} - (\mathbf{k}.\mathbf{E})\mathbf{E}\}$$

Tutorial 4

We can use eqn. (12.2.11) to calculate the electric field in a light wave. At a distance of 10 m from a 100 W lamp, the intensity is 7.96×10^{-2} W/m^2. Calculate the amplitude of the electric field and the magnetic field.

A 1 mW laser has a beam diameter of 2 mm. Assuming the divergence of the beam to be negligible, calculate the electric and magnetic field amplitudes.

Tutorial 5

In this tutorial, we shall look at the phenomenon of total internal reflection in slightly greater detail. Recall that this occurs when the angle of incidence $\theta_1 \geq \theta_c$ where θ_c is called the critical angle and satisfies the relation $\sin\theta_c = n_2/n_1$. Consider eqn. (12.5.18)

$$\frac{E_{01}'}{E_{01}} = \frac{n_2 \cos\theta_1 - n_1 \cos\theta_2}{n_2 \cos\theta_1 + n_1 \cos\theta_2} = \frac{\dfrac{n_2}{n_1}\cos\theta_1 - \cos\theta_2}{\dfrac{n_2}{n_1}\cos\theta_1 + \cos\theta_2}$$

$$= \frac{\dfrac{n_2}{n_1}\cos\theta_1 - \sqrt{1-\sin^2\theta_2}}{\dfrac{n_2}{n_1}\cos\theta_1 + \sqrt{1-\sin^2\theta_2}} = \frac{\dfrac{n_2}{n_1}\cos\theta_1 - \sqrt{1-\dfrac{n_1^2}{n_2^2}\sin^2\theta_1}}{\dfrac{n_2}{n_1}\cos\theta_1 + \sqrt{1-\dfrac{n_1^2}{n_2^2}\sin^2\theta_1}}$$

$$\therefore \quad \frac{E_{01}'}{E_{01}} = \frac{\left(\dfrac{n_2}{n_1}\right)^2 \cos\theta_1 - \sqrt{\left(\dfrac{n_2}{n_1}\right)^2 - \sin^2\theta_1}}{\left(\dfrac{n_2}{n_1}\right)^2 \cos\theta_1 + \sqrt{\left(\dfrac{n_2}{n_1}\right)^2 - \dfrac{n_1^2}{n_2^2}\sin^2\theta_1}}$$

If $\sin\theta_1 > \dfrac{n_2}{n_1}$, the quantity under the square root is negative and so we have

$$\frac{E_{01}'}{E_{01}} = \frac{\left(\dfrac{n_2}{n_1}\right)^2 \cos\theta_1 - i\sqrt{\sin^2\theta_1 - \left(\dfrac{n_2}{n_1}\right)^2}}{\left(\dfrac{n_2}{n_1}\right)^2 \cos\theta_1 + i\sqrt{\sin^2\theta_1 - \left(\dfrac{n_2}{n_1}\right)^2}}$$

This is of the form

$$\frac{E_{01}'}{E_{01}} = \frac{A - iB}{A + iB} \qquad \therefore \quad \left|\frac{E_{01}'}{E_{01}}\right|^2 = 1$$

The reflectance is unity. Let us consider the transmitted wave.

$$\mathbf{E}_2 = \tilde{\mathbf{E}}_{02} e^{i(\mathbf{k}_2.\mathbf{r} - \omega t)}$$

$$\mathbf{k}_2.\mathbf{r} = k_2 z \cos\theta_2 + k_2 x \sin\theta_2$$

$$= k_2 z\sqrt{1-\sin^2\theta_2} + k_2 x \sin\theta_2$$

$$= k_2 z\sqrt{1-\frac{n_1^2}{n_2^2}\sin^2\theta_1} + k_2 x \sin\theta_2$$

$$= ik_2 z\sqrt{\frac{n_1^2}{n_2^2}\sin^2\theta_1 - 1} + k_2 x \sin\theta_2$$

$$\therefore \qquad \mathbf{E}_2 = E_{02}\, e^{-k_2 z\sqrt{\frac{n_1^2}{n_2^2}\sin^2\theta_1 - 1}}\; e^{1(k_2 x \sin\theta_2 - \omega t)}$$

$\therefore$ The transmitted wave is strongly attenuated along the z axis and propagates along the surface in the x direction. This wave is called an *evanescent wave.*

Problems

1. The intensity of sunlight hitting the earth is about 1300 W/m^2. Find the amplitude of electric and magnetic fields.

   (University of Calcutta, 2002)

   [Ans. $E_0 = 0.99 \times 10^3$ N/C. $B_0 = 0.33 \times 10^{-5}$ T]

2. Find the wavelength and propagation speed in copper for radio waves at 1 MHz. Given the conductivity of copper $\sigma = 58 \times 10^6\ (\Omega\text{m})^{-1}$. Choose $\mu = \mu_0$, $\varepsilon = \varepsilon_0$. *(University of Calcutta, 2003)*

   [Ans. $v = 414.9$ m/s., $\lambda = 0.415$ mm]

3. Calculate the time average of the Poynting vector for the elliptically polarized wave

$$\mathbf{E} = (E_{0x} e^{i\pi/2}\,\hat{\mathbf{i}} + E_{0y}\hat{\mathbf{j}})\, e^{i(kz-\omega t)}$$

$$\mathbf{B} = \left(\frac{-E_{0y}}{c}\hat{\mathbf{i}} + \frac{E_{0x}e^{i\pi/2}}{c}\hat{\mathbf{j}}\right) e^{i(kz-\omega t)}$$

   considered in the text. Consider propagation in free space.

4. Using Snell's law show that eqns. (12.5.18) and (12.5.19) can be (rewritten as). (Show also that eqns. (12.5.29) and (12.5.30) can be rewritten as respectively).

$$\frac{E_{01}'}{E_{01}} = \frac{\tan(\theta_1 - \theta_2)}{\tan(\theta_1 + \theta_2)}$$

$$\frac{E_{02}}{E_{01}} = \frac{2\cos\theta_1 \sin\theta_2}{\sin(\theta_1+\theta_2)\cos(\theta_1-\theta_2)}$$

$$\frac{E_{01}'}{E_{01}} = \frac{\sin(\theta_2 - \theta_1)}{\sin(\theta_2 + \theta_1)}$$

$$\frac{E_{02}}{E_{01}} = \frac{2\cos\theta_1 \sin\theta_2}{\sin(\theta_2 + \theta_1)}$$

respectively.

5. Consider an electromagnetic wave propagating along the z axis in a conductor.

$$\mathbf{E} = \tilde{\mathbf{E}}_0 e^{i(kz - \omega t)},\ \mathbf{B} = \tilde{\mathbf{B}}_0 e^{i(kz - \omega t)} \text{ with } \tilde{\mathbf{E}}_0 = E_{0x} e^{i\phi}\,\hat{\mathbf{i}} + E_{0y}\hat{\mathbf{j}},$$

$$\tilde{\mathbf{B}}_0 = \tilde{B}_{0x}\hat{\mathbf{i}} + \tilde{B}_{0y}\hat{\mathbf{j}}.$$

Use eqn. (12.6.11) to obtain $\tilde{B}_{0x}$ and $\tilde{B}_{0y}$ in terms of E_{0x} and E_{0y}. Show that if $\tilde{n} = |\tilde{n}| e^{i\alpha}$,

$$Re\ \tilde{\mathbf{E}}_0 . Re\ \tilde{\mathbf{B}}_0 = -\frac{|\tilde{n}|}{c} \sin\phi \sin\alpha$$

Hence prove that the real electric and magnetic fields are perpendicular to each other only for linearly polarized waves.

6. Find the skin depth at a frequency 1.6 MHz in aluminium where $\sigma = 38.2 \times 10^6\ (\Omega\text{m})^{-1}$ and $\mu \cong \mu_0$. **[Ans.** $d = 0.065$ mm**]**

7. Consider a linearly polarized wave in a conductor propagating along the z axis:

$$\mathbf{E} = E_0 \hat{\mathbf{i}}\, e^{-k_2 z} e^{i(k_1 z - \omega t)}$$

$$\mathbf{B} = \tilde{\mathbf{B}}_0 e^{-k_2 z} e^{i(k_1 z - \omega t)}$$

Take $\tilde{n} = |\tilde{n}| e^{i\alpha}$. Calculate the energy density taking $\mu \cong \mu_0$ and permittivity ε. Find the average energy density and show that for a good conductor, the energy in the magnetic field dominates.

13

Electromagnetic Radiation

13.1 RETARDED POTENTIALS

In this chapter, we shall briefly consider radiation from accelerated charges. Recall that in the Lorentaz guage, the scalar and vector potentials satisfy the following equations in the presence of sources:

$$\nabla^2 \phi - \frac{1}{c^2}\frac{\partial^2 \phi}{\partial t^2} = \frac{-\rho}{\varepsilon_0} \tag{11.6.16}$$

$$\nabla^2 \mathbf{A} - \frac{1}{c^2}\frac{\partial^2 \mathbf{A}}{\partial t^2} = -\mu_0 \mathbf{J} \tag{11.6.17}$$

The solutions of these equations are given by

$$\phi(\mathbf{r}, t) = \frac{1}{4\pi\varepsilon_0}\int \frac{\rho(\mathbf{r}', t_r)}{r_1} dv' \tag{13.1.1}$$

$$\mathbf{A}(\mathbf{r}, t) = \frac{\mu_0}{4\pi}\int \frac{\mathbf{J}(\mathbf{r}', t_r)}{r_1} dv' \tag{13.1.2}$$

In these equations, $\rho(\mathbf{r}', t_r)$ is the charge density at the position specified by $\mathbf{r}'$ and time t_r. The time t_r is called the *retarded time*. It is related to the time t when the potential is calculated, by

$$t_r = t - \frac{r_1}{c} \tag{13.1.3}$$

This retarded time is used because electromagnetic waves travel with speed c in vacuum. So r_1/c is the time taken by the wave to travel from the source point $\mathbf{r}'$ to the field point $\mathbf{r}$.

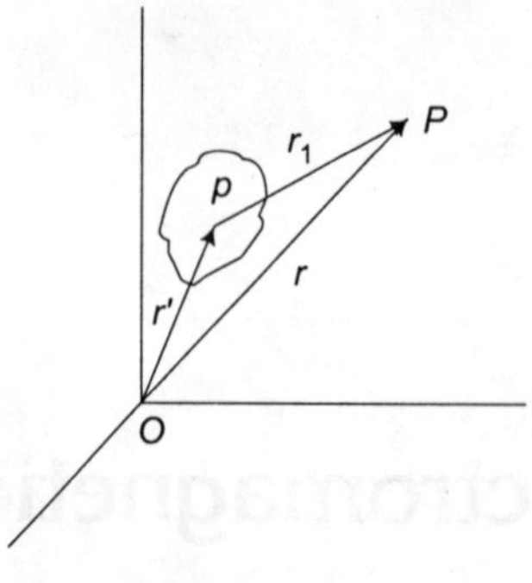

Fig. 13.1

13.2 THE LIENARD-WIECHERT POTENTIALS

We will now calculate the potential due to a point charge moving with a velocity **v**. It may appear, at first sight, that the potential is the same as the Coulomb potential with the distace now being replaced by the distance corresponding to the time when the signal left the charge (the retarded distance). But this is incorrect. The correct procedure is to start with a volume charge distribution and take the limit as the volume tends to zero. The integration in eqn. (13.1.1) is not well defined since the retarded time is different for different parts of the charge distribution. To calculate the potential due to a point charge, we first consider a charge distribution with a finite but small volume. Let us suppose that P (Fig. 13.2) is the field point, i.e. we wish to evaluate potentials at this point. For this purpose, we imagine a sphere centered at P and contracting towards P with speed c. This sphere is called an information-collecting sphere by Panofsky and Phillips.[17] It sweeps across the charge distribution as it converges towards P. It carries information about the charge in an infinitesimal volume element $dv' = dS'dr_1$ as it crosses the element. Let us suppose that this volume element has a velocity **v** (Fig. 13.2). If **v** has a component towards P, the charge measured by P at the retarded distance will be more than the true charge there. This is because the approaching charge stays longer within the information-collecting sphere. ∴ The infinitesimal element of charge observed at P is

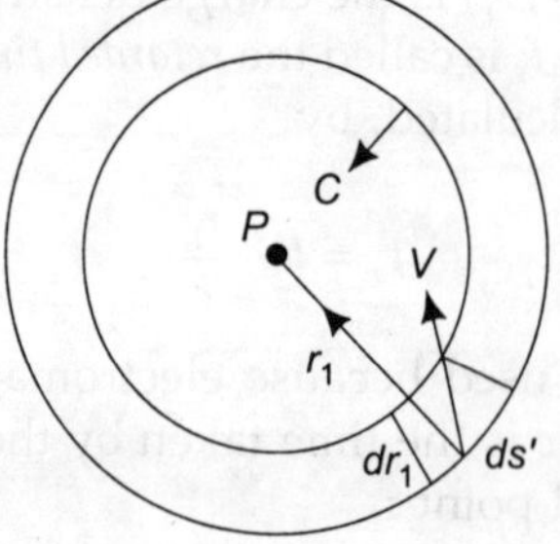

Fig. 13.2

$$dq = \rho(\mathbf{r}', t_r)dv' = \rho(\mathbf{r}', t_r)\,\frac{\mathbf{v}.\mathbf{r}_1}{r_1}dSdt$$

as the information-collecting sphere contracts in time dt. Now $dt = dr_1/c$ and $dS'dr_1 = dv'$

$$\therefore \qquad dq = \rho(\mathbf{r}', t_r)dv' - \rho(\mathbf{r}', t_r)dv'\frac{\mathbf{v}.\mathbf{r}_1}{cr_1}$$

or,

$$\rho(\mathbf{r}', t_r)dv' = \frac{dq}{1-\dfrac{\mathbf{v}.\hat{\mathbf{r}}_1}{c}}$$

$\therefore$ The retarded potential is

$$\phi(\mathbf{r}, t) = \frac{1}{4\pi\varepsilon_0}\int \frac{dq}{\left(\mathbf{r}_1 - \dfrac{\mathbf{r}_1.\mathbf{v}}{c}\right)}$$

Note that r_1 is the retarded distance from P. In other words, it is the distance of the infinitesimal element of charge when the information-collecting sphere crossed it. In the limit of a point charge, the distance-dependent terms are slowly varying. Since $\int dq = q$, the total charge,

$$\phi(\mathbf{r}, t) = \frac{q}{4\pi\varepsilon_0\left(\mathbf{r}_1 - \dfrac{\mathbf{r}_1.\mathbf{v}}{c}\right)} \tag{13.2.1}$$

The vector potential is

$$\mathbf{A}(\mathbf{r}, t) = \frac{\mu_0}{4\pi}\frac{q\mathbf{v}}{\left(\mathbf{r}_1 - \dfrac{\mathbf{r}_1.\mathbf{v}}{c}\right)} \tag{13.2.2}$$

The potentials given by eqns. (13.2.1) and (13.2.2) are the Lienard-Wiechert potentials due to a point charge.

13.3 ELECTRIC DIPOLE RADIATION

A point charge moving with a uniform velocity does not radiate. An accelerated charge does radiate. We will now consider an example of an accelerated charge. Two small spheres are separated by a distance 's' and connected by a fine wire.[14] The system as a whole is electrically neutral (Fig. 13.3). If the charge on the upper sphere at time t is $q(t)$, that on the lower sphere is $-q(t)$. There is a mechanism for moving the charge back and forth through the wire.

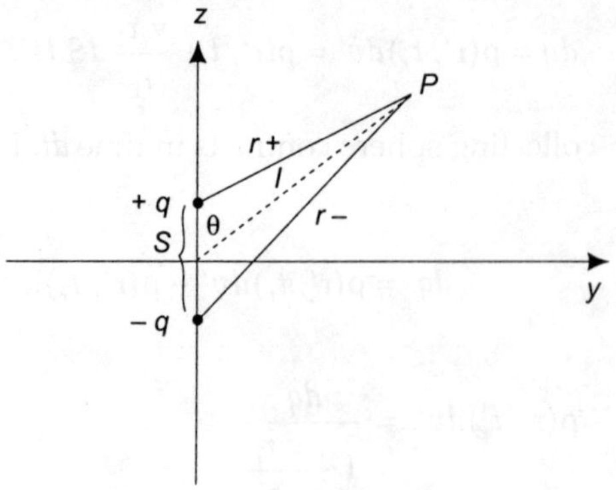

Fig. 13.3

$$\therefore \qquad q(t) = q_0 \cos \omega t$$

This represents an electric dipole with a moment

$$\mathbf{p}(t) = q(t)\mathbf{s} = p_0 \cos \omega t \ \hat{\mathbf{k}}$$

where $p_0 = q_0 s$. The retarded potential at P is given by

$$\phi(\mathbf{r}, t) = \frac{1}{4\pi\varepsilon_0}\left[\frac{q_0 \cos \omega(t - r_+/c)}{r_+} - \frac{q_0 \cos \omega(t - r_-/c)}{r_-}\right] \tag{13.3.1}$$

$$r_+ = \left(r^2 + \frac{s^2}{4} - rs \cos \theta\right)^{1/2}$$

$$r_- = \left(r^2 + \frac{s^2}{4} + rs \cos \theta\right)^{1/2}$$

For a perfect dipole, we have $s << r$. $\therefore$ We can rewrite

$$r_+ = r\left(1 + \frac{s^2}{4r^2} - \frac{s \cos \theta}{r}\right)^{1/2}$$

and expand binomially to first order in s/r:

$$r_+ \cong r\left(1 - \frac{s \cos \theta}{2r}\right) \tag{13.3.2}$$

Similarly, we get,

$$r_- \cong r\left(1 + \frac{s \cos \theta}{2r}\right) \tag{13.3.3}$$

$$\therefore \qquad \frac{1}{r_+} \cong \frac{1}{r}\left(1+\frac{s\cos\theta}{2r}\right) \tag{13.3.4}$$

and

$$\frac{1}{r_-} \cong \frac{1}{r}\left(1-\frac{s\cos\theta}{2r}\right) \tag{13.3.5}$$

Now let us expand the cosine

$$\cos\left(\omega t-\frac{\omega r_+}{c}\right) \cong \cos\left[\left(\omega t-\frac{\omega r}{c}\right)+\frac{\omega s\cos\theta}{2c}\right]$$

$$= \cos\left(\omega t-\frac{\omega r}{c}\right)\cos\left(\frac{\omega s\cos\theta}{2c}\right) - \sin\left(\omega t-\frac{\omega r}{c}\right)\sin\left(\frac{\omega s\cos\theta}{2c}\right)$$

In the limit of a perfect dipole, $s << c/\omega$

$$\cos\left(\omega t-\frac{\omega r_+}{c}\right) \cong \cos\left(\omega t-\frac{\omega r}{c}\right) - \frac{\omega s\cos\theta}{2c}\sin\left(\omega t-\frac{\omega r}{c}\right) \tag{13.3.6}$$

and similarly,

$$\cos\left(\omega t-\frac{\omega r_-}{c}\right) \cong \cos\left(\omega t-\frac{\omega r}{c}\right) + \frac{\omega s\cos\theta}{2c}\sin\left(\omega t-\frac{\omega r}{c}\right) \tag{13.3.7}$$

Substituting eqns. (13.3.4)-(13.3.7) in eqn.(13.3.1), we finally get

$$\phi(\mathbf{r}, t) = \frac{q_0}{4\pi\varepsilon_0 r}\left[\frac{-\omega s}{c}\cos\theta\sin\left(\omega t-\frac{\omega r}{c}\right)+\frac{s\cos\theta}{r}\cos\left(\omega t-\frac{\omega r}{c}\right)\right]$$

$$\therefore \qquad \phi(\mathbf{r}, t) = \frac{p_0\cos\theta}{4\pi\varepsilon_0 r}\left[\frac{-\omega}{c}\sin\omega\left(t-\frac{r}{c}\right)+\frac{1}{r}\cos\omega\left(t-\frac{r}{c}\right)\right] \tag{13.3.8}$$

At large distance from the source, i.e. $r >> c/\omega$, called the *radiation zone*,

$$\phi(\mathbf{r}, t) = \frac{-p_0\omega\cos\theta}{4\pi\varepsilon_0 cr}\sin\omega\left(t-\frac{r}{c}\right) \tag{13.3.9}$$

To find vector potential, we have to calculate the current,

$$\mathbf{I}(t) = \frac{dq}{dt}\hat{\mathbf{k}} = -q_0\omega\sin\omega t\,\hat{\mathbf{k}}$$

The vector potential is then

$$\mathbf{A}(\mathbf{r}, t) = \frac{\mu_0}{4\pi}\hat{\mathbf{k}}\int_{-s/2}^{s/2}\frac{(-q_0\omega)\sin\omega\left(t-\frac{r_1}{c}\right)}{r}dz$$

On integrating, we get a factor of s. So, to first order in s, we can replace the sine function by its value at $s = 0$.

$$\therefore \qquad \mathbf{A}(\mathbf{r}, t) \cong \frac{-\mu_0}{4\pi}\hat{\mathbf{k}}\frac{q_0 s\omega}{r}\sin\omega\left(t - \frac{r}{c}\right)$$

$$\text{or,} \qquad \mathbf{A}(\mathbf{r}, t) = \frac{-\mu_0 p_0 \omega}{4\pi r}\sin\omega\left(t - \frac{r}{c}\right)\hat{\mathbf{k}} \tag{13.3.10}$$

Let us now calculate the electric and magnetic fields. We will have to use spherical polar co-ordinates and so recall that

$$\hat{\mathbf{k}} = \hat{\mathbf{r}}\cos\theta - \hat{\theta}\sin\theta$$

$$\nabla\phi = \frac{\partial\phi}{\partial r}\hat{\mathbf{r}} + \frac{1}{r}\frac{\partial\phi}{\partial\theta}\hat{\theta}$$

$$\frac{\partial\phi}{\partial r} = \frac{-p_0\omega\cos\theta}{4\pi\varepsilon_0 c}\left\{\frac{-1}{r^2}\sin\omega\left(t - \frac{r}{c}\right) - \frac{\omega}{cr}\cos\omega\left(t - \frac{r}{c}\right)\right\}$$

In the radiation zone,

$$\frac{\partial\phi}{\partial r} \cong \frac{p_0\omega^2}{4\pi\varepsilon_0 c^2 r}\cos\theta\cos\omega\left(t - \frac{r}{c}\right)$$

$$\frac{1}{r}\frac{\partial\phi}{\partial\theta} = \frac{p_0\omega\sin\theta}{4\pi\varepsilon_0 c r^2}\sin\omega\left(t - \frac{r}{c}\right)$$

$$\therefore \qquad \nabla\phi \cong \frac{p_0\omega^2}{4\pi\varepsilon_0 c^2 r}\cos\theta\cos\omega\left(t - \frac{r}{c}\right)\hat{\mathbf{r}}$$

$$\frac{\partial\mathbf{A}}{\partial t} = \frac{-\mu_0 p_0\omega^2}{4\pi r}\cos\omega\left(t - \frac{r}{c}\right)(\hat{\mathbf{r}}\cos\theta - \sin\theta\,\hat{\theta})$$

$$\mathbf{E} = -\nabla\phi - \frac{\partial\mathbf{A}}{\partial t}$$

$$\therefore \qquad \mathbf{E} = \frac{-\mu_0 p_0\omega^2}{4\pi r}\sin\theta\cos\omega\left(t - \frac{r}{c}\right)\hat{\theta} \tag{13.3.11}$$

$$\nabla\times\mathbf{A} = \frac{1}{r}\left[\frac{\partial}{\partial r}(rA_\theta) - \frac{\partial A_r}{\partial\theta}\right]\hat{\varphi}$$

$$A_r = \frac{-\mu_0 p_0\omega}{4\pi r}\cos\theta\sin\omega\left(t - \frac{r}{c}\right)$$

$$A_\theta = \frac{\mu_0 p_0 \omega}{4\pi r} \sin\theta \sin\omega\left(t - \frac{r}{c}\right)$$

∴ In the radiation zone,

$$\mathbf{B} \cong \frac{-\mu_0 p_0 \omega^2}{4\pi c r} \sin\theta \sin\omega\left(t - \frac{r}{c}\right)\hat{\varphi} \qquad (13.3.12)$$

Eqns. (13.3.11) and (13.3.12) represent monochromatic spherical waves traveling with the speed of light and having angular frequency ω. They are called spherical waves because their amplitude is not constant but falls off as $1/r$.

Tutorial

In this tutorial, we shall calculate the electric and magnetic fields of a point charge moving with a velocity **v**. We shall use the Lienard-Wiechert potentials. We shall follow ref. 14. Let us first consider the scalar potential ϕ (13.2.1).

$$\phi(\mathbf{r}, t) = \frac{qc}{4\pi\varepsilon_0 (cr - \mathbf{r}_1.\mathbf{v})}$$

Here $\qquad \mathbf{r}_1 = \mathbf{r} - \mathbf{w}(t_r),\ \mathbf{v} = \dot{\mathbf{w}}(t_r) \qquad (1)$

$\mathbf{w}(t_r)$ is the position of the particle at retarded time t_r and **v** is the velocity there. The electric field is given by

$$\mathbf{E} = -\nabla\phi - \frac{\partial \mathbf{A}}{\partial t}$$

$$\nabla\phi = \frac{-qc}{4\pi\varepsilon_0} \cdot \frac{1}{(r_1 c - \mathbf{r}_1.\mathbf{v})^2} \nabla(r_1 c - \mathbf{r}_1.\mathbf{v}) \qquad (2)$$

Now $\qquad r_1 = c(t - t_r) \quad \therefore \nabla r_1 = -c\nabla t_r \qquad (3)$

$$\nabla(\mathbf{r}_1.\mathbf{v}) = \mathbf{r}_1 \times (\nabla \times \mathbf{v}) + \mathbf{v} \times (\nabla \times \mathbf{r}_1) + (\mathbf{r}_1.\nabla)\mathbf{v} + (\mathbf{v}.\nabla)\mathbf{r}_1 \qquad (4)$$

$$\nabla \times \mathbf{v} = \hat{\mathbf{i}}\left(\frac{\partial v_z}{\partial y} - \frac{\partial v_y}{\partial z}\right) + \hat{\mathbf{j}}\left(\frac{\partial v_x}{\partial z} - \frac{\partial v_z}{\partial x}\right) + \hat{\mathbf{k}}\left(\frac{\partial v_y}{\partial x} - \frac{\partial v_x}{\partial y}\right)$$

$$= \hat{\mathbf{i}}\left(\frac{dv_z}{dt_r}\frac{\partial t_r}{\partial y} - \frac{dv_y}{dt_r}\frac{\partial t_r}{\partial z}\right) + \hat{\mathbf{j}}\left(\frac{dv_x}{dt_r}\frac{\partial t_r}{\partial z} - \frac{dv_z}{dt_r}\frac{\partial t_r}{\partial x}\right)$$

$$+ \hat{\mathbf{k}}\left(\frac{dv_y}{dt_r}\frac{\partial t_r}{\partial x} - \frac{dv_x}{dt_r}\frac{\partial t_r}{\partial y}\right)$$

The acceleration of the particle at the retarded time t_r is

$$\mathbf{a} = \hat{\mathbf{i}}\frac{dv_x}{dt_r} + \hat{\mathbf{j}}\frac{dv_y}{dt_r} + \hat{\mathbf{k}}\frac{dv_z}{dt_r}$$

$$\therefore \qquad \nabla \times \mathbf{v} = -\mathbf{a} \times \nabla t_r \tag{5}$$

$$\nabla \times \mathbf{r}_1 = \nabla \times \mathbf{r} - \nabla \times \mathbf{w} = -\nabla \times \mathbf{w}, \text{ since } \nabla \times \mathbf{r} = 0$$

In analogy with eqn.(5), $\nabla \times \mathbf{w} = -\mathbf{v} \times \nabla t_r$

$$\therefore \qquad \nabla \times \mathbf{r}_1 = \mathbf{v} \times \nabla t_r \tag{6}$$

$$(\mathbf{r}_1.\nabla)\mathbf{v} = \left\{(x-x')\frac{\partial}{\partial x} + (y-y')\frac{\partial}{\partial y} + (z-z')\frac{\partial}{\partial z}\right\}\mathbf{V}$$

$$= \left\{(x-x')\frac{\partial t_r}{\partial x}\frac{d\mathbf{v}}{dt_r} + (y-y')\frac{\partial t_r}{\partial y}\frac{d\mathbf{v}}{dt_r} + (z-z')\frac{\partial t_r}{\partial z}\frac{d\mathbf{v}}{dt_r}\right\} \tag{7}$$

$$\therefore \qquad (\mathbf{r}_1.\nabla)\mathbf{v} = (\mathbf{r}_1.\nabla t_r)\,\mathbf{a}$$

$$(\mathbf{v}.\nabla)\mathbf{r}_1 = (\mathbf{v}.\nabla)\mathbf{r} - (\mathbf{v}.\nabla)\mathbf{w}$$

$(\mathbf{v}.\nabla)\mathbf{r} = \mathbf{v}$ and in analogy with eqn.(7)

$$(\mathbf{v}.\nabla)\mathbf{w} = \mathbf{v}(\mathbf{v}.\nabla t_r) \tag{8}$$

Substituting eqns. (5) to (8) into (4), we get,

$$\nabla(\mathbf{r}_1.\mathbf{v}) = -\mathbf{r}_1 \times (\mathbf{a} \times \nabla t_r) + \mathbf{v} \times \mathbf{v} \times \nabla t_r) + \mathbf{a}(\mathbf{r}_1.\nabla t_r) + \mathbf{v} - \mathbf{v}(\mathbf{v}.\nabla t_r)$$

which on simplifying, gives

$$\nabla(\mathbf{r}_1.\mathbf{v}) = (\mathbf{r}_1.\mathbf{a})\nabla t_r - v^2 \nabla t_r + \mathbf{v} \tag{9}$$

Substituting eqn.(9) and eqn.(3) into eqn.(2), we get,

$$\nabla\phi = \frac{qc}{4\pi\varepsilon_0(r_1 c_1 - \mathbf{r}_1.\mathbf{v})^2}\{(c^2 - v^2 + \mathbf{r}_1.\mathbf{a})\nabla t_r + \mathbf{v}\} \tag{10}$$

Let us now evaluate ∇t_r. Starting from eqn.(3), we get,

$$-c\nabla t_r = \nabla r_1 = \nabla(\mathbf{r}_1.\mathbf{r}_1)^{1/2} = \frac{1}{2(\mathbf{r}_1.\mathbf{r}_1)^{1/2}}\nabla(\mathbf{r}_1.\mathbf{r}_1)$$

Now using a vector identity,

$$\nabla(\mathbf{r}_1.\mathbf{r}_1) = 2\mathbf{r}_1 \times (\nabla \times \mathbf{r}_1) + 2(\mathbf{r}_1.\nabla)\mathbf{r}_1$$

$$(\mathbf{r}_1.\nabla)\mathbf{r}_1 = (\mathbf{r}_1.\nabla)\mathbf{r} - (\mathbf{r}_1.\nabla)\mathbf{w} = \mathbf{r}_1 - \mathbf{v}(\mathbf{r}_1.\nabla t_r)$$

Using eqn.(6) we thus get,

$$\nabla(\mathbf{r}_1.\mathbf{r}_1) = 2\{(\mathbf{r}_1 \times (\mathbf{v} \times \nabla t_r) + \mathbf{r}_1 - \mathbf{v}(\mathbf{r}_1.\nabla t_r))\}$$

$$= 2\{r_1 - (\mathbf{r}_1.\mathbf{v})\nabla t_r\}$$

$$\therefore \qquad -c\nabla t_r = \frac{1}{r_1}\{\mathbf{r}_1 - (\mathbf{r}_1.\mathbf{v})\nabla t_r\}$$

Rearranging terms, we get,

$$\nabla t_r = -\frac{\mathbf{r}_1}{(r_1 c - \mathbf{r}_1.\mathbf{v})} \tag{11}$$

Substituting this into eqn. (10), we get

$$\nabla\phi = \frac{qc}{4\pi\varepsilon_0 (r_1 c - \mathbf{r}_1.\mathbf{v})^3}\{(r_1 c - \mathbf{r}_1.\mathbf{v})\dot{\mathbf{v}} - (c^2 - v^2 + \mathbf{r}_1.\mathbf{a})\mathbf{r}_1\} \tag{12}$$

Let us now consider the vector potential. Using $\mu_0 = \dfrac{1}{\varepsilon_0 c^2}$ in eqn. (13.2.2), we have,

$$\mathbf{A} = \frac{q\mathbf{v}}{4\pi\varepsilon_0 c(r_1 c - \mathbf{r}_1.\mathbf{v})}$$

$$\frac{\partial \mathbf{A}}{\partial t} = \frac{q}{4\pi\varepsilon_0 c}\left\{\frac{1}{(r_1 c - \mathbf{r}_1.\mathbf{v})}\frac{\partial \mathbf{v}}{\partial t} - \frac{\mathbf{v}}{(r_1 c - \mathbf{r}_1.\mathbf{v})^2}\frac{\partial}{\partial t}(r_1 c - \mathbf{r}_1.\mathbf{v})\right\}$$

$$= \frac{q}{4\pi\varepsilon_0 c}\left\{\frac{1}{(r_1 c - \mathbf{r}_1.\mathbf{v})}\frac{d\mathbf{v}}{dt_r}\frac{\partial t_r}{\partial t} - \frac{\mathbf{v}}{(r_1 c - \mathbf{r}_1.\mathbf{v})^2}\left[\frac{c\,dr_1}{dt_r} - \frac{\partial(\mathbf{r}_1.\mathbf{v})}{\partial t_r}\frac{\partial t_r}{\partial t}\right]\right\}$$

$$\therefore \frac{\partial \mathbf{A}}{\partial t} = \frac{q}{4\pi\varepsilon_0 c}\left\{\frac{\mathbf{a}}{(r_1 c - \mathbf{r}_1.\mathbf{v})}\frac{\partial t_r}{\partial t} - \frac{\mathbf{v}}{(r_1 c - \mathbf{r}_1.\mathbf{v})^2}\left(\frac{c\,dr_1}{dt_r} - \frac{d\mathbf{r}.\mathbf{v} - \mathbf{r}_1.\mathbf{a}}{dt_r}\right)\frac{\partial t_r}{\partial t}\right\} \tag{13}$$

Let us now evaluate $\dfrac{\partial t_r}{\partial t}$. To do that, consider $\dfrac{dr_1}{dt_r}$.

$$\frac{dr_1}{dt_r} = \frac{d(\mathbf{r}_1.\mathbf{r}_1)^{1/2}}{dt_r} = \frac{\mathbf{r}_1}{\sqrt{\mathbf{r}_1.\mathbf{r}_1}}.\frac{d\mathbf{r}_1}{dt_r} = \frac{\mathbf{r}_1}{r_1}\frac{d\mathbf{r}_1}{dt_r}$$

$$\frac{d\mathbf{r}_1}{dt_r} = -\dot{\mathbf{w}} = -\mathbf{v} \qquad \therefore \frac{dr_1}{dt_r} = \frac{-\mathbf{r}_1.\mathbf{v}}{r} \tag{14}$$

$$r_1 = c(t - t_r) \qquad \therefore \frac{dr_1}{dt_r} = c\left(\frac{\partial t}{\partial t_r} - 1\right) \tag{15}$$

Using eqn.(14) in eqn.(15), we get,

$$\frac{\partial t_r}{\partial t} = \frac{r_1 c}{r_1 c - \mathbf{r}_1.\mathbf{v}} \tag{16}$$

Using eqn.(14) and eqn.(16) in eqn.(13), we get,

$$\frac{\partial \mathbf{A}}{\partial t} = \frac{q}{4\pi\varepsilon_0 c}\left\{\frac{\mathbf{a}r_1 c}{(r_1 c - \mathbf{r}_1.\mathbf{v})^2} - \frac{\mathbf{v}r_1 c}{(r_1 c - \mathbf{r}_1.\mathbf{v})^3}\left(\frac{-c\mathbf{r}_1.\mathbf{v}}{r_1} + v^2 - \mathbf{r}_1.\mathbf{a}\right)\right\}$$

$$= \frac{q}{4\pi\varepsilon_0 c(r_1 c - \mathbf{r}_1.\mathbf{v})^3}\left\{\mathbf{a}r_1(r_1 c - \mathbf{r}_1.\mathbf{v}) - \mathbf{v}r_1\left(\frac{-c\mathbf{r}_1.\mathbf{v}}{r_1} + v^2 - \mathbf{r}_1.\mathbf{a}\right)\right\}$$

$$= \frac{qc}{4\pi\varepsilon_0 c(r_1 c - \mathbf{r}_1.\mathbf{v})^3}\left\{\frac{\mathbf{a}r_1}{c}(r_1 c - \mathbf{r}_1.\mathbf{v}) + \mathbf{v}(\mathbf{r}_1.\mathbf{v}) - \frac{\mathbf{v}v^2 r_1}{c} + \frac{\mathbf{v}r_1}{c}(\mathbf{r}_1.\mathbf{a})\right\}$$

Adding and subtracting a term $r_1 c\mathbf{v}$ within the brackets on the right hand side, we finally get

$$\frac{\partial \mathbf{A}}{\partial t} = \frac{qc}{4\pi\varepsilon_0 c(r_1 c - \mathbf{r}_1.\mathbf{v})^3}\left\{(r_1 c - \mathbf{r}_1.\mathbf{v})\left(-\mathbf{v} + \frac{\mathbf{a}r_1}{c}\right) + \frac{r_1}{c}[c^2 - v^2 + \mathbf{r}_1.\mathbf{a}]\mathbf{v}\right\}$$

Let us now introduce a vector $\mathbf{u} = c\,\hat{\mathbf{r}}_1 - \mathbf{v}$. In terms of this vector, we may rewrite

$$\nabla\phi = \frac{qc}{4\pi\varepsilon_0 (\mathbf{r}_1.\mathbf{u})^3}\{(\mathbf{r}_1.\mathbf{u})\mathbf{v} - (c^2 - v^2 + \mathbf{r}_1.\mathbf{a})\}$$

and

$$\frac{\partial \mathbf{A}}{\partial t} = \frac{qc}{4\pi\varepsilon_0 (\mathbf{r}_1.\mathbf{u})^3}\left\{(\mathbf{r}_1.\mathbf{u})\left(-\mathbf{v} + \frac{\mathbf{a}r_1}{c}\right) + \frac{r_1}{c}[c^2 - v^2 + \mathbf{r}_1.\mathbf{a}]\mathbf{v}\right\}$$

The electric field is then given by

$$\mathbf{E} = \frac{-q}{4\pi\varepsilon_0 (\mathbf{r}_1.\mathbf{u})^3}\{(c^2 - v^2)\,(r_1\mathbf{v} - \mathbf{r}_1 c - (\mathbf{r}_1.\mathbf{a})(\mathbf{r}_1 c - r_1\mathbf{v}) + \mathbf{a}(\mathbf{r}_1.\mathbf{u})r_1\}$$

$$= \frac{qr}{4\pi\varepsilon_0 (\mathbf{r}_1.\mathbf{u})^3}\{(c^2 - v^2)\mathbf{u} + (\mathbf{r}_1.\mathbf{a})\mathbf{u} - (\mathbf{r}_1.\mathbf{u})\mathbf{a}\}$$

$$\therefore \qquad \mathbf{E} = \frac{q}{4\pi\varepsilon_0 (\mathbf{r}_1.\mathbf{u})^3}\{(c^2 - v^2)\mathbf{u} + \mathbf{r}_1 \times (\mathbf{u}\times\mathbf{a}) \tag{17}$$

Let us now calculate the magnetic field. We can write the vector potential as

$$\mathbf{A} = \frac{\mathbf{v}\phi}{c^2} \tag{18}$$

$$\nabla\times\mathbf{A} = \frac{1}{c^2}\{\phi(\nabla\times\mathbf{v}) - \mathbf{v}\times\nabla\phi\}$$

Using eqns. (5) and (11), we get

$$\nabla \times \mathbf{v} = \frac{\mathbf{a} \times \mathbf{r}_1}{(r_1 c - \mathbf{r}_1.\mathbf{v})} \tag{19}$$

Using eqns.(13) and (19), we get,

$$\phi(\nabla \times \mathbf{v}) - \mathbf{v} \times \nabla\phi$$

$$= \frac{qc}{4\pi\varepsilon_0 (r_1 c - \mathbf{r}_1.\mathbf{v})}\left[\frac{\mathbf{a} \times \mathbf{r}_1}{(r_1 c - \mathbf{r}_1.\mathbf{v})} - \mathbf{v} \times \left\{\frac{\mathbf{v}}{(r_1 c - \mathbf{r}_1.\mathbf{v})} - \frac{(c^2 - v^2 + \mathbf{r}_1.\mathbf{a})\mathbf{r}_1}{(r_1 c - \mathbf{r}_1.\mathbf{v})^2}\right\}\right]$$

$$= \frac{qc}{4\pi\varepsilon_0 (r_1 c - \mathbf{r}_1.\mathbf{v})^2}\left[\mathbf{a} \times \mathbf{r}_1 + \frac{(c^2 - v^2 + \mathbf{r}_1.\mathbf{a})(\mathbf{v} \times \mathbf{r}_1)}{(r_1 c - \mathbf{r}_1.\mathbf{v})}\right]$$

$$= \frac{qc}{4\pi\varepsilon_0 (r_1 c - \mathbf{r}_1.\mathbf{v})^3}[(\mathbf{a} \times \mathbf{r}_1)(r_1 c - \mathbf{r}_1.\mathbf{v}) + (c^2 - v^2 + \mathbf{r}_1.\mathbf{a})(\mathbf{v} \times \mathbf{r}_1)]$$

$$= \frac{qc}{4\pi\varepsilon_0 (\mathbf{r}_1.\mathbf{u})^3}[\mathbf{a}(\mathbf{r}_1.\mathbf{u}) + (c^2 - v^2)\mathbf{v} + (\mathbf{r}_1.\mathbf{a})\mathbf{v}] \times \mathbf{r}_1$$

$$\therefore \quad \mathbf{B} = \frac{q}{4\pi\varepsilon_0 c(\mathbf{r}_1.\mathbf{u})^3}\,[\mathbf{a}(\mathbf{r}_1.\mathbf{u}) + (c^2 - v^2)\mathbf{v} + (\mathbf{r}_1.\mathbf{a})\mathbf{v}] \times \mathbf{r}_1 \tag{20}$$

The first term in the electric field falls off as the inverse square of the distance from the particle. If the velocity and acceleration are both zero, $\mathbf{r}_1.\mathbf{u} = cr_1$, $\mathbf{u} = c\hat{\mathbf{r}}_1$ and so we have

$$\mathbf{E} = \frac{q}{4\pi\varepsilon_0 r_1^2}\hat{\mathbf{r}}_1$$

This is why the first term in E is referred to as the *generalized Coulomb field*. It is also known as the *velocity field*. The second term falls off as the inverse first power of r_1 and therefore dominates at large distances. It is this term that is responsible for electromagnetic radiation. It is called the *radiation field* or the *acceleration field*.

Let us now calculate the electric field of a charge moving with constant velocity. From eqn. (17), we find that it is

$$\mathbf{E} = \frac{q(c^2 - v^2)(c\mathbf{r}_1 - \mathbf{v}r_1)}{4\pi\varepsilon_0 (c r_1 - \mathbf{r}_1.\mathbf{v})^3}$$

Recall that $\mathbf{r}_1 = \mathbf{r} - \mathbf{w}(t_r) = \mathbf{r} - \mathbf{v}t_r$ in this case and $r_1 = c(t - t_r)$.

$$\therefore \quad \mathbf{r}_1.\mathbf{v} = \mathbf{r}.\mathbf{v} - v^2 t_r \tag{21}$$

$$c\mathbf{r}_1 - \mathbf{v}r_1 = c(\mathbf{r} - \mathbf{v}t_r) - \mathbf{v}c(t - t_r) = c(\mathbf{r} - \mathbf{v}t) \tag{22}$$

Also, $\quad (\mathbf{r} - \mathbf{v}t_r)^2 = c^2(t - t_r)^2$

The resulting quadratic equation for t_r has the solution

$$t_r = \frac{(c^2 t - \mathbf{r.v}) \pm \{(\mathbf{r.v} - c^2 t)^2 + (c^2 - v^2)(r^2 - c^2 t^2)\}^{1/2}}{(c^2 - v^2)} \tag{23}$$

Using eqn. (21), we have,

$$cr_1 - \mathbf{r}_1.\mathbf{v} = c^2(t - t_r) - (\mathbf{r.v} - v^2 t_r) = c^2\mathbf{t} - \mathbf{r.v} - (c^2 - v^2)t_r$$

$$\therefore \quad cr_1 - \mathbf{r}_1.\mathbf{v} = \{(\mathbf{r.v} - c^2 t)^2 + (c^2 - v^2)(r^2 - c^2 t^2)\}^{1/2} \tag{24}$$

taking the negative sign before the square root in eqn. (23). ∴ Using eqns.(22) and (24), we find the electric field to be

$$\mathbf{E} = \frac{qc(c^2 - v^2)(\mathbf{r} - \mathbf{v}t)}{4\pi\varepsilon_0 \{(\mathbf{r.v} - c^2 t)^2 + (c^2 - v^2)(r^2 - c^2 t^2)\}^{3/2}}$$

We can rewrite this by defining a vector $\mathbf{R} = \mathbf{r} - \mathbf{v}t$ which is the vector from the present position of the particle to P. Let θ be the angle between $\mathbf{R}$ and $\mathbf{v}$ (Fig. 13.4). ∴ $\mathbf{R.v} = Rv \cos\theta$. Also $\mathbf{R.v} = \mathbf{r.v} - v^2 t$. Squaring the right hand sides of both equations, we get,

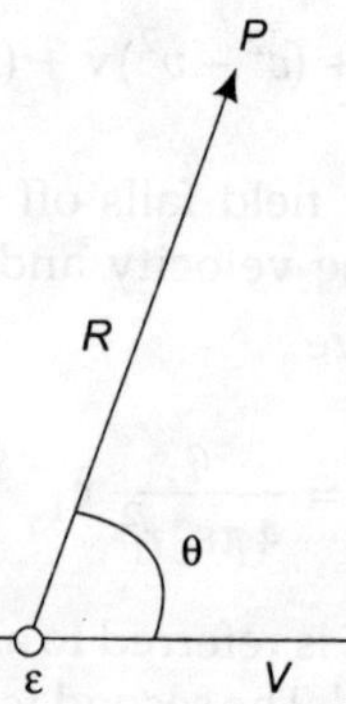

Fig. 13.4

$$(\mathbf{r.v})^2 + v^4 t^2 - 2v^2 t\mathbf{r.v} = R^2 v^2 \cos^2\theta = R^2 v^2 (1 - \sin^2\theta) \tag{25}$$

Now $(\mathbf{r.v} - c^2 t^2) + (c^2 - v^2)(r^2 - c^2 t^2) = (\mathbf{r.v})^2 - 2\,\mathbf{r.v}c^2 t + c^2 r^2 - v^2 r^2 + v^2 c^2 t^2$ (26)

Also

$$R^2 = r^2 - 2\,\mathbf{r.v}t + v^2 t^2$$

$$\therefore \quad R^2 v^2 = r^2 v^2 - 2\,(\mathbf{r.v})tv^2 + v^4 t^2 \tag{27}$$

From eqns. (25) and (27), we get,

$$R^2 v^2 \sin^2\theta = r^2 v^2 - (\mathbf{r.v})^2$$

Using this in eqn. (26), we get,

$$(\mathbf{r.v} - c^2 t)^2 + (c^2 - v^2)(r^2 - c^2 v^2) = R^2 c^2 - R^2 v^2 \sin^2\theta$$

Substituting this and the definition of **R** in the expression for the electric field, we finally get,

$$\mathbf{E} = \frac{q\left(1-\dfrac{v^2}{c^2}\right)\mathbf{R}}{4\pi\varepsilon_0 R^3\left(1-\dfrac{v^2}{c^2}\sin^2\theta\right)^{3/2}} \tag{28}$$

The electric field is much more intense at right angles to the direction of motion of the charge. Moreover, the electric field points from the present position of the charge.

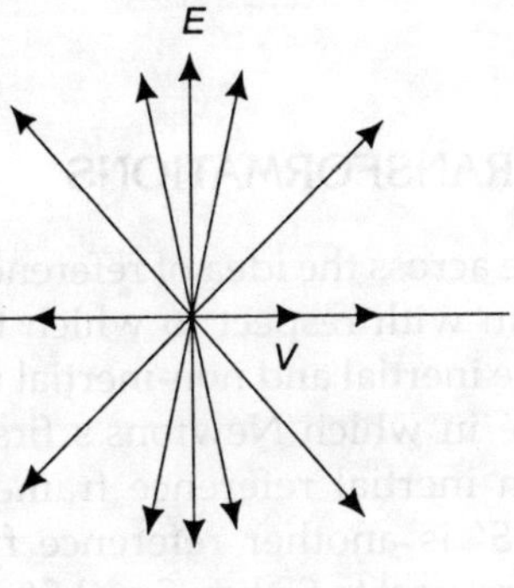

Fig. 13.5

14

Special Relativity: An Introduction

14.1 GALILEAN TRANSFORMATIONS

In mechanics, we come across the idea of reference frames. A reference frame is a co-ordinate system with respect to which the equations of motion are written down. We have inertial and non-inertial reference frames. An inertial reference frame is one in which Newtons's first law is valid. We shall be concerned solely with inertial reference frames in this chapter. If S is a reference frame and S' is another reference frame which moves with a constant velocity with respect to S, then S and S' are a pair of inertial reference frames. But if S' is accelerating with respect to S, then S and S' are a pair of non-inertial frames.

Suppose two particles A and B undergo a collision at a point P in S. This is an example of an *event*. It is specified by (x, y, z, t) where (x, y, z) are the spatial co-ordinates of P and t is the time at which the collision occurs. Consider another frame S' which moves relative to S with a constant speed v along the common x-x' axis (Fig. 14.1).

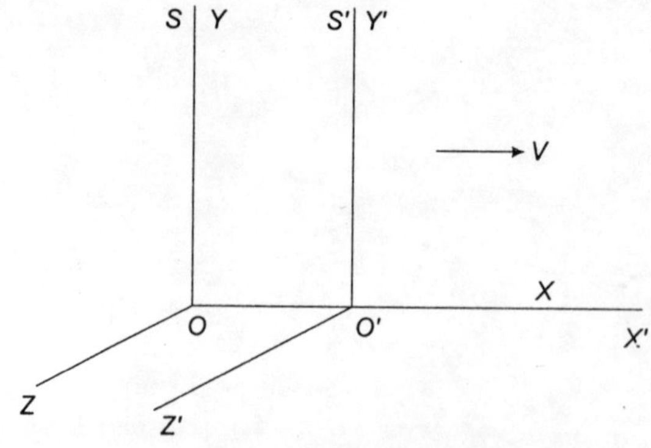

Fig. 14.1

The event occurring at P is specified by (x', y', z', t') in S'. The two sets of co-ordinates are related by the Galilean transformations:

$$x' = x - vt \tag{14.1.1}$$

$$y' = y \tag{14.1.2}$$

$$z' = z \tag{14.1.3}$$

$$t' = t \tag{14.1.4}$$

The last equation (14.1.4) describes the universal nature of the time as was assumed prior to Einstein's special relativity theory.

Let us now consider a particle moving with velocity **u** and acceleration **a** with respect to a frame S. The components of the velocity are u_x, u_y and u_z along the x, y and z axes respectively. The velocity of the particle with respect to S' is $\mathbf{u}'$ with components u_x', u_y' and u_z' respectively. Let us find the transformation between **u** and $\mathbf{u}'$. Note that

$$u_x = \frac{dx}{dt}, u_y = \frac{dy}{dt}, u_z = \frac{dz}{dt}$$

$$u_x' = \frac{dx'}{dt'} = \frac{dx'}{dt} \qquad \text{because of (14.1.4)}$$

$$= \frac{dx}{dt} - v, \qquad \text{differentiating (14.1.1)}$$

$$\therefore \quad u_x' = u_x - v \tag{14.1.5}$$

$$u_y' = \frac{dy'}{dt'} = \frac{dy}{dt} = u_y \tag{14.1.6}$$

$$u_z' = \frac{dz'}{dt'} = \frac{dz}{dt} = u_z \tag{14.1.7}$$

We can write, in general,

$$\mathbf{u}' = \mathbf{u} - \mathbf{v} \tag{14.1.8}$$

This equation is the classical velocity addition theorem. The acceleration of the particle with respect to S has components

$$a_x = \frac{du_x}{dt}, a_y = \frac{du_y}{dt}, a_z = \frac{du_z}{dt}$$

To find the components of acceleration $\mathbf{a}'$ of the particle with respect to S', we differentiate eqns. (14.1.5)–(14.1.7) with respect to t'. We get

$$a_x' = \frac{du_x'}{dt'} = \frac{du_x}{dt} = a_x \qquad \text{since } v \text{ is constant} \tag{14.1.9}$$

$$a_y' = \frac{du_y'}{dt'} = \frac{du_y}{dt} = a_y \tag{14.1.10}$$

$$a_z' = \frac{du_z'}{dt'} = \frac{du_z}{dt} = a_z \tag{14.1.11}$$

Thus, in general

$$\mathbf{a}' = \mathbf{a} \tag{14.1.12}$$

So the acceleration of a particle as measured by S and S' is the same. If we write Newton's second law in the form $\mathbf{F} = m\,\mathbf{a}$, where $\mathbf{F}$ is the force and m is the inertial mass, then the force $\mathbf{F}'$ with respect to S' is $\mathbf{F}' = m\mathbf{a}'$. But because of eqn. (14.1.12), we find $\mathbf{F}' = \mathbf{F}$ (the inertial mass $m' = m$ being the same for all frames). Thus Newton's laws of motion are the same in all inertial frames. Since the principle of conservation of momentum and energy can be deduced from Newton's laws, we can say that they are satisfied for all inertial frames. Thus the laws of mechanics are the same in all inertial reference frames. One of the important consequences of this result is that all ineetial frames are equivalent. By performing experiments in only one inertial frame, we cannot determine whether that frame is in motion with respect to another inertial frame or at rest with respect to it. Only by comparing measurements between the two frames S and S' can we determine their relative velocity. So there is no absolute rest frame. This is sometimes called Newtonian relativity.

We have seen that Newton's laws of mechanics have the same form in a pair of inertial reference frames. We say that the laws of mechanics are invariant under Galilean transformations. But it turns out that Maxwell's equations are not invariant under Galilean transformations. Moreover the classical velocity addition theorem suggests that there is an absolute reference frame in which the speed of light is c. In another inertial frame, moving with speed v with respect to this absolute frame (either in the same or opposite direction to the light beam), the speed of light is $c + v$ or $c - v$. It was believed (see Tutorial 1 of chapter 12) that light waves propagated through the ether frame with speed c. Attempts were thus made to determine the properties of this ether. This medium had to be sufficiently tenuous to allow the motion of planets with no detectable change in their speed and yet have sufficiently large restoring forces to account for the extremely large speed of light.

Example 1: Show that the homogeneous wave equation [eqn. (11.6.16) with $\rho = 0$] is not invariant under Galilean transformations.

Solution: The homogeneous wave equation satisfied by the scalar potential ϕ is

$$\nabla^2\phi - \frac{1}{c^2}\frac{\partial^2\phi}{\partial t^2} = 0$$

In Cartesian co-ordinates,

$$\nabla^2\phi = \frac{\partial^2\phi}{\partial x^2} + \frac{\partial^2\phi}{\partial y^2} + \frac{\partial^2\phi}{\partial z^2}$$

To relate the derivatives with respect to the pair of co-ordinates (x, y, z, t) and (x', y', z', t'), we have to use the chain rule for partial derivatives.

$$\frac{\partial\phi}{\partial x} = \frac{\partial\phi}{\partial x'}\frac{\partial x'}{\partial x} + \frac{\partial\phi}{\partial y'}\frac{\partial y'}{\partial x} + \frac{\partial\phi}{\partial z'}\frac{\partial z'}{\partial x} + \frac{\partial\phi}{\partial t'}\frac{\partial t'}{\partial x} = \frac{\partial\phi}{\partial x'}$$

using eqns.(14.1.1)–(14.1.4)

$$\frac{\partial\phi}{\partial y} = \frac{\partial\phi}{\partial x'}\frac{\partial x'}{\partial y} + \frac{\partial\phi}{\partial y'}\frac{\partial y'}{\partial y} + \frac{\partial\phi}{\partial z'}\frac{\partial z'}{\partial y} + \frac{\partial\phi}{\partial t'}\frac{\partial t'}{\partial y} = \frac{\partial\phi}{\partial y'}$$

Similarly, $\frac{\partial\phi}{\partial z} = \frac{\partial\phi}{\partial z'}$

$$\frac{\partial\phi}{\partial t} = \frac{\partial\phi}{\partial x'}\frac{\partial x'}{\partial t} + \frac{\partial\phi}{\partial y'}\frac{\partial y'}{\partial t} + \frac{\partial\phi}{\partial z'}\frac{\partial z'}{\partial t} + \frac{\partial\phi}{\partial t'}\frac{\partial t'}{\partial x} = -\frac{v\partial\phi}{\partial x} + \frac{\partial\phi}{\partial t'}$$

$$\frac{\partial^2\phi}{\partial x^2} = \frac{\partial}{\partial x}\left(\frac{\partial\phi}{\partial x}\right)$$

$$= \frac{\partial}{\partial x'}\left(\frac{\partial\phi}{\partial x'}\right)\frac{\partial x'}{\partial x} + \frac{\partial}{\partial y'}\left(\frac{\partial\phi}{\partial x'}\right)\frac{\partial y'}{\partial x} + \frac{\partial}{\partial z'}\left(\frac{\partial\phi}{\partial x'}\right)\frac{\partial z'}{\partial x}$$

$$+ \frac{\partial}{\partial t'}\left(\frac{\partial\phi}{\partial x'}\right)\frac{\partial t'}{\partial x} = \frac{\partial^2\phi}{\partial x'^2}$$

Similarly, $\frac{\partial^2\phi}{\partial y^2} = \frac{\partial^2\phi}{\partial y'^2}, \frac{\partial^2\phi}{\partial z^2} = \frac{\partial^2\phi}{\partial z'^2}$

$$\frac{\partial^2\phi}{\partial t^2} = \frac{\partial}{\partial t}\left(-\frac{v\partial\phi}{\partial x'} + \frac{\partial\phi}{\partial t'}\right)$$

$$= \frac{\partial}{\partial x'}\left(-\frac{v\partial\phi}{\partial x'} + \frac{\partial\phi}{\partial t'}\right)\frac{\partial x'}{\partial t} + \frac{\partial}{\partial y'}\left(-\frac{v\partial\phi}{\partial x'} + \frac{\partial\phi}{\partial t'}\right)\frac{\partial y'}{\partial t}$$

$$+ \frac{\partial}{\partial z'}\left(-\frac{v\partial\phi}{\partial x'} + \frac{\partial\phi}{\partial t'}\right)\frac{\partial z'}{\partial t} + \frac{\partial}{\partial t'}\left(-\frac{v\partial\phi}{\partial x'} + \frac{\partial\phi}{\partial t'}\right)\frac{\partial t'}{\partial t}$$

$$= \frac{v^2\partial^2\phi}{\partial x'^2} - \frac{2v\partial^2\phi}{\partial x'\partial t'} + \frac{\partial^2\phi}{\partial t'^2}$$

∴ The homogeneous wave equation becomes

$$\left(1-\frac{v^2}{c^2}\right)\frac{\partial^2\phi}{\partial x'^2}+\frac{2v}{c^2}\frac{\partial^2\phi}{\partial x'\partial t'}+\frac{\partial^2\phi}{\partial y'^2}+\frac{\partial^2\phi}{\partial z'^2}-\frac{1}{c^2}\frac{\partial^2\phi}{\partial t'^2}=0$$

which is of a different form.

14.2 THE MICHELSON–MORLEY EXPERIMENT

One of the most famous attempts to find the "ether" frame was by A.A. Michelson and E.W. Morley in 1887. A schematic diagram of the apparatus is shown in Fig. 14.2[18a] S is a source of light. P is a glass plate with a semi-transparent metal coating on the face towards S. It is a called a beam splitter. M_1 and M_2 are two mirrors. C is a compensating plate. Light from S is incident on P. Part of it is transmitted through the glass and is incident on M_1. It is reflected from M_1 back towards P where a part of it is reflected and moves back through the plate towards the telescope T. The other part of the light incident from S towards P is reflected at P, passes through C and is reflected at M_2. It then passes through C again and is partly transmitted back through the glass plate P and moves back towards the telescope. In absence of the compensating plate C, the beam moving towards M_1 passes thrice through the glass of the glass of the plate P whereas the beam moving towards M_2 passes only once through P. With the compensating plate C, both beams move thrice through the glass. The optical path through glass is then the same for both beams. Hence the name of C. The mirrors M_1 and M_2 are almost at right angles to each other and plate P is at 45° to either. This setup is called a Michelson interferometer. If $l_1 = l_2$ and the mirrors are not exactly at right angles to each other, interference fringes are observed in the telescope just as we observe interference due to a wedge-shaped air film between glass plates.

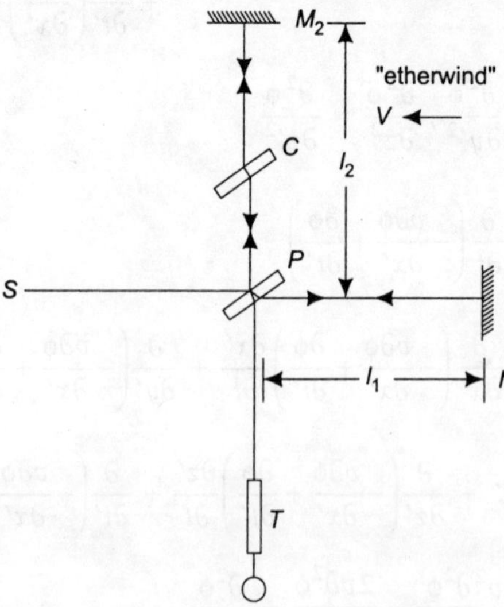

Fig. 14.2

Suppose that the whole apparatus is moving to the right through the ether with a speed v. Then, in the laboratory frame there is an ether wind moving in the opposite direction with speed v. Then the speed of light in the lab frame is $c - v$ when it goes from P to M_1 and $c + v$ when it moves from M_1 to P. The time t_1 taken by the light to go from P to M_1 and back is

$$t_1 = \frac{l_1}{c-v} + \frac{l_1}{c+v} = \frac{2l_1 c}{c^2 - v^2}$$

or,
$$t_1 = \frac{2l_1/c}{1-\frac{v^2}{c^2}} \tag{14.2.1}$$

Let t_2 be the time required for the beam to go from P to M_2 and back. Let us consider the process from the point of view of the ether frame (Fig. 14.3). Then, from the diagram

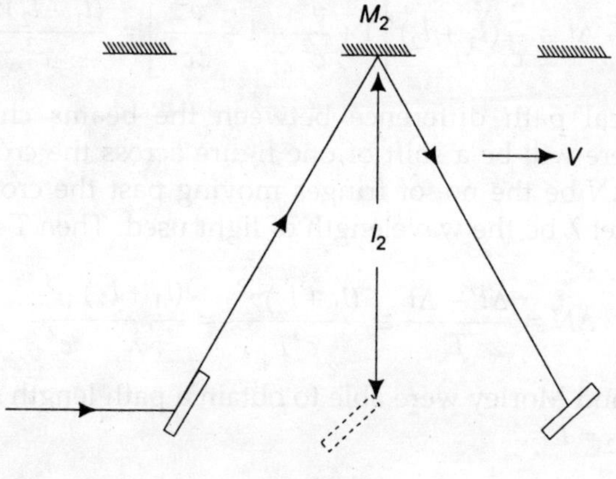

Fig. 14.3

$$2\left\{l_2^2 + \left(\frac{vt_2}{2}\right)^2\right\}^{1/2} = ct_2$$

or,
$$t_2 = \frac{2l_2}{\sqrt{c^2 - v^2}} = \frac{2l_2}{c\sqrt{1-\frac{v^2}{c^2}}} \tag{14.2.2}$$

The difference in transit times is

$$\Delta t = t_2 - t_1 = \frac{2}{c}\left[\frac{l_2}{\sqrt{1-\frac{v^2}{c^2}}} - \frac{l_1}{1-\frac{v^2}{c^2}}\right] \tag{14.2.3}$$

Suppose now that the apparatus is rotated through 90°. Then l_1 and l_2 interchange positions and we have, for the difference in transit times,

$$\Delta t' = t_2' - t_1' = \frac{2}{c}\left[\frac{l_2}{1-\dfrac{v^2}{c^2}} - \frac{l_1}{\sqrt{1-\dfrac{v^2}{c^2}}}\right] \tag{14.2.4}$$

$$\therefore \qquad \Delta t' - \Delta t = \frac{2}{c}\left[\frac{l_1+l_2}{1-\dfrac{v^2}{c^2}} - \frac{(l_1+l_2)}{\sqrt{1-\dfrac{v^2}{c^2}}}\right]$$

Using the binomial expansion and keeping terms only upto v^2/c^2, we get

$$\Delta t' - \Delta t \cong \frac{2}{c}(l_1+l_2)\left[1+\frac{v^2}{c^2}-1-\frac{v^2}{2c^2}\right] = \frac{(l_1+l_2)v^2}{c^3}$$

If the optical path difference between the beams changes by one wavelength, there will be a shift of one figure across the cross hairs of the telescope. Let ΔN be the no. of fringes moving past the cross hairs as the pattern shifts. Let λ be the wavelength of light used. Then $T = \lambda/c$.

$$\therefore \qquad \Delta N = \frac{\Delta t' - \Delta t}{T} \cong \frac{(l_1+l_2)v^2}{c^3 T} = \frac{(l_1+l_2)}{\lambda}\frac{v^2}{c^2}$$

Michelson and Morley were able to obtain a path length $l_1 + l_2$ of about 22m. Also $l_1 \cong l_2 = l$

So

$$\Delta N = \frac{2lv^2}{\lambda c^2}$$

If we use for v the orbital speed of the earth about the sun (~ 30 km/s.), then $v/c \cong 10^{-4}$ and $\Delta N = 0.4$. Michelson and Morley rotated the apparatus through 360° and took readings during day and night and in all the seasons but did not observe any fringe shift at all. Thus the experiment did not find any evidence for an absolute (ether) frame.

14.3 THE POSTULATES OF SPECIAL RELATIVITY

In 1905, Einstein put forward his two postulates of special relatively.

Postulate 1: The laws of physics are the same in all inertial reference frames.

Postulate 2: The speed of light in free space always has the same value c.

Postulate 1 is the principle of relativity and asserts that there is no preferred inertial frame. All inertial frames are equivalent. Postulate 2 is the principle of constancy of the speed of light. The Michelson-Morley experiment provided a proof of this postulate. This was also confirmed by other experiments which we shall not go into. See p. 73-74 of reference 18a for details.

14.4 THE LORENTZ TRANSFORMATION

It turns out that the solution to the difficulties mentioned earlier lies in the postulates of special relativity and is embodied in a new set of transformations. This new set of transformations is the Lorentz transformation. We shall follow ref. 18b in deriving these transformations.

We shall use the principle of homogeneity of space and time and also the isotropy of space, in addition to the postulates above. The homogeneity of space and time implies that the results of a measurement of a length or a time interval should not depend on where and when it is performed. The transformations should thus be linear. The most general linear transformation between the set of co-ordinates (x, y, z, t) and (x', y', z', t') is

$$x' = a_{11}x + a_{12}y + a_{13}z + a_{14}t \tag{14.4.1}$$

$$y' = a_{21}x + a_{22}y + a_{23}z + a_{24}t \tag{14.4.2}$$

$$z' = a_{31}x + a_{32}y + a_{33}z + a_{34}t \tag{14.4.3}$$

$$t' = a_{41}x + a_{42}y + a_{43}z + a_{44}t \tag{14.4.4}$$

Look back at Fig. 14.1. The x and x' axes always coincide. The equation representing the x-axis is $y = 0$, $z = 0$. $\therefore$ $y = 0$, $z = 0$ should imply $y' = 0$, $z' = 0$. Then eqns. (14.4.2) and (14.4.3) simplify to

$$y' = a_{22}y + a_{23}z \tag{14.4.2a}$$

$$z' = a_{32}y + a_{33}z \tag{14.4.3a}$$

The XY plane goes over into the $X'Y'$ plane. Similarly, the XZ plane goes over into the $X'Y'$ plane. In other words, $z = 0$ implies $z' = 0$ and $y = 0$ implies $y' = 0$. $\therefore$ Eqns. (14.4.2a) and (14.4.3a) become

$$y' = a_{22}y \tag{14.4.2b}$$

$$z' = a_{33}z \tag{14.4.3b}$$

Consider a rod of unit length at rest in the S frame, lying along the y-axis. $\therefore$ From eqn. (14.4.2b), its length as measured by the S' observer is $y' = a_{22}$. Now suppose the rod is at rest in the S' frame, lying along the y'-axis. The S' observer will now measure its length to be unity. According to eqn. (14.4.2b), its length, as measured by the S observer is $y = 1/a_{22}$. Since these measurements are reciprocal in nature and the postulate of relativity (postulate 1) requires these measurements to be identical, we must have $a_{22} = 1/a_{22}$, or $a_{22} = 1$. Similarly, $a_{33} = 1$. $\therefore$ Eqns. (14.4.2b) and (14.4.3b) reduce to

$$y' = y \tag{14.4.5}$$

$$z' = z \tag{14.4.6}$$

Let us now come to eqn. (14.4.4). This transformation should not depend on y and z. If it did, then clocks placed symmetrically with respect to the x-axis, i.e. at $+y$ and $-y$ would disagree with each other. Similarly, clocks placed at $+z$ and $-z$ would show different values for t'. Thus $a_{42} = 0$ and $a_{43} = 0$ and eqn. (14.4.4) becomes

$$t' = a_{41}x + a_{44}t \tag{14.4.4a}$$

Let us now consider eqn. (14.4.1) The origin of the S' frame coincides with the origin of the S frame at $t = 0$. Thereafter the origin O' moves along the common x-x' axis with speed v. ∴ With respect to the S frame, its equation is $x = vt$. This means that $x' = 0$ is identical with $x = vt$. ∴ Eqn.(14.4.1) becomes

$$x' = a_{11}(x - vt) \tag{14.4.1a}$$

To determine the constants a_{11}, a_{41} and a_{44}, we consider a spherical electromagnetic wave expanding radially outwards from the origin O at $t = 0$. The equation of the wave in the two reference frames S and S' have the same form because of the principle of constancy of the speed of light:

$$x'^2 + y'^2 + z'^2 = c^2t'^2 \tag{14.4.7}$$

$$x^2 + y^2 + z^2 = c^2t^2 \tag{14.4.8}$$

Substituting eqns. (14.4.1a), (14.4.4a), (14.4.5) and (14.4.6) into eqn. (14.4.7), we get

$$a_{11}^2(x - vt)^2 + y^2 + z^2 = c^2(a_{41}x + a_{44}t)^2$$

or, $$(a_{11}^2 - c^2a_{41}^2)x^2 + y^2 + z^2 - 2(va_{11}^2 + c^2a_{41}a_{44})xt = (c^2a_{44}^2 - v^2a_{11}^2)t^2$$

This equation must agree with eqn. (14.4.8) and so we must have

$$a_{11}^2 - c^2a_{41}^2 = 1 \tag{14.4.9a}$$

$$c^2a_{44}^2 - v^2a_{11}^2 = c^2 \tag{14.4.9b}$$

$$va_{11}^2 + c^2a_{41}a_{44} = 0 \tag{14.4.9c}$$

From eqn. (14.4.9c), we get, $a_{41} = \dfrac{-va_{11}^2}{c^2a_{44}}$, Substituting this into eqn. (14.4.9a), we get

$$a_{11}^2\left(1 - \frac{v^2a_{11}^2}{c^2a_{44}^2}\right) = 1 \tag{14.4.10}$$

Dividing eqn. (14.4.9b) throughout by $c^2a_{44}^2$, we get,

$$1 \quad \frac{v^2a_{11}^2}{c^2a_{44}^2} = \frac{1}{a_{44}^2} \tag{14.4.11}$$

Substituting this in eqn. (14.4.10), we get,

$$a_{11}^2 = a_{44}^2 \tag{14.4.12}$$

Putting this in eqn. (14.4.11), we get,

$$a_{44} = \frac{1}{\sqrt{1-\frac{v^2}{c^2}}} = a_{11} \tag{14.4.13}$$

where we have taken the positive sign throughout. And lastly, we get,

$$a_{41} = \frac{-v}{c^2}.\frac{1}{\sqrt{1-\frac{v^2}{c^2}}} \tag{14.4.14}$$

The transformation eqns. (14.4.1a), (14.4.5), (14.4.6) and (14.4.4a), are finally,

$$x' = \frac{x - vt}{\sqrt{1-\frac{v^2}{c^2}}} \tag{14.4.15}$$

$$y' = y \tag{14.4.5}$$

$$z' = x \tag{14.4.6}$$

$$t' = \frac{t - \frac{v^2}{c^2}}{\sqrt{1-\frac{v^2}{c^2}}} \tag{14.4.16}$$

These equations are knows as the *Lorentz transformation*. They give the transformation from S to S'. Let us now find the transformation from S' to S, i.e. the inverse transformation. We have to solve for x, y, z, t in terms of x', y', z', t'. Rewriting eqn. (14.4.15), making x the subject of the equation, we get,

$$x = x'\sqrt{1-\frac{v^2}{c^2}} + vt$$

Similarly, rewriting eqn. (14.4.16), making t the subject of the equation, we get,

$$t = t'\sqrt{1-\frac{v^2}{c^2}} + \frac{vx}{c^2}$$

Substituting for t in the above equation for x, we get,

$$x = x'\sqrt{1-\frac{v^2}{c^2}} + vt'\sqrt{1-\frac{v^2}{c^2}} + \frac{v^2x}{c^2}$$

or, simplifying, we get,

$$x = \frac{x' + vt'}{\sqrt{1 - \frac{v^2}{c^2}}} \tag{14.4.17}$$

Now, substituting for x in the above equation for t, we get,

$$t = t'\sqrt{1 - \frac{v^2}{c^2}} + \frac{vx'}{c^2}\sqrt{1 - \frac{v^2}{c^2}} + \frac{v^2 t}{c^2}$$

or, simplifying, we get,

$$t = \frac{t' + \frac{vx'}{c^2}}{\sqrt{1 - \frac{v^2}{c^2}}} \tag{14.4.18}$$

These two equation, (14.4.17) and (14.4.18) together with

$$y = y' \tag{14.4.5}$$

$$z = z' \tag{14.4.6}$$

are the inverse Lorentz transformations. Note that S moves with speed $-v$ with respect to S'. So we expect to obtain the inverse transformation by replacing v by $-v$ in the Lorentz transformation. This agrees with what we have obtained.

14.5 THE RELATIVITY OF SIMULTANEITY

One of the new ideas introduced by special relativity is that events which are simultaneous in one inertial frame are not simultaneous in another inertial frame. This has come about from the finite speed of light and thus the absence of a universal time. In order to record an event, one needs an observer and a clock to record its time and location. When two events ar simultaneous, it means that they occur at the same time. If they occur at the same location, the clock at this location records the time of occurrence. However, if the two events occur at different locations, then they will be called simultaneous if the clocks at those two locations record the same time. For this conclusion to be valid, the clocks must be synchronized in a given reference frame. Let A and B be two locations in a given reference frame, with a clock stationed at each location. Let there be a source of light at A and a mirror at B. Let t_0 be the time required by light to go from A to B and back to A. The observer at A sets his clock to 0 when the light signal starts out from A and observer B sets his clock to $t_0/2$ when the light signal reaches him. Then the clocks at A and B are said to be synchronized.

Let us now consider an observer O in the S frame and an observer O' in the S' frame. Suppose two events occur at two points C and D in S such that the light signals from them reach O at the same time. Clearly O observes the events to be simultaneous and also by nothing the locations of C and D, he observes himself to be at the midpoint between C and D. Observer O′ is moving with speed v towards the location of D, say. So, clearly, he has moved to the right by a certain distance by the time the light signal from D reaches him. The light signal from D reaches O′ earlier than that from C because O′ is moving away from C. ∴ To O′, the event at D occurs earlier than the event at C. Thus the events are not simultaneous to O′.

14.6 LENGTH CONTRACTION AND TIME DILATION

We will now look at two consequences of the Lorentz transformations. Let us first consider length contraction. Let a rod of length l be at rest in the S frame, lying along the x-axis. the end co-ordinates of the rod as measured in the S frame are x_2 and x_1 so that

$$l = x_2 - x_1 \tag{14.6.1}$$

To measure the length l' of the rod in the S' frame, the observer in this frame must note the co-ordinates x_2' and x_1' of the rod at the same time t'. From eqn. (14.4.17), we get,

$$x_2 = \frac{x_2' + vt_2'}{\sqrt{1 - \frac{v^2}{c^2}}}$$

$$x_1 = \frac{x_1' + vt_1'}{\sqrt{1 - \frac{v^2}{c^2}}} \qquad \therefore \quad x_2 - x_1 = \frac{x_2' + x_1'}{\sqrt{1 - \frac{v^2}{c^2}}}$$

Let us call $l' = x_2' - x_1'$

$$\therefore \qquad l' = l\sqrt{1 - \frac{v^2}{c^2}} = \frac{l}{\gamma} \tag{14.6.2}$$

$$\gamma = \sqrt{1 - \frac{v^2}{c^2}}$$

∴ The length of the rod as measured in the S' frame is shorter than the length of the rod as measured in the rest frame of the rod, i.e. the S frame. ∴ The rod has its greatest length in the S frame. The length of the rod in a frame in which it is at rest is called its *proper length*.

Now let us look at the phenomenon of time dilation. Let $\Delta t = t_2 - t_1$ be the time interval between two events occurring at the same point in the S frame. This time interval is measured by the same clock in S, i.e. the clock at the location of the event in S. What is the time interval $\Delta t'$ between these events as measured in S'? Since S' is moving with respect to S with a speed v, the events occur at different places is S' and are timed by two different clocks in S'. These clocks in S' are, however synchronized with each other in S'. $\Delta t' = t_2' - t_1'$. So the readings of the two moving clocks (in S') must be compared with the reading of one stationary clock (in S). From eqn. (14.4.16), we get,

$$t_2' = \gamma\left(t_2 - \frac{vx}{c^2}\right)$$

$$t_1' = \gamma\left(t_1 - \frac{vx}{c^2}\right)$$

$$\therefore \quad t_2' - t_1' = \gamma(t_2 - t_1)$$

$$\therefore \quad \Delta t' = \gamma \Delta t \tag{14.6.3}$$

∴ The time interval between the events is larger by a factor γ as measured in the S' frame. The time interval Δt between the events as measured by a clock in S is called the *proper time* τ. The time interval between the two events is thus shortest in a reference frame in which the clock is at rest. The single clock in this example appears moving in the S' frame and the elapsed time between the two events appears shorter than the elapsed time as measured by clocks in S'. The scale of time as measured by the moving clock appears stretched or dilated.[18a]

Example 2: A 100 MeV electron for which $v = 0.999975c$ moves along the axis of an evacuated tube which has a length l' of 3.00 m as measured by a laboratory observer S' with respect to whom the tube is at rest. An observer S moving with the electron would see the tube moving past at a speed v. What length would observer S measure for this tube?

Solution: In the S frame, the length of the tube l would appear to be shorter. It would be

$$l = l'\sqrt{1-\left(\frac{v}{c}\right)^2} = 3.00 \times \sqrt{1-0.99995} = 2.12 \times 10^2 \text{ m.}$$

Example 3: (a) If the average (proper) lifetime of a μ-meson is 2.3×10^{-6} sec., what average distance would it travel in vacuum before dying as measured in reference frames in which its velocity is 0.00c, 0.60c, 0.90c and 0.99c? (b) Compare each of these distances with the distance the meson sees itself travelling through.

Solution: (a) For $v = 0$, $d = 0$. For a reference frame in which the speed of the μ-meson is $v = 0.60c$, the lifetime is measured to be

$$\Delta t = \frac{2.3 \times 10^{-6}}{\sqrt{1-(0.6)^2}} = 2.875 \times 10^{-6} s.$$

∴ In this frame, it travels a distance $d = 0.6c\ \Delta t = 518$ m.

For $v = 0.90c$, $\Delta t = 5.275 \times 10^{-6}$ s, $d = 0.9c\ \Delta t = 1.42$ km.

For $v = 0.90c$, $\Delta t = 16.3 \times 10^{-6}$ s, $d = 0.99c\ \Delta t = 4.84$ km.

(b) In the frame of the muon, its lifetime is 2.3×10^{-6} s. $d = 0$ in the first case. In the second case, the muon sees the reference frame move past with a speed $0.60c$, and it travels a distance $d' = d\sqrt{1-(0.6)^2} = 518 \times 0.8 = 414$ m.

For $v = 0.90c$, $d' = d\sqrt{1-(0.9)^2} = 1.42 \times 10^3 \times 0.436 = 619$ m.

For $v = 0.99c$, $d' = d\sqrt{1-(0.99)^2} = 4.84 \times 10^3 \times 0.141 = 682$ m.

As far as lengths perpendicular to the direction of motion are concerned, they are measured to be the same by both observers. A rod of length 1 m lying along the y axis in the S frame would be measured to have a length of 1m by the observer in the S' frame moving along the common x-x' axis. Consider a rod of length 1m lying along the y axis of the S frame with one end at the origin O. Consider another rod of length 1m lying along the y' axis of the S' frame with one end at the origin O'. Let us suppose that each has a pointer attached to the other end which puts a mark on the other rod as they pass each other. The observers agree on the simultaneity of the measurement and the result of the measurement. After the rods pass each other, there are two possibilities. Either each observer finds his pointer marked by the other. Or, one observer finds his rod marked at a place below the pointer and the other observer finds his rod unmarked. Then they both must agree that one rod is shorter than the other. If that is the case then one reference frame is an absolute one in the sense that a rod gets stretched in that frame or gets contracted (in the opposite direction). This would contradict the principle of relativity. Thus the first possibility is the correct one. Both observers measure the length of the other rod to be 1 m.

14.7 EINSTEIN'S VELOCITY ADDITION RULE

Let us now consider the transformation of velocities using the Lorenz transformations. Let (u_x, u_y, u_z) be the components of velocity $\mathbf{u}$ of a particle in the S frame and (u_x', u_y', u_z') be the components of velocity $\mathbf{u}'$ of the particle in the S' frame. Note that

$$u_x = \frac{dx}{dt},\ u_y = \frac{dy}{dt},\ u_z = \frac{dz}{dt},\ u_x' = \frac{dx'}{dt'},\ u_y' = \frac{dy'}{dt'},\ u_z' = \frac{dz'}{dt'}$$

Take differentials of both sides of eqn. (14.4.15).

$$dx' = \frac{dx - vdt}{\sqrt{1-\dfrac{v^2}{c^2}}}$$

or, $$u_x'dt = \frac{u_x dt - v dt}{\sqrt{1-\frac{v^2}{c^2}}} = \frac{(u_x - v)dt}{\sqrt{1-\frac{v^2}{c^2}}} \tag{14.7.1}$$

Take differentials of both sides of eqn. (14.4.16)

$$dt' = \frac{dt - \frac{v dx}{c^2}}{\sqrt{1-\frac{v^2}{c^2}}} = \frac{dt\left(1-\frac{u_x v}{c^2}\right)}{\sqrt{1-\frac{v^2}{c^2}}} \tag{14.7.2}$$

Substituting for dt' from (14.7.2) into (14.7.1), we get,

$$\frac{u_x' dt\left(1-\frac{u_x v}{c^2}\right)}{\sqrt{1-\frac{v^2}{c^2}}} = \frac{(u_x - v)dt}{\sqrt{1-\frac{v^2}{c^2}}}$$

or, $$u_x' = \frac{u_x - v}{\left(1-\frac{u_x v}{c^2}\right)} \tag{14.7.3}$$

Taking differentials of both sides of eqn. (14.4.5), we get

$$dy' = dy$$

$$\therefore \quad \frac{dy'}{dt'} = \frac{dy}{dt'}$$

or, $$u_y' = \frac{dy}{dt}\frac{\sqrt{1-\frac{v^2}{c^2}}}{\left(1-\frac{u_x v}{c^2}\right)},$$ using eqn. (14.7.2)

$$u_y' = u_y \frac{\sqrt{1-\frac{v^2}{c^2}}}{\left(1-\frac{u_x v}{c^2}\right)} \tag{14.7.4}$$

Similarly, the transformation of the z component of velocity is

$$u_z' = u_z \frac{\sqrt{1-\frac{v^2}{c^2}}}{\left(1-\frac{u_x v}{c^2}\right)} \tag{14.7.5}$$

Thus eqns. (14.7.3)–(14.7.5) are the relativistic transformation of velocities. Note that if u_x, $v << c$, then eqns. (14.7.3)–(14.7.5) reduce to the classical velocity addition rule. Let us consider an example where $u_x = c$, $u_y = 0$, $u_z = 0$. Then eqn. (14.7.3) becomes.

$$u_x' = \frac{c - v}{1 - \frac{v}{c}} = c$$

This is in clear contradiction to the classical velocity addition rule where $u_x' = c - v$. To derive the inverse transformations, consider eqn. (14.7.3)

$$u_x' - u_x'\frac{u_x v}{c^2} = u_x - v$$

or, $$u_x = u_x' + v - \frac{u_x u_x' v}{c^2}$$

or, $$u_x = \frac{u_x' + v}{1 + \frac{u_x' v}{c^2}} \tag{14.7.6}$$

This is what we expect to get if we replace v by $-v$ in eqn. (14.7.3). From eqn. (14.7.6),

$$1 - \frac{u_x v}{c^2} = 1 - \left(\frac{u_x' v + v^2}{c^2 + u_x' v}\right) = \frac{c^2 - v^2}{c^2 + u_x' v}$$

∴ $$1 - \frac{u_x v}{c^2} = \frac{1 - \frac{v^2}{c^2}}{1 + \frac{u_x' v}{c^2}} \tag{14.7.7}$$

From eqn. (14.7.4), we get,

$$u_y = \frac{u_y'\left(1 - \frac{u_x v}{c^2}\right)}{\sqrt{1 - \frac{v^2}{c^2}}} = \frac{u_y'\left(1 - \frac{v^2}{c^2}\right)}{\left(1 + \frac{u_x' v}{c^2}\right)\sqrt{1 - \frac{v^2}{c^2}}} \text{ using eqn. (14.7.7)}$$

∴ $$u_y = \frac{u_y'\sqrt{1 - \frac{v^2}{c^2}}}{\left(1 + \frac{u_x' v}{c^2}\right)} \tag{14.7.8}$$

Similarly,
$$u_z = \frac{u_z' \sqrt{1 - \frac{v^2}{c^2}}}{\left(1 + \frac{u_x' v}{c^2}\right)} \tag{14.7.9}$$

This is what we expect to get when we replace v by $-v$ in eqns. (14.7.4) and (14.7.5). If we put $u_x' = c$, $v = c/2$ in eqn. (14.7.6), we get $u_x = c$. Classically, we would get $u_x = 3c/2$. But c is the ultimate speed which can be attained by a body. This is thus clear from eqn. (14.7.6)

14.8 DOPPLER EFFECT IN RELATIVITY

We are all familiar with the Doppler effect in classical physics. It is the apparent increase or decrease in frequency of a wave (e.g. a sound wave) as the source moves towards or away from an observer. We will now study the phenomenon when relativistic effects come into play. Let us suppose that a plane electromagnetic wave of unit amplitude is emitted from the origin O' in the S' frame. The wave normal makes an angle θ' with the x' axis (Fig. 14.4) and lies in the x' y' plane. The S' frame moves with a speed v with respect to the S frame. The displacement of the wave in the S' frame is given by

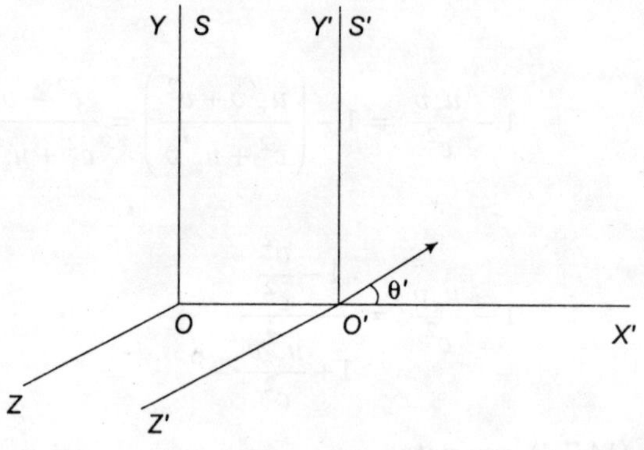

Fig. 14.4

$$\xi' = \cos\left(\frac{x'\cos\theta' + y'\sin\theta'}{\lambda'} - \nu' t'\right) \tag{14.8.1}$$

Since the Lorentz transformation is linear, a plane is transformed into a plane. ∴ The displacement of the wave in the S frame is of the same form

$$\xi = \cos\left(\frac{x\cos\theta + y\sin\theta}{\lambda} - \nu t\right) \tag{14.8.2}$$

We have $c = \nu\lambda = \nu'\lambda'$, and θ is the angle which the wave normal makes with the x axis. We must have $\xi = \xi'$. Let us apply the Lorentz transformation (14.4.15), (14.4.16), (14.4.5) to eqn. (14.8.1). We get

$$\xi' = \cos\left\{\frac{\cos\theta'}{\lambda'}\gamma(x - vt) + \frac{\sin\theta'}{\lambda'}y - \nu'\gamma\left(t - \frac{vx}{c^2}\right)\right\}$$

or, $$\xi' = \cos\left\{x\left(\frac{\gamma\cos\theta'}{\lambda'} + \frac{\gamma\nu' v}{c^2}\right) + y\frac{\sin\theta'}{\lambda'} - t'\left(\frac{\gamma v\cos\theta'}{\lambda'} + \nu'\gamma\right)\right\} \quad (14.8.3)$$

Equating coefficients of x, y and t on the right hand side of eqn. (14.8.2) and (14.8.3), we get,

$$\frac{\cos\theta}{\lambda} = \gamma\left(\frac{\cos\theta'}{\lambda'} + \frac{\nu' v}{c^2}\right)$$

or, $$\frac{\cos\theta}{\lambda} = \frac{\gamma}{\lambda'}\left(\cos\theta' + \frac{v}{c}\right) \quad (14.8.4)$$

$$\frac{\sin\theta}{\lambda} = \frac{\sin\theta'}{\lambda'} \quad (14.8.5)$$

and $$\nu = \gamma\left(\frac{v\cos\theta'}{\lambda'} + \nu'\right)$$

or, $$\nu = \gamma\nu'\left(\frac{v}{c}\cos\theta' + 1\right) \quad (14.8.6)$$

Dividing eqns. (14.8.5) by (14.8.4), we get

$$\tan\theta = \frac{\sin\theta'}{\gamma\left(\cos\theta' + \frac{v}{c}\right)} \quad (14.8.7)$$

We will consider this equation again in Tutorial 3. Eqn. (14.8.6) is the relativistic equation for the Doppler effect. The inverse transformation is given by replacing v by $-v$ in this equation and ν by ν', θ by θ':

$$\nu' = \gamma\nu\left(1 - \frac{v}{c}\cos\theta\right) \quad (14.8.8)$$

If we consider nonrelativistic speeds, $v/c << 1$. Then $\gamma = \dfrac{1}{\sqrt{1 - \dfrac{v^2}{c^2}}} \cong 1$, expanding binomially and keeping terms only upto first order in v/c.

$$\therefore \qquad \nu' \cong \nu\left(1 - \frac{v}{c}\cos\theta\right)$$

$$\text{or,} \qquad \nu = \frac{\nu'}{1 - \frac{v}{c}\cos\theta} \cong \nu'\left(1 + \frac{v}{c}\cos\theta\right)$$

For $\theta = 0$, which corresponds to the source moving towards the observer or observer moving towards the source, $\nu = \nu'\left(1 + \frac{v}{c}\right)$. $\therefore$ The observed frequency is greater than the proper frequency. For $\theta = 180°$, we have the source moving away from the observer, or the observer moving away from the source. $\nu = \nu'\left(1 - \frac{v}{c}\right)$. $\therefore$ The observed frequency is less than the proper frequency. For $\theta = 90°$, $\nu = \nu'$. Thus, upto first order in v/c, there is no Doppler effect transverse to the direction of motion. This is true classically also.

Now let us suppose that v is not small compared to c. In this case, for $\theta = 0$ in eqn.(14.8.8), we get,

$$\nu' = \frac{\nu\left(1 - \frac{v}{c}\right)}{\sqrt{1 - \frac{v^2}{c^2}}} = \nu\sqrt{\frac{c - v}{c + v}}$$

$$\text{or,} \qquad \nu = \nu'\sqrt{\frac{c + v}{c - v}} \qquad (14.8.9)$$

For $\theta = 180°$, we get,

$$\nu' = \nu\frac{1 + \frac{v}{c}}{\sqrt{1 - \frac{v^2}{c^2}}} = \nu\sqrt{\frac{c + v}{c - v}}$$

$$\therefore \qquad \nu = \nu'\sqrt{\frac{c - v}{c + v}} \qquad (14.8.10)$$

These two correspond to the longitudinal Doppler effect. The source and observer are either moving away from each other or towards each other. For $\theta = 90°$, we get, from eqn. (14.8.8),

$$\nu' = \nu\gamma$$

$$\text{or} \qquad \nu = \frac{\nu'}{\gamma} = \nu'\sqrt{1 - \frac{v^2}{c^2}} \qquad (14.8.11)$$

This is the *transverse Doppler effect*. This effect is absent classically. At right angles to the direction of relative motion, the frequency is lower than the proper frequency of the source.

14.9 RELATIVISTIC DYNAMICS

Let us now consider an elastic collision between two particles A and B in two frames S and S'. We will see that if conservation of linear momentum in one frame is to imply conservation of linear momentum in the other frame, then we must redefine linear momentum. Here we shall follow the approach of ref. 18a. Let A and B have the same mass. Let $\mathbf{u}_A'$ and $\mathbf{u}_B'$ be the velocities of A and B before the collision and $\mathbf{U}_A'$ and $\mathbf{U}_B'$ be the velocities after collision (in the S' frame). We choose $\mathbf{u}_A' = -\mathbf{u}_B'$. $\therefore$ $\mathbf{u}_{yA}' = -\mathbf{u}_{yB}'$, $\mathbf{u}_{xA}' = -\mathbf{u}_{xB}'$. Since the collision is elastic and the bodies have the same mass, we must have, after the collision, $\mathbf{U}_{yA}' = -\mathbf{U}_{yB}'$, $\mathbf{U}_{xA}' = -\mathbf{U}_{xB}'$. Further, $\mathbf{u}_{yA}' = -\mathbf{U}_{yA}'$, $\mathbf{u}_{yB}' = -\mathbf{U}_{yB}'$ and $\mathbf{u}_{xA}' = \mathbf{U}_{xA}'$, $\mathbf{u}_{xB}' = \mathbf{U}_{xB}'$. $\therefore$ The x components of velocity are unchanged. The

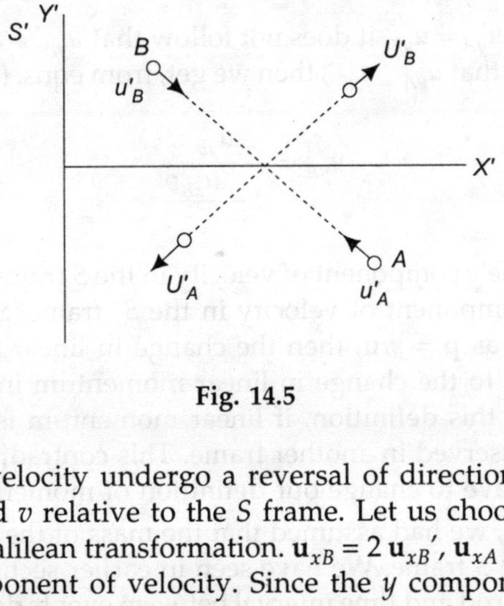

Fig. 14.5

y components of velocity undergo a reversal of direction. The S' frame is moving with speed v relative to the S frame. Let us choose $\mathbf{v} = \mathbf{u}_{xB}'$. Then, according to the Galilean transformation. $\mathbf{u}_{xB} = 2\,\mathbf{u}_{xB}'$, $\mathbf{u}_{xA} = 0$. The particle A has only a y component of velocity. Since the y component of velocity is unaffected by a Galilean transformation. $\mathbf{u}_{yA} = \mathbf{u}_{yA}'$, $\mathbf{u}_{yB} = \mathbf{u}_{yB}'$. Also $\mathbf{u}_{yA}$

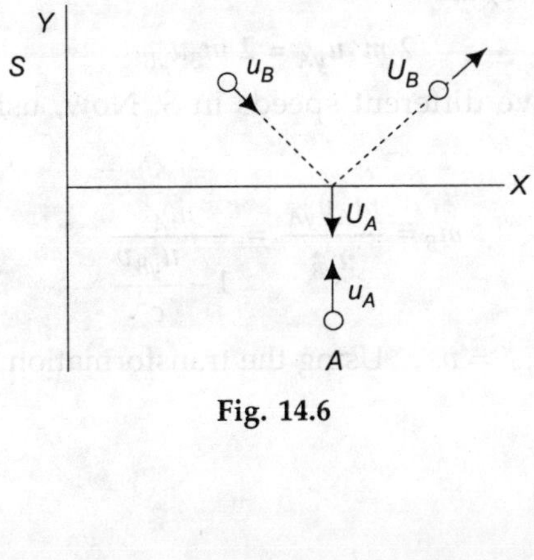

Fig. 14.6

$= -\mathbf{U}_{yA}$, $\mathbf{u}_{yB} = -\mathbf{U}_{yB}$. The loss in momentum of A equals the gain in momentum by B.

$$2\,mu_{yA} = 2\,mu_{yB} \tag{14.9.1}$$

or,

$$u_{yA} = u_{yB} \tag{14.9.2}$$

Now let us use the Lorentz transformation instead of the Galilean one. Then we get, from eqn. (14.7.4),

$$u_{yB}' = \frac{u_{yB}\sqrt{1-\dfrac{v^2}{c^2}}}{1-\dfrac{u_{xB}v}{c^2}} \tag{14.9.3}$$

and for particle A,

$$u_{yA}' = u_{yA}\sqrt{1-\frac{v^2}{c^2}} \tag{14.9.4}$$

since $u_{xA} = 0$. So if $u_{yA} = u_{yB}$, it does not follow that $u_{yA}' = u_{yB}'$ and vice versa. Thus if we assume that $u_{yA'} = u_{yB}'$, then we get, from eqns. (14.9.3) and (14.9.4).

$$u_{yA} = \frac{u_{yB}}{1-\dfrac{u_{xB}v}{c^2}} \tag{14.9.5}$$

∴ The change in the y component of velocity in the S frame is not equal to the change in the y component of velocity in the S' frame. So if we define the linear momentum as $\mathbf{p} = m\mathbf{u}$, then the change in linear momentum in one frame is not equal to the change in linear momentum in another frame. In other words, with this definition, if linear momentum is conserved in one frame, it is not conserved in another frame. This contradicts the principle of relativity. So we have to change our definition of momentum.

In eqn. (14.9.1), we had assumed that the mass of the bodies is the same as measured in the S frame. We have seen in earlier sections of this chapter that the length of a rod and time interval between events depend on the speed of the rod, etc. So let us consider the possibility of mass depending on speed and rewrite eqn (14.9.1) as

$$2\,m_A u_{yA} = 2\,m_B u_{yB} \tag{14.9.6}$$

because A and B have different speeds in S. Now, using eqn. (14.9.5) in (14.9.6), we get,

$$m_B = \frac{m_A u_{yA}}{u_{yB}} = \frac{m_A}{1-\dfrac{u_{yB}v}{c^2}} \tag{14.9.7}$$

We have taken $u_{xB}' = v$. ∴ Using the transformation (14.7.3), we get

$$v = \frac{u_{xB} - v}{1 - \frac{u_{xB} v}{c^2}}$$

Solving for v, we get,

$$v = \frac{c^2}{u_{xB}}\left(1 - \sqrt{1 - \frac{u_{xB}^{\ 2}}{c^2}}\right)$$

If we substitute this expression for v into eqn. (14.9.7), we get,

$$m_B = \frac{m_A}{\sqrt{1 - \frac{u_{xB}^{\ 2}}{c^2}}}$$

In order to find how mass varies with speed, let us consider the case when the y and y' components of velocity are set equal to zero. In that case, the collision is a grazing one as seen in the S' frame. In the S frame, A is at rest and B grazes past it. Since A is at rest, its mass m_A is called the rest mass m_0. Particle B has a mass m_B, which we call the relativistic mass m. Let us put $u_{xB} = u$.

$$\therefore \qquad m = \frac{m_0}{\sqrt{1 - \frac{u^2}{c^2}}} \qquad (14.9.8)$$

Eqn. (14.9.8) tells us how the relativistic mass varies with speed u.

Example 4: How many times m_0 is the mass m when u = 0.5c, 0.8c, 0.9c, 0.99c?

Solution: For $u = 0.5c,\ m = \frac{m_0}{\sqrt{\frac{3}{4}}} = 1.15\ m_0$

For $u = 0.8c,\ m = \frac{m_0}{\sqrt{1-(0.8)^2}} = 1.67\ m_0$

For $u = 0.9c,\ m = \frac{m_0}{\sqrt{1-(0.9)^2}} = 2.29\ m_0$

For $u = 0.99c,\ m = \frac{m_0}{\sqrt{1-(0.99)^2}} = 7.09\ m_0$

The linear momentum is defined as

$$\mathbf{p} = \frac{m_0 \mathbf{u}}{\sqrt{1 - \frac{u^2}{c^2}}} \tag{14.9.9}$$

The force on a particle of rest mass m_0 and moving with a velocity **u** is then

$$\mathbf{F} = \frac{d\mathbf{p}}{dt} = \frac{d}{dt}\left(\frac{m_0 \mathbf{u}}{\sqrt{1 - \frac{u^2}{c^2}}}\right) \tag{14.9.10}$$

The kinetic energy K of the particle is equal to the work done on it by a force **F** in increasing its velocity from 0 to **u**.

$$\therefore \qquad K = \int_0^u \mathbf{F}.d\mathbf{l} = \int_0^u F\,dx \qquad \text{in one dimension}$$

$$= \int_0^u \frac{d}{dt}(mu)\,dx = \int_0^u d(mu)u$$

$$= \int_0^u (u^2 dm + mudu)$$

Squaring eqn. (14.9.8), we get,

$$m^2 = \frac{m_0^2}{1 - \frac{u^2}{c^2}}$$

or, $$m^2c^2 - m^2u^2 = m_0{}^2c^2$$

Taking differentials of both sides of this equation, we get,

$$2\,mc^2 dm = 2\,u^2 mdm - m^2.2\,udu = 0$$

or $$mudu + u^2 dm = c^2 dm$$

$$\therefore \qquad K = \int_0^u c^2 dm = c^2 \int_{m_0}^{m} dm = c^2(m - m_0)$$

or, using eqn. (14.9.8), we get,

$$K = m_0c^2 \left[\frac{1}{\sqrt{1 - \frac{u^2}{c^2}}} - 1\right] \tag{14.9.11}$$

If we take $mc^2 = E$, the total energy of the particle, then we can write,

$$E = m_0c^2 + K \tag{14.9.12}$$

where m_0c^2 is called the rest energy of the particle. It is the energy of the particle when it is at rest. The expression for kinetic energy above can be expanded binomially for $u << c$.

$$K = m_0c^2\left[\left(1+\frac{u^2}{c^2}\right)^{-1/2} - 1\right]$$

$$\cong m_0c^2\left[1+\frac{1}{2}\left(\frac{u}{c}\right)^2+\frac{3}{8}\left(\frac{u}{c}\right)^4-1\right]$$

$$K = \frac{1}{2}\, m_0u^2$$

taking the first two terms. This is the classical result.

We shall now derive a few useful relations between the kinetic energy, linear momentum and total energy. Squaring eqn. (14.9.9), we get,

$$p^2 = \frac{m_0^2u^2c^2}{c^2-u^2} \tag{14.9.13}$$

Again, $$E^2 = m^2c^4 = \frac{m_0^2c^6}{c^2-u^2}$$

$\therefore$ $$p^2 = \frac{m_0^2u^2c^4}{c^2-u^2} \tag{14.9.14}$$

$\therefore$ $$E^2 - p^2c^2 = m_0^2c^4$$

or, $$E^2 = p^2c^2 + m_0^2c^4 \tag{14.9.15}$$

This is one useful result. Squaring eqn. (14.9.12), we get,

$$E^2 = K^2 + 2\, m_0c^2K + m_0^2c^4 \tag{14.9.16}$$

$\therefore$ Comparing the right hand sides of eqns. (14.9.15) and (14.9.16), we get,

$$p^2c^2 = K^2 + 2\, m_0c^2K \tag{14.9.17}$$

This is another useful result.

Let us now consider an inelastic collision. Let two identical particles A and B of rest mass m_0 move with equal speeds towards each other along the x' axis in the S' frame. After the collision, they stick together and form a composite particle C (Fig. 14.6). This particle of mass M_0, say, must be at rest, from the conservation of linear momentum. The S' frame moves with a speed $v = u'$ with respect to the S frame. $\therefore$ In this frame, A is at rest and B moves with speed u (Fig. 14.7) before the collision. From the transformation of velocities, eqn.(14.7.6), we get,

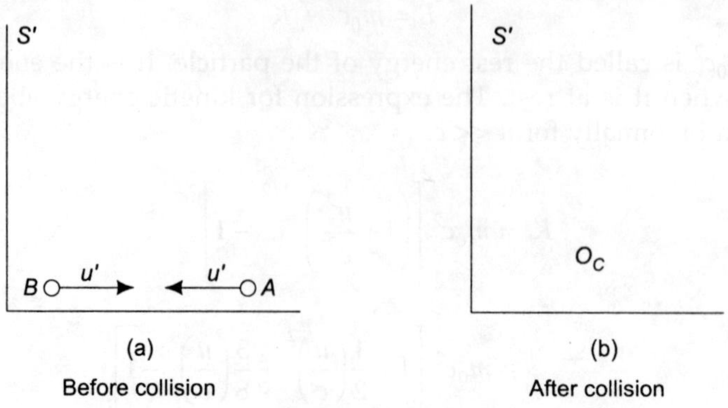

Fig. 14.7

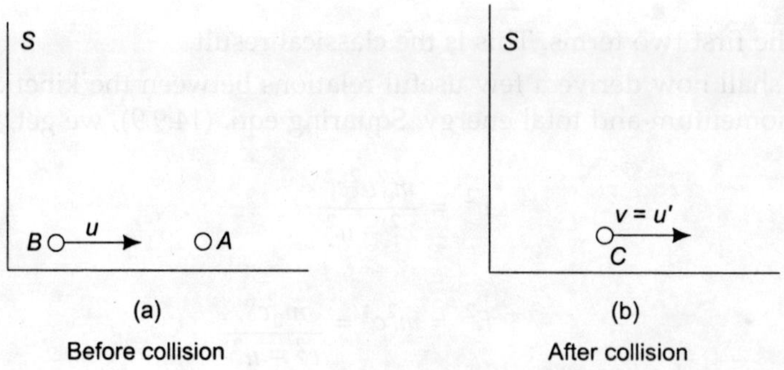

Fig. 14.8

$$u = \frac{u' + v}{1 + \dfrac{u'v}{c^2}} = \frac{2\,u'}{1 + \dfrac{u'^2}{c^2}} \tag{14.9.18}$$

After collision, the composite particle C must move with a speed $v = u'$. From the conservation of linear momentum in the S frame, we get,

$$\frac{m_0 u}{\sqrt{1 - \dfrac{u^2}{c^2}}} = \frac{M_0 u'}{\sqrt{1 - \dfrac{u'^2}{c^2}}}$$

Using eqn. (14.9.18), we get,

$$\sqrt{1 - \frac{u^2}{c^2}} = \frac{1 - \dfrac{u'^2}{c^2}}{1 + \dfrac{u'^2}{c^2}} \tag{14.9.19}$$

$$\therefore \qquad \frac{m_0 u\left(1+\frac{u'^2}{c^2}\right)}{\sqrt{1-\frac{u'^2}{c^2}}} = M_0 u'$$

or, using (14.9.18) again,

$$M_0 = \frac{2\,m_0}{\sqrt{1-\frac{u'^2}{c^2}}} \tag{14.9.20}$$

Since this is an inelastic collision, kinetic energy is not conserved. In the S' frame, the total kinetic energy before the collision is the sum of the kinetic energies of A and B. It is thus

$$K = 2\,m_0 c^2\left[\frac{1}{\sqrt{1-\frac{u^2}{c^2}}} - 1\right]$$

After the collision, the total kinetic energy is zero because C is at rest. Thus all the kinetic energy goes into rest energy of particle C. Let us calculate the total energy before the collision. It is

$$E_i' = \frac{2\,m_0 c^2}{\sqrt{1-\frac{u'^2}{c^2}}} \tag{14.9.21}$$

After the collision, it is

$$E_f' = M_0 c^2 \tag{14.9.22}$$

Using eqn. (14.9.20) in eqn. (14.9.22) and comparing with eqn. (14.9.21), we find that $E_i' = E_f'$

$\therefore$ Total energy is conserved. Let us now look at the total mass before the collision. It is

$$\frac{2\,m_0}{\sqrt{1-\frac{u'^2}{c^2}}}$$

and after collision it is M_0. $\therefore$ From eqn. (14.9.20), we see that the total mass is conserved. Thus conservation of total energy and total mass are equivalent.

Let us now consider the situation in the S frame. The total energy before the collision is

$$E_i = \frac{m_0 c^2}{\sqrt{1 - \frac{u^2}{c^2}}} + m_0 c^2 \tag{14.9.23}$$

After the collision, it is

$$E_f = \frac{M_0 c^2}{\sqrt{1 - \frac{u'^2}{c^2}}} \tag{14.9.24}$$

Using eqn. (14.9.19)in eqn. (14.9.22), we find

$$E_i = m_0 c^2 \left[\frac{1 + \frac{u'^2}{c^2}}{1 - \frac{u'^2}{c^2}} - 1 \right] = \frac{2 m_0 c^2}{1 - \frac{u'^2}{c^2}}$$

$$= E_f \qquad \text{using eqn. (14.9.20)}$$

∴ Total energy is conserved in S. Let us now look at the total mass. Before the collision, it is

$$\frac{m_0}{\sqrt{1 - \frac{u^2}{c^2}}} + m_0$$

$$= m_0 \left[\frac{1 + \frac{u'^2}{c^2}}{1 - \frac{u'^2}{c^2}} - 1 \right] = \frac{2 m_0}{1 - \frac{u'^2}{c^2}}$$

After the collision, the total mass is

$$\frac{M_0}{\sqrt{1 - \frac{u'^2}{c^2}}}$$

This is equal to the intial mass because of eqn. (14.9.20). Thus the relation $E = mc^2$ expresses the equivalence of mass and energy. Mass m can also be expressed in terms of energy units.

Example 5: The average lifetime of μ-mesons at rest is 2.3×10^{-6}s. A laboratory measurement on μ-mesons yields an average lifetime of 6.9×10^{-6}s. (a) What is the speed of the mesons in the laboratory? (b) the rest mass of a μ-meson is 207 m_e. What is the effective mass of such a meson when moving at this speed? (c) What is its kinetic energy? What is its momentum?

Solution: (a) Let u be the speed of the mesons in the laboratory. The proper (average) lifetime is $\Delta\tau = 2.3 \times 10^{-6}$s. The lifetime Δt as measured in the laboratory is

$$\therefore \quad \Delta t = \frac{\Delta\tau}{\sqrt{1-\frac{u^2}{c^2}}}$$

$$6.9 \times 10^{-6} = \frac{2.3\times 10^{-6}}{\sqrt{1-\frac{u^2}{c^2}}}$$

$$\therefore \quad u = 0.943c$$

(b) Effective mass $m = \dfrac{m_0}{\sqrt{1-\frac{u^2}{c^2}}} = 3\, m_0 = 621\, m_c$

(c) $$K = m_0c^2\left[\frac{1}{\sqrt{1-\frac{u^2}{c^2}}} - 1\right] = m_0c^2[3-1] = 2\, m_0c^2$$

$$= 2 \times 207\, m_ec^2 = 414\, m_ec^2 = 211.6 \text{ MeV}.$$

Momentum $p = \dfrac{m_0u}{\sqrt{1-\frac{u^2}{c^2}}} = 621\, m_e \times 0.943c$

$$= 621 \times 9.11 \times 10^{-31} \times 0.943 \times 3 \times 10^8$$

$$p = 1.6 \times 10^{-19} \text{ kgm/s}.$$

Momentum can also be expressed as

$$p = 621 \times 0.511\, \frac{\text{MeV}}{c^2} \times 0.943c = 299 \text{ MeV/c}$$

14.10 FOUR-VECTORS AND MINKOWSKI DIAGRAMS

Let us now determine how linear momentum and total energy transform in going from one frame to the other. Let (p_x, p_y, p_z) be the components of linear momentum and E be the energy of a particle of rest mass m_0 in the S frame and moving with a speed u. The corresponding quantities in the S' frame are (p_x', p_y', p_z'), E' and u'. Thus from eqn (14.9.9).

$$p_x = \frac{m_0 u_x}{\sqrt{1 - \frac{u^2}{c^2}}} \tag{14.10.1a}$$

$$p_y = \frac{m_0 u_y}{\sqrt{1 - \frac{u^2}{c^2}}} \tag{14.10.1b}$$

$$p_z = \frac{m_0 u_z}{\sqrt{1 - \frac{u^2}{c^2}}} \tag{14.10.1c}$$

In the S' frame, we have,

$$p_x' = \frac{m_0 u_x'}{\sqrt{1 - \frac{u'^2}{c^2}}} \tag{14.10.2a}$$

$$p_y' = \frac{m_0 u_y'}{\sqrt{1 - \frac{u'^2}{c^2}}} \tag{14.10.2b}$$

$$p_z' = \frac{m_0 u_z'}{\sqrt{1 - \frac{u'^2}{c^2}}} \tag{14.10.2c}$$

The velocity component transform according to eqns. (14.7.3), (14.7.4) and (14.7.5). Squaring and adding these equations, we get

$$u'^2 = u_x'^2 + u_y'^2 + u_z'^2 = \frac{u^2 - 2\,u_x v + v^2 - \dfrac{u_y^2 v^2}{c^2} - \dfrac{u_z^2 v^2}{c^2}}{\left(1 - \dfrac{u_x v}{c^2}\right)^2}$$

where $\quad u^2 = u_x^2 + u_y^2 + u_z^2$

$$\therefore \quad 1 - \frac{u'^2}{c^2} = \frac{1 + \dfrac{u^2 v^2}{c^4} - \dfrac{u^2}{c^2} - \dfrac{v^2}{c^2}}{\left(1 - \dfrac{u_x v}{c^2}\right)^2}$$

$$\therefore \quad 1-\frac{u'^2}{c^2} = \frac{\left(1-\frac{u^2}{c^2}\right)\left(1-\frac{v^2}{c^2}\right)}{\left(1-\frac{u_x v}{c^2}\right)^2}$$

$$\frac{1}{\sqrt{1-\frac{u'^2}{c^2}}} = \frac{1-\frac{u_x v}{c^2}}{\sqrt{\left(1-\frac{u^2}{c^2}\right)\left(1-\frac{v^2}{c^2}\right)}} \tag{14.10.3}$$

Substituting eqn. (14.10.3) and eqn. (14.7.3) into eqn. (14.10.2a), we get

$$p_x' = \frac{m_0(u_x - v)}{\sqrt{\left(1-\frac{u^2}{c^2}\right)\left(1-\frac{v^2}{c^2}\right)}} \tag{14.10.4}$$

The total energy in the S frame is

$$E = \frac{m_0 c^2}{\sqrt{1-\frac{u^2}{c^2}}}$$

$$\therefore \quad \frac{E}{c^2} = \frac{m_0}{\sqrt{1-\frac{u^2}{c^2}}} \tag{14.10.5}$$

$\therefore$ Using (14.10.1a) and (14.10.5), we get,

$$p_x - \frac{Ev}{c^2} = \frac{m_0(u_x - v)}{\sqrt{1-\frac{u^2}{c^2}}}$$

Using this in eqn. (14.10.4), we get,

$$p_x' = \frac{1}{\sqrt{1-\frac{v^2}{c^2}}}\left(p_x - \frac{Ev}{c^2}\right) \tag{14.10.6}$$

Let us now consider eqn. (14.10.2b). Using eqn. (14.7.4) and (14.10.3), we get,

$$p_y' = \frac{m_0 u_y}{\sqrt{1-\frac{u^2}{c^2}}}$$

$\therefore$ $$p_y' = p_y \qquad (14.10.7)$$

Similarly, $$p_z' = p_z \qquad (14.10.8)$$

Let us now consider the transformation of the energy. In the S' frame, the energy of the particle is

$$E' = \frac{m_0 c^2}{\sqrt{1-\frac{u'^2}{c^2}}}$$

$$= \frac{m_0 c^2 - m_0 u_x v}{\left(1-\frac{u^2}{c^2}\right)\left(1-\frac{v^2}{c^2}\right)} \qquad \text{using eqn. (14.10.3)}$$

$$E' = \frac{1}{\sqrt{1-\frac{v^2}{c^2}}}(E - vp_x) \qquad (14.10.9)$$

We can rewrite this by dividing throughout by c^2:

$$\frac{E'}{c^2} = \frac{1}{\sqrt{1-\frac{v^2}{c^2}}}\left(\frac{E}{c^2} - \frac{vp_x}{c^2}\right) \qquad (14.10.9a)$$

Eqns. (14.10.6)–(14.10.9) are the transformation of energy and linear momentum according to special relativity.

Example 6: A particle as observed in a certain reference frame has a total energy of 5 GeV and a momentum of 3 GeV/c. (a) What is its enegy in a frame in which its momentum is equal to 4 GeV/c? (b) What is the relative velocity of the two reference frames?

Solution: Assume the particle to be moving along the common x-x' axis, with its y and z components (and y' and z' components) of momentum being zero.

$\therefore$ $$E' = \frac{c^2 p'}{u'}, \; E = \frac{c^2 p}{u}$$

Let $E' = 5$ GeV, $cp' = 3$ GeV, $cp = 4$ GeV

Since $E'^2 - (cp')^2 \; E^2 - (cp)^2$, we get $E = 5.66$ GeV

$$\frac{u}{c} = \frac{cp}{E} = \frac{4}{5.66} = 0.707$$

$$\frac{u'}{c} = \frac{cp'}{E'} = \frac{3}{5} = 0.6$$

$$u' = \frac{u-v}{1-\frac{uv}{c^2}} \quad \therefore \quad \frac{u'}{c} = \frac{\frac{u}{c}-\frac{v}{c}}{1-\frac{uv}{c^2}}$$

$$\therefore \quad 0.6 = \frac{0.707 - \frac{v}{c}}{1-\frac{0.707\,v}{c^2}}$$

Solving for v/c, we get,

$$\frac{v}{c} = 0.186$$

Let us come back to eqn. (14.4.15) and (14.4.16) and rewrite them as

$$x' = \gamma(x - \beta ct) \tag{14.10.10}$$

$$ct' = \gamma(ct - \beta x) \tag{14.10.11}$$

with $\gamma = \frac{1}{\sqrt{1-\beta^2}}$. The quantities x, y, z, ct form a four-dimensional spacetime.

We can plot a graph of x vs. ct, taking ct along the vertical axis (Fig. 14.8) in two dimensions. Such a diagram is called a Minkowski diagram. The path of a light ray is along the dotted line, making an angle of 45° with the x axis. Its equation is $x = ct$. Light can be described in the particle picture in terms of photons. The energy E and momentum p of a photon satisfy the relation

$$E = cp \tag{14.10.12}$$

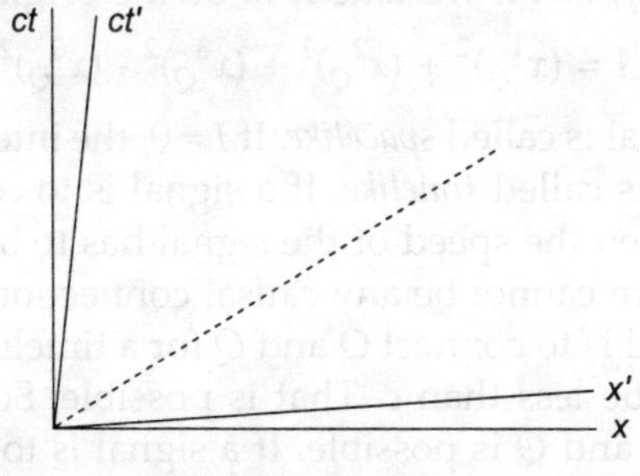

Fig. 14.9

This is true for all frequencies of the electromagnetic spectrum. Photons have zero rest mass. The dotted line in the diagram is called the world line of photon.

Squaring eqns. (14.10.10) and (14.10.11), we get,

$$x'^2 = \gamma^2(x - \beta ct)^2$$

$$c^2t^2 = \gamma^2(ct - \beta x)^2$$

$$x'^2 - c^2t'^2 = \gamma^2\{x^2 - 2\beta ctx + \beta^2c^2t^2 - (c^2t^2 - 2\beta ctx + \beta^2x^2)\}$$

$$= \gamma^2(1 - \beta^2)(x^2 - c^2t^2), \qquad \text{on simplifying}$$

$$\therefore \quad x'^2 - c^2t'^2 = x^2 - c^2t^2 \qquad (14.10.13)$$

$\therefore$ The quantity $x^2 - c^2t^2$ is invariant under Lorentz transformations. If we put $x^2 - c^2t^2 = 1$, it represents a hyperbola in the Minkowski diagram. The equation $ct = 0$ defines the x axis. Then $x = 1$ defines unit distance along the x axis. Because of (14.10.3), we must also have $x'^2 - c^2t'^2 = 1$. The equation $ct' = 0$ defines the x' axis. Note that the x' axis is tilted in the diagram because its equation is $x = vt$ in the S frame. $\therefore$ The equation $x' = 1$ also defines unit distance along the x'axis.

Let us now introduce a new notation. Let $x^0 = ct, x^1 = x, x^2 = y, x^3 = z$. Here the super scripts on x (i.e. 0, 1, 2, 3) are not powers of x. They are merely indices. In this notation, the Lorentz transformations (14.10.10), (14.4.5), (14.4.6), (14.4.11) become

$$x^{0\prime} = \gamma(x^0 - \beta x^1) \qquad (14.10.14)$$

$$x^{1\prime} = \gamma(x^1 - \beta x^0) \qquad (14.10.15)$$

$$x^{2\prime} = x^2 \qquad (14.10.16)$$

$$x^{3\prime} = x^3 \qquad (14.10.14)$$

The quantities (x^0, x^1, x^2, x^3) and similarly $(x^{0\prime}, x^{1\prime}, x^{2\prime}, x^{3\prime})$ are called *four-vectors*. The transformation of the four vectors from reference frame S to S' is given by eqn. (14.10.14)-(14.10.17). The interval between two points $P(x^0{}_P, x^1{}_P, x^2{}_P, x^3{}_P)$ and $Q(x^0{}_Q, x^1{}_Q, x^2{}_Q, x^3{}_Q)$ is given by

$$I = (\Delta x^1)^2 + (\Delta x^2)^2 + (\Delta x^3)^2 - (\Delta x^0)^2$$

where $\Delta x^1 = x^1{}_Q - x^1{}_P$, etc. If we take P to be the origin, then

$$I = (x^1{}_Q)^2 + (x^2{}_Q)^2 + (x^3{}_Q)^2 - (x^0{}_Q)^2$$

If $I > 0$, the interval is called *spacelike*. If $I = 0$, the interval is called *lightlike*. If $I < 0$, the interval is called *timelike*. If a signal is to connect O and Q for a spacelike interval, then the speed of the signal has to be greater than c. That is impossible. So there cannot be any causal connection between two events at O and Q. If a signal is to connect O and Q for a timelike interval, the speed of the signal has to be less than c. That is possible. So a causal connection between events at O and Q is possible. If a signal is to connect O and Q for a lightlike interval, the signal must be a light signal. So there can be a causal connection between events O and Q.

Another example of a four-vector is the energy and momentum four-vector. Let $p^0 = E/c, p^1 = p_x, p^2 = p_y, p^3 = p_z$. Then eqns. (14.10.6)-(14.10.9) give

$$p^{0\prime} = \gamma(p^0 - \beta p^1) \tag{14.10.18}$$

$$p^{1\prime} = \gamma(p^1 - \beta p^0) \tag{14.10.19}$$

$$p^{2\prime} = p^2 \tag{14.10.20}$$

$$p^{3\prime} = p^3 \tag{14.10.21}$$

Note that $\dfrac{E^2}{c^2} - p^2 = \dfrac{E'^2}{c^2} - p'^2 = m_0c^2$ which is an invariant.

Let us now find the transformation properties of the force. The forces on a particle in the S and S' frames are $\mathbf{F} = d\mathbf{p}/dt$ and $\mathbf{F}' = d\mathbf{p}'/dt'$ respectively. The x component of $\mathbf{F}'$ is

$$F_x{}' = \frac{dp_x{}'}{dt} = \frac{\dfrac{dp_x{}'}{dt}}{\dfrac{dt'}{dt}} = \frac{\gamma\dfrac{dp_x}{dt} - \dfrac{v\,dE}{c^2\,dt}}{\gamma\left(1 - \dfrac{v\,dx}{c^2\,dt}\right)}$$

using eqns. (14.4.16) and (14.10.6)

$$\therefore \quad F_x{}' = \frac{F_x - \dfrac{v\,dE}{c^2\,dt}}{1 - \dfrac{u_x v}{c^2}} \tag{14.10.22}$$

Eqn. (14.9.15) can be written as

$$E^2 = c^2\,(\mathbf{p.p}) + m_0{}^2c^4$$

Differentiating with respect to time t, we get,

$$\frac{E\,dE}{dt} = \frac{c^2\mathbf{p}.d\mathbf{p}}{dt}$$

$$\therefore \quad \frac{dE}{dt} = \frac{c^2\mathbf{p.F}}{E} = \frac{\mathbf{p.F}}{m} = \mathbf{u.F}$$

Substituting this in eqn. (14.10.22), we get,

$$F_x{}' = \frac{F_x - \dfrac{v\,(\mathbf{F.u})}{c^2}}{1 - \dfrac{u_x v}{c^2}} \tag{14.10.23}$$

The y component of $\mathbf{F}'$ is

$$F_y' = \frac{dp_y'}{dt'} = \frac{\frac{dp_y'}{dt}}{\frac{dt'}{dt}} = \frac{\frac{dp_y}{dt}}{\gamma\left(1-\frac{v\,dx}{c^2\,dt}\right)}$$

$$F_y' = \frac{F_y}{\gamma\left(1-\frac{u_x v}{c^2}\right)} \tag{14.10.24}$$

If $u = 0$, then

$$F_x' = F_x \tag{14.10.25}$$

$$F_y' = \frac{F_y}{\gamma} \tag{14.10.26}$$

14.11 TRANSFORMATION OF ELECTRIC AND MAGNETIC FIELDS

Let us now study how electric and magnetic fields transform under Lorentz transformations. We shall take a specific case to deduce the transformation. if the electric and magnetic fields are to have a unique meaning, it should be possible to predict E' and B' in a reference frame S' from a knowledge of E and B in a reference frame S, at the same spacetime point. A knowledge of the sources of E and B should be unnecessary for this. So the result of the derivation should be valid in general.

Let us consider a frame S where there are a pair of long charged sheets One of them carries positive charge with surface density σ and other carries negative charge with surface density $-\sigma$ (Fig. 14.10a). The separation between them is small compared to their dimensions. The electric field between them is given by

$$E_z = \frac{\sigma}{\varepsilon_0} \tag{14.11.1}$$

Consider another reference frame S' which is moving relative to S with speed v along the positive x-x' axis.[⊕] The electric field between the plates in this frame is given by

$$E_z' = \frac{\sigma'}{\varepsilon_0} \tag{14.11.2}$$

Due to length contraction, the sides of the sheet along the x-x' axis appear shorter in the S' frame by a factor $\gamma = 1/\sqrt{1-\frac{v^2}{c^2}}$. Since charge is invariant, the

⊕ ∴ In S' the plates appear to move in the direction of the negative x-x' axis with speed v.

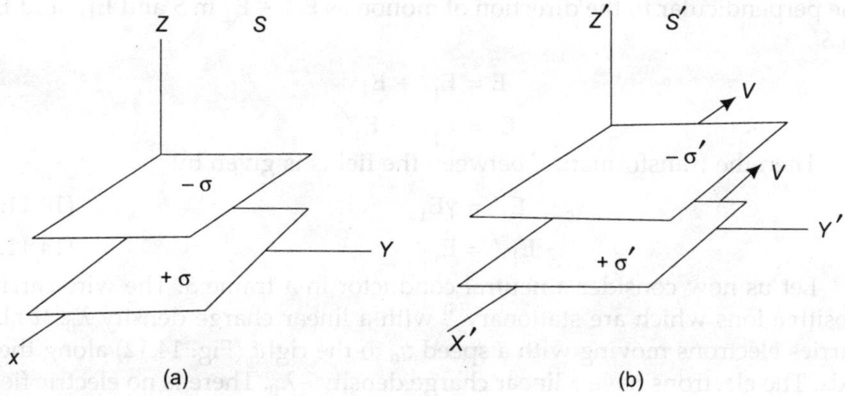

Fig. 14.10

surface charge density is increased by the same factor. Thus,

$$\sigma' = \gamma\sigma \tag{14.11.3}$$

Substituting this in eqn. (14.11.2) and comparing with (14.11.1), we get,

$$E_z{}' = \gamma E_z \tag{14.11.4}$$

Now suppose the sheets lie in the YZ plane (Fig. 14.11) The S' frame is moving as usual, relative to the S frame along the x-x' axis. In the S' frame, the distance between the sheets appears to be shorter. But the electric field is independent of the distance between the sheets. So the electric field in the S

$$E_x = \frac{\sigma}{\varepsilon_0} = E_x{}' \tag{14.11.5}$$

and S' frames are the same. Eqns. (14.11.4) and (14.11.5) give the transformation of the electric fields due to static charge distribution in S.

Consider two reference frames S and S' with S' moving with a velocity v relative to S. We can resolve the electric fields **E** and **E′** in the two frames along

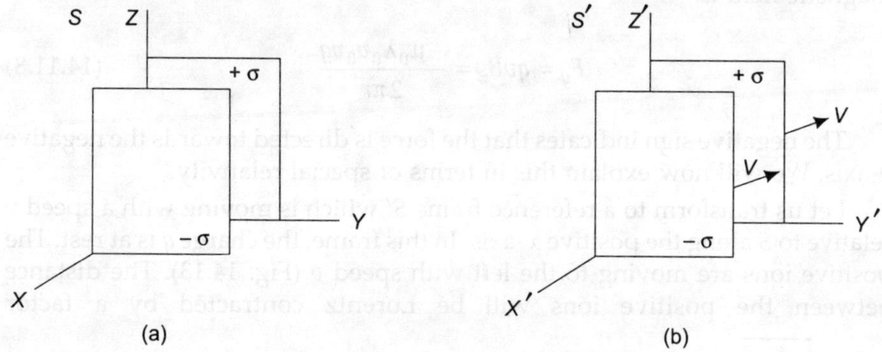

Fig. 14.11

the perpendicular to the direction of motion as $\mathbf{E}_{||} + \mathbf{E}_{\perp}$ in S and $\mathbf{E}_{||}'$ and $\mathbf{E}_{\perp}'$ in S'.

$$\mathbf{E} = \mathbf{E}_{||} + \mathbf{E}_{\perp}$$

$$\mathbf{E}' = \mathbf{E}_{||}' + \mathbf{E}_{\perp}'$$

Then the transformation between the fields is given by

$$\mathbf{E}_{\perp}' = \gamma \mathbf{E}_{\perp} \tag{14.11.6}$$

$$\mathbf{E}_{||}' = \mathbf{E}_{||} \tag{14.11.7}$$

Let us now consider a neutral conductor in a frame S. The wire carries positive ions which are stationary[4] with a linear charge density λ_0. It also carries electrons moving with a speed v_0 to the right (Fig. 14.12) along the x axis. The electrons have a linear charge density $-\lambda_0$. There is no electric field outside the wire. Consider a test charge q moving with a speed v along the x-axis. The wire carries a current $I = -\lambda_0 v_0$. The negative sign implies that the current is directed towards the negative x-axis. So there is a magnetic field

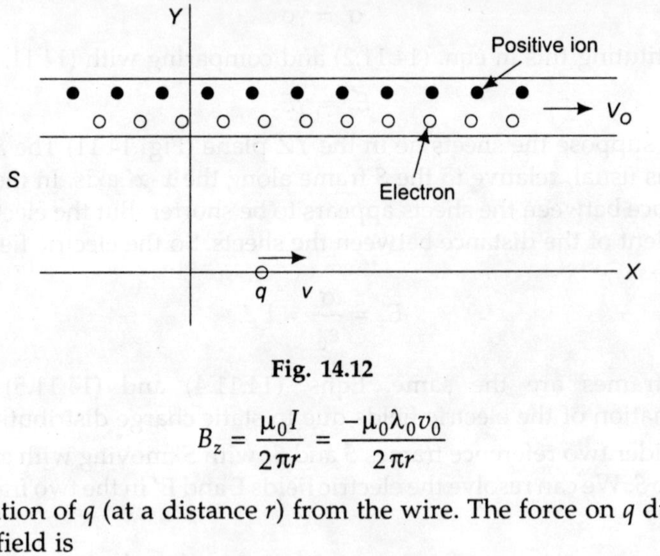

Fig. 14.12

$$B_z = \frac{\mu_0 I}{2\pi r} = \frac{-\mu_0 \lambda_0 v_0}{2\pi r}$$

at the location of q (at a distance r) from the wire. The force on q due to this magnetic field is

$$F_y = qvB_z = \frac{-\mu_0 \lambda_0 v_0 v q}{2\pi r} \tag{14.11.8}$$

The negative sign indicates that the force is directed towards the negative y-axis. We will now explain this in terms of special relativity.

Let us transform to a reference frame S' which is moving with a speed v relative to S along the positive x-axis. In this frame, the charge q is at rest. The positive ions are moving to the left with speed v (Fig. 14.13). The distance between the positive ions will be Lorentz contracted by a factor $\gamma = 1/\sqrt{1-\frac{v^2}{c^2}}$. So their linear charge density increases by the same factor. The

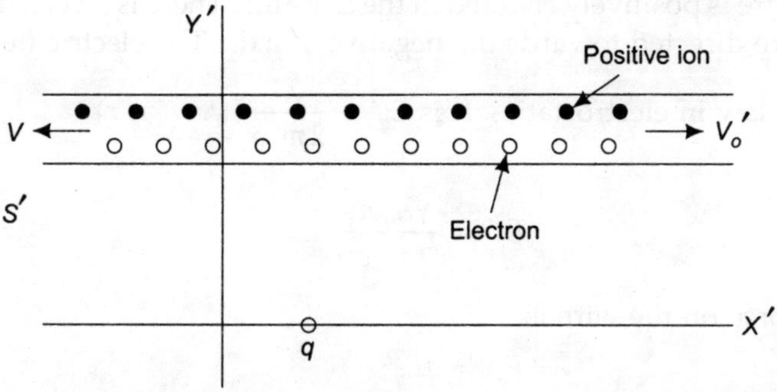

Fig. 14.13

charge density of the positive ions is thus $\gamma\lambda_0$. The distance between the electrons is increased because the electrons were already moving in the S frame. The linear charge density of electrons in a frame in which the electrons are at rest would be $-\lambda_0/\gamma_0$, where $\gamma_0 = 1/\sqrt{1-\frac{v^2}{c^2}}$. Let the speed of the electrons in S' be v_0'. $\therefore$ The linear charge density of the electrons in the S' frame is $-\gamma'\lambda_0/\gamma_0$ where $\gamma' = 1/\sqrt{1-\frac{v_0'^2}{c^2}}$. From the Lorentz transformation of velocities, we have

$$v_0' = \frac{v_0 - v}{1-\frac{vv_0}{c^2}} \tag{14.11.9}$$

After some algebra, we get

$$\frac{1}{\sqrt{1-\frac{v'^2_0}{c^2}}} = \frac{1-\frac{vv_0}{c^2}}{\sqrt{\left(1-\frac{v_0^2}{c^2}\right)\left(1-\frac{v^2}{c^2}\right)}}$$

$$\therefore \qquad \gamma_0' = \gamma_0\gamma\left(1-\frac{vv_0}{c^2}\right) \tag{14.11.10}$$

$\therefore$ The total charge on the wire in the S' frame is

$$\lambda' = \gamma\lambda_0 - \frac{\gamma_0'\lambda_0}{\gamma_0}$$

Substituting for γ_0' from eqn. (14.11.10), we get

$$\lambda' = \frac{\gamma\lambda_0 vv_0}{c^2} \tag{14.11.11}$$

$\therefore$ The wire is positively charged in the S' frame. There is an electric field $E_y{}'$ on the wire directed towards the negative y' axis. The electric field is given by Gauss' law in electrostatics. It is $E_y{}' = \dfrac{\lambda'}{2\pi\varepsilon_0 r}$ $(\because r' = r)$

$$\therefore \qquad E_y{}' = \frac{\gamma\lambda_0 v v_0}{2\pi\varepsilon_0 c^2 r}$$

The force on the wire is

$$F_y{}' = qE_y{}'$$

$$\therefore \qquad F_y{}' = \frac{q\gamma\lambda_0 v v_0}{2\pi\varepsilon_0 c^2 r} \tag{14.11.12}$$

Transforming this force to the frame S, we find

$$F_y = \frac{F_y{}'}{\gamma} = \frac{q\lambda_0 v v_0}{2\pi\varepsilon_0 c^2 r}$$

or,

$$F_y = \frac{q\mu_0\lambda_0 v v_0}{2\pi r} = qvB \tag{14.11.13}$$

Thus the magnetic field is a convenient way to calculate the force on a moving charge due to other moving charges.

Let us now consider the transformation of electric and magnetic fields between two frames S and S'. Let there be a pair of long, parallel plane sheets, one carrying a charge density $+\sigma$ and the other, a charge density $-\sigma$. They are moving with a speed v_0 in the direction of the positive x-axis in the S frame (Fig. 14.14). The sheets constitute a positive current $K_x = \sigma v_0$. The electric field is normal to the plates and has magnitude

$$E_y = \frac{\sigma}{\varepsilon_0} \tag{14.11.14}$$

The magnetic field is along the z-axis and has magnitude

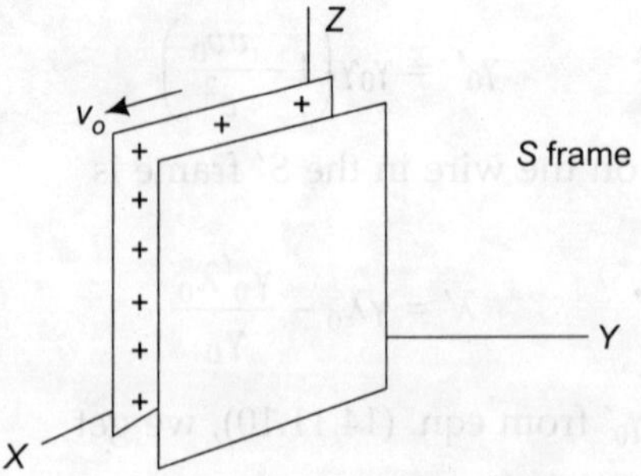

Fig. 14.14

$$B_z = \mu_0 K_x = \mu_0 \sigma v_0 \tag{14.11.15}$$

Consider another frame S' which is moving relative to the frame S with a speed v. In this frame, the speed of the charge sheet v_0' is given by

$$v_0' = \frac{v_0 - v}{1 - \dfrac{v v_0}{c^2}} \tag{14.11.16}$$

The charge density in the rest frame of the charges is σ/γ_0. $\therefore$ The charge density in the rest frame S' is

$$\sigma' = \frac{\sigma \gamma_0'}{\gamma_0} \tag{14.11.17}$$

where
$$\gamma_0' = \frac{1}{\sqrt{1 - \dfrac{v_0'^2}{c^2}}}$$

$\therefore$ Using eqn. (14.11.10), we get

$$\sigma' = \sigma\gamma\left(1 - \frac{v v_0}{c^2}\right) \tag{14.11.18}$$

The current density in the reference frame S' is

$$K_x = \sigma' v_0'$$

$$= \sigma\gamma\left(1 - \frac{v v_0}{c^2}\right)\frac{(v_0 - v)}{\left(1 - \dfrac{v v_0}{c^2}\right)}$$

$$\therefore \quad K_x' = \sigma\gamma(v_0 - v) \tag{14.11.19}$$

The electric field in the S' frame is

$$E_y' = \frac{\sigma'}{\varepsilon_0} = \frac{\sigma\gamma}{\varepsilon_0}\left(1 - \frac{v v_0}{c^2}\right)$$

$$\therefore \quad E_y' = \gamma\left(\frac{\sigma}{\varepsilon_0} - \mu_0 \sigma v v_0\right) \text{ using } \mu_0 = \frac{1}{\varepsilon_0 c^2}$$

$$\therefore \quad E_y' = \gamma\,(E_y - vB_z) \tag{14.11.20}$$

The magnetic field in the S' frame is

$$B_z' = \mu_0 \sigma v_0'$$

$$\therefore \quad B_z' = \mu_0 \sigma \gamma (v_0 - v)$$

$$\therefore \quad B_z' = \gamma\left(\mu_0 \sigma v_0 - \frac{\sigma v}{\varepsilon_0 c^2}\right)$$

$$\therefore \qquad B_z' = \gamma\left(B_z - \frac{vE_y}{c^2}\right) \tag{14.11.21}$$

Now let us suppose that the long plane parallel sheets are oriented parallel to the XY plane, and moving in the direction of the positive x-axis in the S frame with a speed v_0 (Fig. 14.15). The electric field is

$$E_z = \frac{\sigma}{\varepsilon_0} \tag{14.11.22}$$

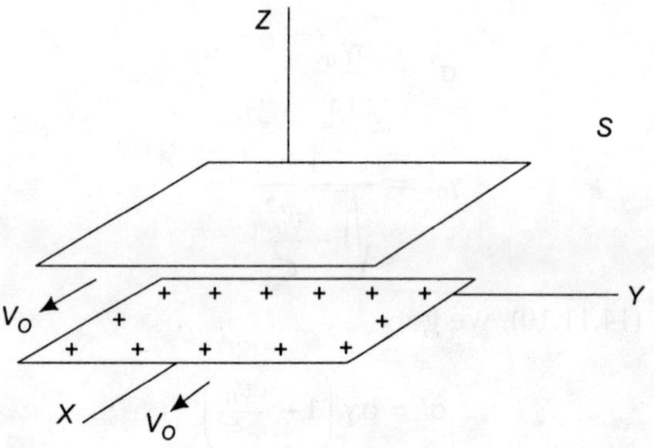

Fig. 14.15

and the magnetic field is in the negative y direction.

$$B_y = -\mu_0\sigma v_0 \tag{14.11.23}$$

In the S' frame, which moves with a speed v relative to the S frame in the direction of the positive x-axis, the speed v_0' of the charged sheet is given by eqn. (14.11.16). The charge density σ' in the S' frame and the charge density σ in the S frame are related by eqn. (14.11.18). The electric field in the S' frame is

$$E_z' = \frac{\sigma'}{\varepsilon_0} = \sigma\gamma\left(1 - \frac{vv_0}{c^2}\right)$$

$$= \gamma\left(\frac{\sigma}{\varepsilon_0} - \frac{\sigma v v_0}{\varepsilon_0 c^2}\right)$$

$$= \gamma\left(\frac{\sigma}{\varepsilon_0} - \mu_0\sigma v v_0\right)$$

$$\therefore \qquad E_z' = \gamma(E_z + vB_y) \tag{14.11.24}$$

The magnetic field in the S' frame is

$$B_y' = \mu_0 \sigma' v_0'$$
$$= -\mu_0 \sigma \gamma (v_0 - v), \quad \text{using (14.11.16) \& (14.11.18)}$$

$$\therefore \quad B_y' = \gamma \left(-\mu_0 \sigma v_0 + \frac{\sigma v}{\varepsilon_0 c^2} \right)$$

$$\therefore \quad B_y' = \gamma \left(B_y + \frac{v E_z}{c^2} \right) \tag{14.11.25}$$

Now let us consider a solenoid with its axis along the x direction, having n turns per unit length and carrying a current I. The magnetic field is in the x direction and is $B_x = \mu_0 nI$. In the S' frame, the solenoid will be Lorentz contracted. So the number of turns per unit length, n' will be greater. The current I' will be reduced by the same factor, because the current is charge per second and the clock in S' is slower.[4]

$$\therefore \quad B_x' = \mu_0 n' I' = B_x.$$

We can now write the transformations compactly by breaking up the electric and magnetic fields parallel to and perpendicular to the direction of motion. Then we have

$$\mathbf{E}_{||}' = \mathbf{E}_{||} \tag{14.11.26}$$

$$\mathbf{E}_{\perp}' = \gamma \{ \mathbf{E}_{\perp} + (\mathbf{v} \times \mathbf{B})_{\perp} \} \tag{14.11.27}$$

And

$$\mathbf{B}_{||}' = \mathbf{B}_{||} \tag{14.11.28}$$

$$\mathbf{B}_{\perp}' = \gamma \left\{ \mathbf{B}_{\perp} - \frac{1}{c^2} (\mathbf{v} \times \mathbf{E})_{\perp} \right\} \tag{14.11.29}$$

Example 7: Calculate the electric and magnetic field of a charge q moving with a uniform velocity $\mathbf{u}$, using the transformation equations for the fields.

Solution: Let the charge be moving with a speed u along the positive x-axis, in the S frame. The electric and magnetic fields due to this charge are $\mathbf{E}$ and $\mathbf{B}$. Consider another frame S' which is moving relative to S with a speed u along the common x-x' axis. In this frame, the charge is at rest.[18b] $\therefore$ In the S' frame, the electric field at the point (x', y', z') at a distance r' from the charge is

$$\mathbf{E}' = \frac{q\mathbf{r}}{4\pi\varepsilon_0 r^3} \tag{1}$$

(with $r' = \sqrt{x'^2 + y'^2 + z'^2}$, assuming that the charge is at the origin of S'). The magnetic field is zero.

$$\mathbf{B}' = 0 \tag{2}$$

In the S frame, the co-ordinates of the charge (X, Y, Z) are $(ut, 0, 0)$. The co-ordinates (x, y, z, t) and (x', y', z', t') of the field point are related by the Lorentz transformation.

$$\therefore \qquad r' = \sqrt{\gamma^2 (x - ut)^2 + y^2 + z^2}$$

with
$$\gamma = \frac{1}{\sqrt{1 - \dfrac{u^2}{c^2}}}$$

To calculate the fields in the S frame, we have to use the inverse transformation. These are

$$E_x = E_x' \tag{3}$$

$$E_y = \gamma(E_y' + uB_z') \tag{4}$$

$$E_z = \gamma(E_z' - uB_y') \tag{5}$$

From eqn (1), we have

$$E_x' = \frac{qx'}{4\pi\varepsilon_0 (x'^2 + y'^2 + z'^2)^{3/2}}$$

$$E_y' = \frac{qy'}{4\pi\varepsilon_0 (x'^2 + y'^2 + z'^2)^{3/2}}$$

$$E_z' = \frac{qz'}{4\pi\varepsilon_0 (x'^2 + y'^2 + z'^2)^{3/2}}$$

$\therefore$ Using (2), (3), (4) and (5), we get

$$E_x = \frac{q\gamma (x - X)}{4\pi\varepsilon_0 \{\gamma^2 (x - X)^2 + y^2 + z^2\}^{3/2}} \tag{6}$$

$$E_y = \frac{q\gamma y}{4\pi\varepsilon_0 \{\gamma^2 (x - X)^2 + y^2 + z^2\}^{3/2}} \tag{7}$$

$$E_z = \frac{q\gamma z}{4\pi\varepsilon_0 \{\gamma^2 (x - X)^2 + y^2 + z^2\}^{3/2}} \tag{8}$$

The inverse transformation for the magnetic field is

$$B_x = B_x' \tag{9}$$

$$B_y = \gamma\left(B_y' - \frac{uE_z'}{c^2}\right) \tag{10}$$

$$B_z = \gamma\left(B_z' + \frac{uE_y'}{c^2}\right) \tag{11}$$

$$\therefore \quad B_x = 0$$

$$B_y = \frac{-\mu_0 q\gamma z u}{4\pi\{\gamma^2(x-X)^2+y^2+z^2\}^{3/2}}$$

$$B_z = \frac{\mu_0 q\gamma y u}{4\pi\{\gamma^2(x-X)^2+y^2+z^2\}^{3/2}}$$

Let us consider the electric field **E** in the S' frame at $t = 0$. It is

$$\mathbf{E} = \frac{q\gamma\mathbf{r}}{4\pi\varepsilon_0\{\gamma^2x^2+y^2+z^2\}^{3/2}} \qquad (\because X = ut = 0)$$

Let **r** make an angle θ with the x-axis.

$$\therefore \quad x = r\cos\theta,\ y^2+z^2 = r^2\sin^2\theta$$

$$\begin{aligned}\gamma^2x^2+y^2+z^2 &= \gamma^2r^2\cos^2\theta + r^2\sin^2\theta \\ &= \gamma^2r^2(1-\sin^2\theta)+r^2\sin^2\theta \\ &= \gamma^2r^2 + r^2\sin^2\theta(1-\gamma^2) \\ &= \gamma^2r^2\left(1-\frac{u^2}{c^2}\sin^2\theta\right)\end{aligned}$$

$$\therefore \quad \gamma^2x^2+y^2+z^2 = \gamma^2r^2(1-\beta^2\sin^2\theta) \qquad \text{where } \beta = \frac{u}{c}$$

$$\therefore \quad \mathbf{E} = \frac{q\mathbf{r}}{\gamma^2 4\pi\varepsilon_0 r^3(1-\beta^2\sin^2\theta)^{3/2}}$$

$$\therefore \quad \mathbf{E} = \frac{q(1-\beta^2)\mathbf{r}}{4\pi\varepsilon_0(1-\beta^2\sin^2\theta)^{3/2}r^3}$$

which is the result we obtained in the Tutorial of Chapter 13

The transformation equations are written in SI and Gaussian units in the table below:

Transformation Equation	*SI*	*Guassian*
(14.11.26)	$\mathbf{E}_{\|\|}' = \mathbf{E}_{\|\|}$	$\mathbf{E}_{\|\|}' = \mathbf{E}_{\|\|}$
(14.11.27)	$\mathbf{E}_\perp' = \gamma\{\mathbf{E}_\perp + (\mathbf{v}\times\mathbf{B})_\perp\}$	$\mathbf{E}_\perp' = \gamma\left\{\mathbf{E}_\perp + \frac{1}{c}(\mathbf{v}\times\mathbf{B})_\perp\right\}$
(14.11.28)	$\mathbf{B}_{\|\|}' = \mathbf{B}_{\|\|}$	$\mathbf{B}_{\|\|}' = \mathbf{B}_{\|\|}$
(14.11.29)	$\mathbf{B}_\perp' = \gamma\left\{\mathbf{B}_\perp - \frac{1}{c^2}(\mathbf{v}\times\mathbf{E})_\perp\right\}$	$\mathbf{B}_\perp' = \gamma\left\{\mathbf{B}_\perp - \frac{1}{c}(\mathbf{v}\times\mathbf{E})_\perp\right\}$

Tutorial 1

Show that the homogeneous wave equation (considered in Example 1 of this chapter) is invariant under Lorentz transformations.

Consider the equations

$$\nabla \times \mathbf{E} = \frac{-\partial \mathbf{B}}{\partial t} \tag{1}$$

$$\nabla . \mathbf{B} = 0 \tag{2}$$

In component form, they read

$$\frac{\partial E_z}{\partial y} - \frac{\partial E_y}{\partial z} = \frac{-\partial B_x}{\partial t} \tag{3a}$$

$$\frac{\partial E_x}{\partial z} - \frac{\partial E_z}{\partial x} = \frac{-\partial B_y}{\partial t} \tag{3b}$$

$$\frac{\partial E_y}{\partial x} - \frac{\partial E_x}{\partial y} = \frac{-\partial B_z}{\partial t} \tag{3c}$$

$$\frac{\partial B_x}{\partial x} + \frac{\partial B_y}{\partial y} + \frac{\partial B_z}{\partial z} = 0 \tag{2}$$

Use the Lorentz transformation and the inverse transformations for the fields to show that the equations in the S' frame are

$$\nabla' \times \mathbf{E}' = \frac{-\partial \mathbf{B}'}{\partial t'} \tag{4}$$

$$\nabla' . \mathbf{B}' = 0 \tag{5}$$

Tutorial 2

Before the special theory of relativity was proposed by Einstein, it was believed that light waves propagated through a medium called the ether. In a material medium which was moving, it was proposed that light was carried along by the medium and by the ether. If the ether was stationary, the light will appear to a stationary observer as though only a part of the velocity of the medium were added to it. This idea was put forward by J.A. Fresnel in 1817. He derived an expression for the speed of light in a medium of refractive index n moving with speed v_w. The speed of light in such a moving medium is

$$v = \frac{c}{n} \pm v_w \left(1 - \frac{1}{n^2}\right) \tag{1}$$

The change in speed is less than v_w. The quantity $1 - \frac{1}{n^2}$ is called the Fresnel drag coefficient. This hypothesis was verified in an experiment by Fizeau in 1851.

We can explain this result using the relativistic addition of velocities. Using the same notation as above, the speed of light v with respect to a stationary observer is

$$v = \frac{\frac{c}{n} + v_w}{1 + \frac{v_w}{nc}} \cong \left(v_w + \frac{c}{n}\right)\left(1 - \frac{v_w}{nc}\right)$$

using the binomial expansion.

$$\therefore \quad v = v_w - \frac{v_w{}^2}{nc} + \frac{c}{n} - \frac{v_w}{n^2}$$

$$\cong v_w + \frac{c}{n} - \frac{v_w}{n^2}$$ where we neglect terms of second order in v_w.

$$\therefore \quad v = \frac{c}{n} + v_w\left(1 - \frac{1}{n^2}\right)$$

which is the same as eqn. (1). (The negative sign is the case of motion of the medium in the opposite direction.)

Tutorial 3

The phenomenon of stellar aberration was first observed by James Bradley in 1725. He was trying to measure the distances of stars by looking for an apparent change in the position of stars as the earth revolves around the sun. If the earth was stationary, a telescope would have to be pointed directly at the observe it. But due to the motion of the earth, the telescope must be tilted at an angle α given by

$$\tan \alpha = \frac{v}{c} \tag{1}$$

It was known that the earth goes around the sun at a speed of about $v = 30$ km/s and so with $c = 3 \times 10^8$ m/s, $\alpha = 20.5''$. This was verified experimentally. At that time, it was believed that the light waves propagated through a medium called the ether. So the above phenomenon of aberration implied that the ether was not dragged along with the earth. Had it been dragged along, it would not have been necessary to tilt the telescope. The ether was undisturbed by the motion of the earth.

We can explain this from the point of view of special relativity. Consider eqn.(14.8.7) and Fig. 14.4.

$$\tan\theta = \frac{\sin\theta'}{\gamma(\cos\theta' + \beta)} \tag{14.8.7}$$

S is a frame in which the star is at rest. *S'* is a frame attached to the earth. We shall consider the inverse of eqn. (14.8.7). It is

$$\tan\theta' = \frac{\sin\theta}{\gamma(\cos\theta - \beta)} \tag{2}$$

If the star is directly overhead in the *S* frame, one receives plane waves whose direction of propagation is along the negative *y*-axis (Fig. 14.16a). $\therefore\ \theta = 3\pi/2$. So eqn. (1) gives

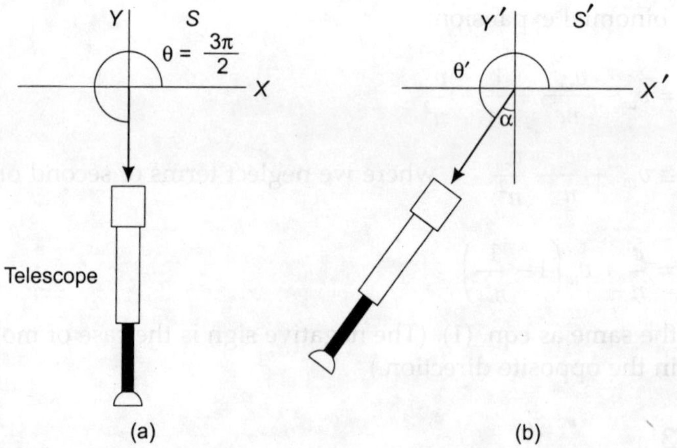

Fig. 14.16

$$\tan\theta' = \frac{1}{\gamma\beta} \tag{3}$$

When $\beta << 1$, $\gamma \cong 1$.

$$\therefore \qquad \tan\theta' = \frac{1}{\beta} = \frac{c}{v} \tag{4}$$

To compare with eqn.(3), note from Fig. 14.16 that

$$\theta' + \alpha = \frac{3\pi}{2} \tag{5}$$

Tutorial 4

In this tutorial, we shall discuss the twin paradox. Consider twins *A* and *B*. *A* remains on the earth. *B* starts out from earth on a spaceship and after some years, returns to earth on the spaceship. On his return, he finds his twin A to be older than him. The paradox lies in the apparent preferred frame of A. If both frames, i.e. the earth and the spaceship are equivalent, then A should find B older than him and vice versa. So there is an apparent contradiction here also.

There are three steps involved in this example. Firstly, B accelerates to a constant velocity. Secondly, B reverses his velocity (since he will have to return to A). Thirdly, B decelerates to come to rest at A's location. These three steps involve accelerated motion. So the situation is not symmetric with respect to A and B. If we wish to apply special relativity, we have to assume that the time of acceleration is very short. Also, we have to assume that B switches from an inertial frame S' (when he is moving away from A) to another inertial frame S'' (when he is moving towards A). Suppose, the constant speed of S' relative to S (the earth frame) is $v = 0.8c$ and similarly that of S'' relative to S is $v' = -0.8c$. Let B be away from A for six years[18b] as measured by his (B's) clock. Furthermore, let each one of them send light signals to the other at intervals of one year as measured by his clock. When B is moving away from A, the frequency of the signals from B will be reduced due to the Doppler effect. It will become, according to eqn. (14.8.10),

$$\nu = \nu' \sqrt{\frac{c - v}{c + v}} = \nu' \sqrt{\frac{1 - 0.8}{1 + 0.8}} = \frac{\nu'}{3} \tag{1}$$

So A receives the first signal from B after three years according to A's clock. In this time interval, A has sent three signals to B. B receives the first signal from A after three years according to B's clock because B's clock is running slow. At this point, B switches over to another reference frame S'' which is moving towards A with speed $v = 0.8c$. The frequency of signals received by A is increased due to the Doppler effect. It will become, according to eqn. (14.8.9),

$$\nu = \nu' \sqrt{\frac{c + v}{c - v}} = \nu' \sqrt{\frac{1 + 0.8}{1 - 0.8}} = 3\nu' \tag{2}$$

In the next three years according to B's clock, B sends three signals. But A will be sending nine signals from the time B reverses his velocity because, according to his (A's) clock, nine years have passed. A receives the three signals send by B in the last year according to A's clock, because of eqn. (2). ∴ According to A's clock, ten years have passed since B left him. According to B, six years have passed since he left A. However, A sends ten signals and B receives ten signals. B sends six signals and A receives six signals.

Problems

1. The half-life of pions at rest is 1.77×10^{-8}s. A collimated pion beam, leaving the accelerator target at a velocity of 0.99c, is found to drop to half its original intensity. Find the distance traveled by the pions in the laboratory. *(University of Calcutta, 1999)*

   [Ans. 37.1m]

2. A space traveler with velocity v synchronizes his clock ($t' = 0$) with the earth friend ($t = 0$). The earth friend then observes both clocks simultaneously, t directly and t' through a telescope. Show that when t' reads one hour, t reads

$$\sqrt{\frac{1+\beta}{1-\beta}} \qquad \left(\beta = \frac{v}{c}\right)$$

   (University of Calcutta, 2002)

3. Two rods of proper length l_0 move lengthwise towards each other parallel to the common axis with the same velocity v relative to the laboratory frame. Show that the length of each rod in the reference frame fixed to the other rod is

$$l = \frac{l_0(1-\beta^2)}{(1+\beta^2)}, \qquad \left(\beta = \frac{v}{c}\right)$$

   (University of Calcutta, 2003)

4. An electron moves in the positive x-direction in frame S at a speed $v = 0.8c$. (a) What are its momentum and its energy in frame S? (b) Frame S' moves to the right at a speed 0.6c with respect to S. Find the momentum and energy of the electron in this frame.

   [Ans. $p_x = 3.64 \times 10^{-22}$ kgm/s, $E = 0.85$ MeV, $p_x' = 1.13 \times 10^{-22}$ kgm/s, $E' = 0.56$ MeV]

5. A K meson traveling through the laboratory breaks up into two π mesons. One of the π mesons is left at rest. What was the energy of the K? What is the energy of the remaining π meson? (Rest mass of K meson = 494 MeV; rest mass of π meson $\cong$ 137 MeV)

   [Ans. Kinetic energy of π meson = 616.6 MeV, of K meson = 396.6 MeV]

6. According to observers in the frame S, the following events occurred in the xy plane. A singly charged positive ion which had been moving with the constant velocity $v = 0.6c$ in the direction of the positive y axis passed through the origin at $t = 0$. At the same time, a similar ion which had been moving with the same speed but in the direction of the negative y axis, passed the point (2, 0, 0) on the x axis. The distances are in cm.

(a) What is the strength and direction of the electric field at $t = 0$ at the point (3, 0, 0)?

(b) What is the strength and direction of the magnetic field at the same place and time?

$$\left[\textbf{Ans. } \text{(a) } \mathbf{E} = \frac{25e}{18}\,\hat{\mathbf{i}},\ \text{(b) } \mathbf{B} = \frac{2e}{3}\,\hat{\mathbf{k}}\right]$$

(**Hint:** Use Gaussian units.)

Indian Contributions in Physics

In this section we shall sketch very briefly some important work by Indian physicists and their contributions to science in India.

Aryabhata

Aryabhata was born in the latter half of the 5th century A.D. A well-known astronomer and mathematician, he made a detailed study of solar and lunar eclipses. He put forward the proposition that the earth rotates about its axis.[19] He also gave the modern approximate value of 3.1416 for π, expressing it in the form of a fraction $\frac{62832}{20000}$[20].

C.V. Raman

Chandrashekhar Venkata Raman was born in 1888. A brilliant student, he stood first in the B.A. examination (1904) winning the Gold Medal. He also stood first in his M.A. examination (1907). In 1906, he published his first scientific paper concerning the diffraction of light. Although he joined the Financial Civil Service in 1907, he had a keen interest in Physics. He was posted in Calcutta. He used to do research at the Indian Association for the Cultivation of Science there, everyday after attending office. In 1917, he was offered the Sir Taraknath Palit Professorship of Physics at Calcutta University by the vice chancellor Sir Ashutosh Mukherjee. Here he pursued an active research and teaching programme. One of his major contributions to physics came during his tenure in Calcutta. He was studying scattering of light by liquids. In 1922, he published a paper. "On the molecular scattering of light in water and the colour of the sea" in the Proceedings of the Royal Society. In 1928, he discovered the effect which now bears his name. Using monochromatic light from a powerful mercury arc, he analyzed the light scattered by liquids, organic vapours and gaseous CO_2 and N_2O with a spectroscope. He and his students, including K.S. Krishnan, found that each incident spectral line at angular frequency ω_L was accompanied by weaker lines at $\omega_L + \omega_1$, $\omega_L - \omega_1$, $\omega_L + \omega_2$, $\omega_L - \omega_2$,.... Raman attributed the additional lines to an energy exchange between the incident photon and the internal excitations of the scattering medium.[21] Raman showed that the effect is also exhibited by crystals and amorphous solids. The new lines are characteristic of the scattering medium. The shift in frequency is independent of the frequency of the incident radiation. Raman was knighted in 1929 and was awarded the Nobel Prize in Physics in 1930, for "his investigations on the scattering of light and the discovery of the effect known after him."

In 1933, Raman moved to the Indian Institute of Science, Bangalore. One of his major contributions there along with Nagendra Nath was the explanation of the diffraction of visible light by high-frequency sound waves generated in fluids by a transducer.

Raman set up a Research Institute in Bangalore on land donated by the Maharaja of Mysore. This Insititute, name after him, started functioning in 1949 and is one of India's best research institutions today. Raman died in 1970.

H.J. Bhabha

Homi Jehangir Bhabha, who is regarded as the father of Nuclear Energy in India, was born in 1909. In 1930, he obtained a Mechanical Sciences Tripos from Cambridge University. In 1931, he published his first paper on cosmic rays. Cosmic rays are charged particles which come to earth from outer space. This field of research was to occupy him for a major part of his life. In 1935, he calculated the cross section for electron-position scattering. This came to be known as Bhabha scattering. The positron is the antiparticle of the electron. It has positive charge and a mass equal to that of the electron. The cross section of a reaction is a measure of the probability of occurrence of the reaction. In his calculation, Bhabha emphasized the importance of the one-photon intermediate state

$$e^+ + e^- \to \gamma \to e^+ + e^-$$

By 1936 it was known that cosmic ray showers usually consisted of photons, electrons and positrons. Bhabha, in collaboration with W. Heitler, showed that this could be interpreted as a cascade phenomenon. For example, a photon with energy ~ 1 GeV$^{\oplus}$ traversing lead creates and e^+e^- pair within a mean distance $\leq$ 1 cm. These particles suffer Bremsstrahlung,$^{\otimes}$ the newly created photons make pairs again and so on.[22] In 1937, Bhabha suggested that Yukawa's mechanism for β-radioactivity implies that a free meson U^+ will decay spontaneously as

$$U^+ \to e^+ + \nu$$

The ν is a neutrino (a neutral, nearly massless particle).

On his return to India, Bhabha joined the Indian Institute of Science, Bangalore in 1940 as Reader in Physics.[23] In 1941, he was elected to the

---

$^{\oplus}$1 GeV = 10^9 eV

$^{\otimes}$loss of energy by radiation

Royal Society. A couple of years later he become Professor of Cosmic Ray Research. Bhabha belonged to a rich educated family, related to the Tatas. He persuaded the Tatas to set up a research centre of excellence in India. The Tata Institute of Fundamental Research (TIFR) was set up in 1945 and he became the first director.

Bhabha became the first Chairman of the Indian Atomic Energy Commission in 1948. Under his supervision, the Trombay Atomic Energy Establishment was built near Mumbai. The first low power "swimming pool" reactor named APSARA was built and went critical in 1956.[23] He died in a plane crash in 1966.

V.A. Sarabhai

Vikram Ambalal Sarabhai is regarded as the pioneer of India's Space Programme. Born in 1919, he obtained his Tripos in Natural Sciences from Cambridge University in 1939. He completed his Ph.D. from Cambridge in 1947, working on cosmic rays and nuclear fission. He belonged to a wealthy family of industrialists. On his return to India, he set up the Physical Research Laboratory in Ahmedabad.[23] Today, this is one of India's premier research institutes. He was also involved in the establishment of the Indian Institute of Management[23], Ahmedabad. In 1962, Prime Minister Jawaharlal Nehru gave him the responsibility of organizing Space research in India. He established a rocket launching station at Thumba in Kerala. It was later expanded into a full-fledged Space Science and Technology Centre which bears his name today. Sarabhai conceived a ten-year programme for Indian Space technology development which consisted of developing weather rockets and launching of space satellites devoted to spreading educational programmes and the INSAT series.

After the death of H.J. Bhabha in 1966, Sarabhai was asked to take over as the Chairman of the Atomic Energy Commission. The planning of nuclear power plants at Tarapur, Kalpakkam, etc. was done during his tenure. He won the Bhatnagar Memorial Prize for Physics in 1962. He died in 1971.

S.N. Bose

Satyendra Nath Bose was born in 1894. He had an extraordinary talent for mathematics. He stood first in both his B.Sc. and M.Sc. (1915) in "Mixed Mathematics" from Presidency College, Calcutta. In 1916, he was appointed a Lecturer in the newly created physics department at the University College of Science at Calcutta. In 1921, he joined Dacca University as a Reader in Physics. In 1924, he made his most important contribution to physics. He derived the energy density of blackbody radiation in a logical manner introducing the concept of indistinguishability.[24] He considered the radiation to be a gas of photons whose number was not conserved. He sent the paper entitled "Planck's law and light quantum hypothesis" to the journal Philosophical Magazine but it was not accepted. Subsequently he sent it to Einstein with a request to have it translated into German and published in the journal Zeitschrift fur Physik.[24] Einstein was impressed by Bose's paper and translated it and had it published. Einstein later considered a gas of atoms whose number is conserved (unlike photons) and derived the formula for the average number of such atoms with an energy E at a temperature T. This came to be known as Bose-Einstein statistics and is obeyed by particles whose spin is an integral multiple of $\hbar$. In 1945, Bose returned to Calcutta University as Khaira Professor of Physics. He died in 1974.

M.N. Saha

Meghnad Saha was born in 1893. A brilliant student, he completed his B.Sc. (1913) and M.Sc. (1915) from Calcutta University. He was appointed a lecturer in the Physics Department at the University College of Science, Calcutta. In 1917, he published his first paper "On Maxwell's stresses, concerning the electromagnetic theory of radiation" in Philosophical Magazine.[23] In the next two years he published papers on light, elasticity, quantum theory, etc. He was awarded the D.Sc. by Calcutta University in 1919. Thereafter, he turned his attention to the spectra of stars. The central body of the sun emits a continuous spectrum. The atmosphere of the sun is at a lower temperature than the central region but contains elements in gaseous form. These elements absorb some of the radiation emitted by the central body at wavelengths characteristic of them. This gives rise to dark lines in the solar spectrum. These were discovered by Fraunhofer in 1814. But the above explanation was given by Kirchhoff in 1859. Spectroscopy using discharge tubes containing different gases had shown emission lines characteristic of gaseous element. It was found that dark lines in the solar spectrum occurred at the same position as the emission lines of the corresponding element. Another way of confirming this explanation was to observe the solar spectrum during a solar eclipse. The solar atmosphere is laid bare at the time of totality and the dark lines eclipse. The solar atmosphere is laid bare at the time of totality and the dark lines appear as bright lines. This spectrum is called a flash spectrum[25] and was first observed in 1872 by Prof. Young. The intensities and other characteristics of the Fraunhofer lines gave information on the abundance and physical state of elements of the sun. At the time that Saha began his work, about 64 elements had been identified in the sun whereas about 92 elements were known to exist on earth.[25] Moreover, the abundances of the elements did not agree. Saha put forward his theory of thermal ionization. In his theory, he showed that the abundance of the elements depend on the extent of ionization of the element. Na and Mg were present almost completely in ionized form in the solar atmosphere and hence strongly intense lines of these elements were observed. Rb and Cs. Which had much lower ionization potentials, were almost absent in the spectrum of the solar atmosphere.

Saha was appointed as the Khaira Professor of Physics at Calcutta University in 1921. He later joined Allahabad University in 1923. Here he guided many young research students to successful careers as scientists. He was elected a Fellow of the Royal Society in 1927. He founded the U.P. Academy of Sciences in 1930.[23] He returned to Calcutta University in 1938. The Institute of Nuclear Physics was set up in 1950. Today it is called the Saha Institute of Nuclear Physics and is one of India's premier research institutes. Saha died in 1956.

S. Chandrasekhar

Subrahmanyan Chandrasekhar was born in 1910. He did his B.A. (Hons.) form Presidency College, Madras. He published his first paper in the Proceedings of the Royal Society at the age of 18. After graduation, he won an Indian Government Scholarship to Trinity College. He had become interested in astrophysics by this time and he obtained his Ph.D. from Cambridge University in 1933.[23] One of the problems he was working on was to consider what would happen to a star after it ran out of nuclear fuel. Earlier, R.H. Fowler had suggested that although gravity would dominate after this, the star would be prevented from collapsing to a point by degeneracy pressure–a quantum mechanical effect. It would become a *white dwarf*. Fowler had applied quantum statistics since he reasoned that at the high densities in a white dwarf, there would be a degenerate gas of electrons. Chandrasekhar further proposed that at these high densities, relativistic effects also had to be taken into account. After extensive calculations, he concluded that if the initial mass of the star was greater than or equal to 1.4 times the mass of the sun, the star would collapse to a point. This limiting mass is called the Chandrasekhar limit. His theory was opposed vigorously by Sir Arthur Eddington, one of the eminent scientists at that time in England. But later observations proved Chandrasekhar's theory to be correct.

Chandrasekhar joined the faculty of the University of Chicago in 1936, at its Yerkes Observatory campus. Here he further developed his work on the interiors of stars, including degenerate matter, and published a book "An introduction to the study of Stellar Structure" in 1939.[26] He also worked on the gravitational frictional drag on a star passing through a tenuous cloud of stars and wrote a book entitled "Principles of Stellar Dynamics" (1943). Continuing his research on stellar interiors, he studied radiative transfer. He identified the causes of opacity of hydrogen at the typical surface temperature of stars and wrote another book "Radiative Transfer" in 1950. In 1952, he was appointed Morton D. Hall Distinguished Service Professor in Astronomy and Astrophysics at the University of Chicago. He was awarded the Gold Medal of the Royal Astronomical Society of England in 1952.

Thereafter, Chandrasekhar worked on plasma physics and hydrodynamics, investigating the stability of magnetic fluids.[26] He wrote a book on this subject, entitled "Hydrodynamic and hydromagnetic stability" (1961). Chandrasekhar worked on the dynamical properties of rotating black holes and published a book "The Mathematical Theory of Black Holes"(1983). He was awarded the Nobel Prize in Physics in 1983. This delay in the award was primarily due to the opposition by Arthur Eddington. Chandrasekhar died in 1995.

References

1. L.A. Pipes (1946), *Applied Mathematics for Engineers and Physicists,* McGraw-Hill Book Co.
2. J.R. Reitz, F.J. Milford and R.W. Christy (1986), *Foundations of Electromagnetic Theory,* Addison-Wesley/Narosa, 3rd Edition.
3. J.D. Jackson (1999), *Classical Electrodynamics,* John Wiley and Sons, 3rd Edition.
4. E.M. Purcell (1985), *Electricity and Magnetism* (Berkeley Physics Course – Vol. 2) McGraw-Hill, 2nd Edition.
5. D. Halliday and R. Resnick (1985), *Physics,* Part II, Wiley Eastern, 2nd Edition.
6. S.P. Parker ed. (1986), *McGraw-Hill Dictionary of Physics.*
7. F.H. Bower (1960), *Thermistor,* McGraw-Hill Encyclopedia of Science and Technology, vol. 13
8. J.A. Edminister (1987), *Theory and Problems of Electric Circuits,* Schaum's Outline Series, McGraw-Hill, 2nd Edition.
9. J. Millman and C. Halkias (1972), *Integrated Electronics,* McGraw-Hill, International Edition.
10. K.S. Krane (1988), *Introductory Nuclear Physics,* John Wiley.
11. J. Yarwood (1988), *Atomic Physics,* Oxford University Press
12. R.P. Feynman, R. Leighton, M. Sands (1964), The *Feynaman Lectures on Physics,* Addition-Wesley.
13. W.T. Silfvast (1998), *Laser Fundamentals,* Combridge University Press.
14. D.J. Griffiths (1989), *Introduction to Electrodynamics,* Prentice-Hall of India, 2nd Edition.
15. E. Hecht (2002), *Optics,* Pearson Education Asia, 4th Edition.
16. M. Born and E. Wolf (1989), *Principles of Optics,* Pergamom Press, 6th Edition.
17. W.K.H. Panofsky and M. Philips (1962), *Classical Electricity and Magnetism.* Addison-Wesley, 2nd Edition.

18a. A.P French (1972), *Special Relativity,* ELBS/Van Nostrand and Reinhold (UK).

18b. R. Resnick (1972), *Introduction to Special Relativity*, Wiley Eastern.

19. D.N. Kundra and S. D. Kundra (1983), *A new textbook of the History of India*, Part I, Navdeep Publications.

20. A.L. Basham (2001), *The wonder that was India*, Rupa & Co.

21. A. Jayaraman and A.K. Ramdas (April 1988), *C.V. Raman*, Physics Today.

22. A. Pais (1986), *Inward Bound*, Clarendon Press, Oxford.

23. K.V. Gopalakrishnan (2001), *Some eminent Indian scientists and Technologists*, NCERT.

24. G. Venkataraman (1992), *Bose and His Statistics*, Universities Press.

25. M.N. Saha and B.N. Srivastava (1965), *A Treatise on Heat*, The Indian Press, 5th Edition.

26. E.N. Parker (November 1995), *Obituary*, Physics Today.

Index